Lecture Notes in Geoinformation and Cartography

Series Editors: William Cartwright, Georg Gartner, Liqiu Meng, Michael P. Peterson

William Cartwright · Georg Gartner ·
Antje Lehn (Eds.)

Cartography and Art

Springer

Editors

Prof. Dr. Dr. William Cartwright
RMIT University
School of Mathematical and
Geospatial Sciences
Melbourne VIC 3000
Australia
william.cartwright@rmit.edu.au

Prof. Georg Gartner
Vienna University of Technology
Institute of Geoinformation and
Cartography
Erzherzog-Johannplatz 1
A-1040 Vienna
Austria
georg.gartner@tuwien.ac.at

Dr. Antje Lehn
Akademie der Bildenden
Künste Wien
Schillerplatz 3
A-1010 Vienna
Austria
a.lehn@akbild.ac.at

ISBN: 978-3-540-68567-8 e-ISBN: 978-3-540-68569-2

DOI 10.1007/978-3-540-68569-2

Lecture Notes in Geoinformation and Cartography ISSN: 1863-2246

Library of Congress Control Number: 2008942186

Camera-ready by Manuela Schmidt and Felix Ortag

Cover design: deblik, Berlin

Cover illustration: Oliver Maerki, Academy of Fine Arts Vienna

Printed on acid-free paper

9 8 7 6 5 4 3 2 1

springer.com

Preface

This book is the fruition of work from contributors to the *Art and Cartography: Cartography and Art* symposium held in Vienna in February 2008. This meeting brought together cartographers who were interested in the design and aesthetics elements of cartography and artists who use maps as the basis for their art or who incorporate place and space in their expressions.

The outcome of bringing together these like minds culminated in a wonderful event, spanning three evenings and two days in the Austrian capital. Papers, exhibitions and installations provided a forum for appreciating the endeavors of artists and cartographers and their representations of geography.

As well as indulging in an expansive and expressive occasion attendees were able to reflect on their own work and discuss similar elements in each other's work. It also allowed cartographers and artists to discuss the potential for collaboration in future research and development.

To recognise the significance of this event, paper authors were invited to further develop their work and contribute chapters to this book. We believe that this book marks both a significant occasion in Vienna and a starting point for future collaborative efforts between artists and cartographers.

The editors would like to acknowledge the work of Manuela Schmidt and Felix Ortag, who undertook the task of the design and layout of the chapters.

William Cartwright, Melbourne, Australia
Georg Gartner, Vienna, Austria
Antje Lehn, Vienna, Austria

List of Contributors

Andel, Emanuel
ÖIR-Informationsdienste GmbH
Vienna, Austria

Angus, Alice
Proboscis
London, United Kingdom

Benová, Alexandra
Faculty of Natural Sciences
Comenius University in Bratislava
Bratislava, Slovak Republic

Bär, Hans Rudolf
Institute of Cartography
ETH Zurich
Zurich, Switzerland

Caquard, Sébastien
Département de Géographie
Université de Montréal
Montréal, Canada

Cartwright, William
School of Mathematical and Geospatial Sciences
RMIT University
Melbourne, Australia

Christophe, Sidonie
National Geographic Institute
COGIT Laboratory
Saint Mandé Cedex, France

Downton, Peter
School of Architecture and Design
RMIT University
Melbourne, Australia

Dykhuis, Peter
Halifax, Canada

Edquist, Harriet
School of Architecture and Design
RMIT University
Melbourne, Australia

Fairbairn, David
School of Civil Engineering and Geosciences
Newcastle University
Newcastle upon Tyne, United Kingdom

Feranec, Ján
Institute of Geography
Slovak Academy of Sciences
Bratislava, Slovak Republic

Fiel, Wolfgang
Vienna, Austria

Friedmannová, Lucie
Department of Geography
Masaryk University
Brno, Czech Republic

Gartner, Georg
Institute of Geoinformation and Cartography
Vienna University of Technology
Vienna, Austria

Häfele, Theresa Anna
Vienna, Austria

Holter, Maria Christine
Vienna, Austria

Hurni, Lorenz
Institute of Cartography
ETH Zurich
Zurich, Switzerland

Jobst, Markus
Institute of Geoinformation and Cartography
Vienna University of Technology
Vienna, Austria

Kinberger, Michaela
Department of Geography and Regional Research
University of Vienna
Vienna, Austria

Kinman, Edward L.
Department of Natural Sciences
Longwood University
Farmville, VA, USA

Koskinen, Ilpo
University of Art and Design Helsinki
School of Design
Helsinki, Finland

Kriz, Karel
Department of Geography and Regional Research
University of Vienna
Vienna, Austria

Lechthaler, Mirjanka
Institute of Geoinformation and Cartography
Vienna University of Technology
Vienna, Austria

Lehn, Antje
Institute for Art and Architecture
Academy of Fine Arts Vienna
Vienna, Austria

Mourão, Manuela
Department of English
Old Dominion University
Norfolk, VA, USA

Müller-Funk, Sabine
Vienna, Austria

Ortag, Felix
Institute of Geoinformation and Cartography
Vienna University of Technology
Vienna, Austria

Piatti, Barbara
Institute of Cartography
ETH Zurich
Zurich, Switzerland

Plath, Nils
School of Design
University of Applied Sciences
Münster, Germany

Pravda, Ján
Institute of Geography
Slovak Academy of Sciences
Bratislava, Slovak Republic

Reuschel, Anne-Kathrin
Institute of Cartography
ETH Zurich
Zurich, Switzerland

Sarjakoski, Tapani
Finnish Geodetic Institute
Department of Geoinformatics and Cartography
Masala, Finland

Sarjakoski, L. Tiina
Finnish Geodetic Institute
Department of Geoinformatics and Cartography
Masala, Finland

Skupin, André
Department of Geography
San Diego State University
San Diego, CA, USA

Skupin, Marinta
Louisiana State Museum
New Orleans, LA, USA

Spanring, Christian
ÖIR-Informationsdienste GmbH
Vienna, Austria

Tarnai, Titusz
Academy of Fine Arts Vienna
Vienna, Austria

Thun-Hohenstein, Felicitas
Academy of Fine Arts Vienna
Vienna, Austria

Trischak, Evamaria
Vienna, Austria

Vaughan, Laurene
School of Applied Communication
RMIT University
Melbourne, Australia

von Ungern-Sternberg, Armin
Department of Philosophy and Philology
University of Mainz
Mainz, Germany

Vujaković, Peter
Department of Geographical and Life Sciences
Canterbury Christ Church University
Canterbury, United Kingdom

Widorn, Verena
Department of Art History
University of Vienna
Vienna, Austria

Wright, Benjamin
Institute for Comparative Studies in Literature
Carleton University
Ottawa, Canada

Ylirisku, Salu
University of Art and Design Helsinki
School of Design
Helsinki, Finland

Table of Contents

Section I: Theory

1 Maps and Mapping in the Eyes of Artists and Cartographers – Experiences from the International Symposium on Cartography and Art

William Cartwright[1], Georg Gartner[2] and Antje Lehn[3]

[1] School of Mathematical and Geospatial Sciences, RMIT University, Australia
[2] Institute of Geoinformation and Cartography
Vienna University of Technology, Austria
[3] Institute for Art and Architecture, Academy of Fine Arts Vienna, Austria

1.1 Introduction

Contemporary methods for depicting the earth and its cultural and natural attributes use graphic and non-graphic formats, maps and map-related artefacts, for visualising geography and for building virtual landscapes and environments. The discipline area of cartography, traditionally, has applied art (design), science and technology to map-making to design and realise these products. Prior to the mid 1950s cartographic artefacts were built under the theoretical and practical 'umbrella' of this partnership of art, science and technology. However, since then it is argued, the theory and methodology associated with visualising geography has focused on science and technology, and away from art. This 'move' away from art was accelerated by: 1. Computing, computers and complete automated systems; and 2. A 'quest' to gain 'scientific legitimacy' by using scientific visualization as a lodestone for gauging the 'quality' of theories and applications (Cartwright 2008, Gartner 2008). In order to bridge the gap between science and art in cartography initiatives have been taken which led into the organisation of an International Symposium on "Cartography and Art" in Vienna 2008.

1.2 Establishing an Exchange Forum for Cartographers and Artists

The International Symposium "Cartography and Art – Art and Cartography" took place from January 31 to February 2, 2008 in Vienna, Austria. It was backed by the International Cartographic Association (ICA) and organised by the Institute of Geoinformation and Cartography of the Vienna University of Technology, the Institute for Art and Architecture, Academy of Fine Arts Vienna and the School of Mathematical and Geospatial Sciences, RMIT University.

Ideas behind establishing such an event were developed and derived from various discussions of cartographers to topics like ‘aesthetics’, ‘artistic aspects in cartographic communication’ or ‘cartographic heritage’, which finally led into the foundation of a new Working Group of the International Cartographic Association called “Art and Cartography”. The objectives of this working group are primarily to stimulate and establish co-operations and links between cartographers and artists. With this it might be possible, to bring the artistic aspects of maps again as a main focus. The symposium in Vienna was the result of the initiative of William Cartwright (President of the International Cartographic Association, then Chair of the ICA Art and Cartography Working Group, RMIT University Melbourne), Antje Lehn (Academy of Fine Arts Vienna) and Georg Gartner (Vice-President of the International Cartographic Association, Vienna University of Technology) and was meant as a first attempt to establish an exchange forum for cartographers, artists and architects to enable discussions and to encourage interdisciplinary cross-fertilisation of ideas and concepts.

1.3 The Symposium on “Cartography and Art”

Interested participants were invited to present their work on topics related to the artistic aspects of cartography. There were no specifications that limited contributions to look at maps of historic periods only, but a general call was made for contributions that addressed contemporary aspects of cartography and art. Furthermore, interested scientists and artists who attempt to communicate space and place by means of modern media or experimental artefacts and establish alternative options to the use of maps as the main mean of spatial communication were invited to participate. The only limitation was that they needed to keep to the ‘spatial context’.

The call for papers aimed at attracting contributions as scientific papers, posters, installations, videos, web applications and experimental media. It was answered by many contributions from more then 20 countries and various disciplines. A wide spectrum of the submitted contributions was the result. The selection of the contributions was undertaken by William Cartwright, Antje Lehn and Georg Gartner. This was done with respect to criteria of scientific and artistic originality and quality. Finally more then 60 contributions, representing authors of 15 different countries, were selected.

Fig. 1.1. Extract of the work by Ruth Watson – The Reformed World (for Johannes Stabius) Photograph: James Geurts

Due to the heterogenous nature of contributions three different event locations were used to present papers, artefacts and installations in an appropriate way. Venues for the symposium were the main hall of Academy of Fine Arts Vienna (artefacts, paintings, installations, videos), the Kunsthalle Wien project space (Contemporary Art Gallery used for short presentations by artists, works from in the fields of architecture, conceptual art and digital media) and Vienna University of Technology (scientific contributions presented as lectures and posters).

The event began with a vernissage of the exhibition 'zoomandscale' at the Academy of Fine Arts Vienna. Here, works were presented related to the context 'Cartography and Art' from the perspective of artists. These works drew a thematic bow from maps to artistic descriptions of specific places as a starting point for orientation strategies, and maps as metaphors. The exhibition was curated and organised by a group of artists 'collabor.at' together with Antje Lehn. It encouraged visitors to participate in the event through the use of reduced instruments for measuring and surveying like ladders and field glasses, which contained associations to the process of mapping. Large paper banners structured the space and created a backdrop to the artwork. Artists included Wolfgang Fiel, Gabu Heindl, Nicole Six & Paul Petritsch, Christian Mayer (all from Austria), Peter Dykhuis (Canada), Manuela Mourao (USA), Proboscis (UK), Ludo Slagmolen (The Netherlands), Laurene Vaughan (Australia) and Ruth Watson (New Zealand).

The scientific lecture program held at the Vienna University of Technology was organised in themes. Philippe Rekacewicz, cartographer and journalist of the French monthly newspaper 'Le Monde diplomatique' was invited as a keynote speaker. In his lecture he drew a picture of the joint aspects of artistic map production to mapping art. His main argument consisted of the idea that artistic elements in maps have a high significance for the efficiency of the cartographic communication processes. Following this keynote, the opening session, 'Theory' consisted of presentations by William Cartwright (Australia), David Fairbairn (England), Felicitas Thun-Hohenstein (Austria) and Markus Jobst (Austria). As result of this introductory session the theoretical basis of the importance of re-addressing the relationship between art and cartography was established. From the presentations it became clear that it is very useful to compare the various terms, concepts and theories of the diverse disciplines represented at the event. Finally, all presenters accentuated the importance of cooperation between the various disciplines, which provides the ability for increasing the efficiency of the cartographic communication processes by enhancing mapping processes with art elements.

In the following session 'Cartographic Design' detailed suggestions were given to provide support, through examples, the topics previously mentioned. Bob Lilley (England) analysed the artistic aspects in historical and contemporary maps at the United Kingdom's Ordnance Survey, David Forrest (Scotland) featured the aspect of 'composition' as the central element of map design, Jari Korpi (Finland) discussed the design of associative symbols and their application to crisis management applications and Jeroen van den Worm (The Netherlands) presented a general methodology for the standardisation of cartographic symbols.

The first day of the symposium was completed with two sessions entitled 'Integrated Media' and 'Non Graphics'. In the first session Sebastien Caquard (Canada) and Teresa Castro (France) drew parallels between the composition and elements of cinema and cartography. Anna-Lena Kornfeld (Germany) demonstrated with 'Soundslike' the potential of the acoustic channel for the communciation of spatial information, while Robert Edsall (USA) addressed the similarities in composition and design aspects analysed between music and cartography. In the first presentation of the session 'Non Graphics' André Skupin (USA) analysed the role of written text in cartography and art, followed by presentations to the context of literature and cartography by Armin von Ungern-Sternberg (Germany), Nils Plath (Germany) and Harriet Edquist (Australia).

For those participants who were still 'hungry' for more the second part of the exhibit 'zoomandscale' was offered. In the nearby project space of the 'Kunsthalle Wien' various multimedia presentations of conceptual works related to the context of cartography and art were given. The evening session included presentations by Wolfgang Fiel, Sabine Müller-Funk, Waltraud Palme, Christian Spanring, Titusz Tarnai and Evamaria Trischak (all from Austria). The exhibition also contained performance site models and animations by architecture students of the Academy of Fine Arts, an interactive map of Europe, a mapping project describing the city of Vienna by the intersection of the minute points and other multimedia installations focussing on spatial and architectural themes.

The second day of the scientific program at TU Vienna started with poster presentations. Manfred Buchroithner (Germany) analysed 3D-relief representations in the context of aesthetic aspects, Zsolt Török (Hungary) discussed the context of modern technologies and map production, Alexander Wolodtschenko (Germany) presented prehistoric maps as an example of cartosemiotics, Jan Blaha (Czech Republic) analysed aspects of creativity in cartography and Jesus Reyes (Hungary) presented maps of children. Finally, a display of paintings by various artists who had developed their representations of one particular place was presented by Vanessa Parravicini and Klaus Kramer (Austria).

The presentation program began with a session on 'Aesthetics'. The presenters, Ján Feranec (Slovakia), Karel Kriz (Austria) and Alexander Kent (England) analysed the term 'aesthetics' from a holistic point of view and demonstrated, through historic and contemporary examples the aesthetic aspects of maps and cartographic processes. The following session on 'Non-traditional mapping' focused on the use of cartographic methods to represent 'non traditional' information. Stephanie Deitrick (USA) analysed the effective representation of 'Uncertainty', Barbara Piatti and Anne-Kathrin Reuschel (Switzerland) described through a brilliant dialogue concepts of linking cartography and literature to produce an atlas of literature, Christina Ljungberg (Sweden) presented 'Fluid Spaces' and finally 'Informal Geographies' were presented by Michaela Kinberger and Verena Widorn (Austria), highlighting the sacred landscapes of Lahaul.

The final sessions focused on the term 'design'. In the first session the perspective of designers, artists and architects was given and the second session was a forum for cartographers to give their perspective. The 'artist oriented' design session was opened by the internationally awarded designer Angie Rattay (Austria). She displayed her "Planet Earth – Directions for Use", which has been featured in the mass media. Edward Kinman (USA) presented his work on representing places by ceramics, while in the presentations of Laurene Vaughan (Australia) and Peter Downton (Australia) presented conceptual works in the context of architectural artefacts and design principles.

The 'cartography oriented design' session was opened by Sidonie Christophe (France). She presented the outcomes of interesting experiments conducted IGN Paris, to incorporate colour palettes of famous painters into map legends for topographic mapping. A similiar topic was presented by Lucie Friedmannova (Czech Republic). She analysed Claude Monet's colour schemes and demonstrated a method to apply these to cartography. The final presentations by Alexandra Benova (Slovakia) and Mirjanka Lechthaler (Austria) dealt more generally with the definition of map styles and historical aspects of cartography and art.

1.4 Conclusions

Over a glass of *prosecco* the program was later analysed. Ideas for ongoing co-operations were discussed and established, for example the strengthened role of the ICA Working Group on Art and Cartography (http://artcarto.wordpress.com). Stimulated by the enormous response of the participants the organisers began this book as a formal outcome of the symposium.

References

Cartwright W (2008) Cartography and Art. In: ICA News, 50, 2008

Gartner G (2008) Das Symposium „Cartography and Art". In: Kartographische Nachrichten, 58, 4, 2008

2 Art and Cartographic Communication

William Cartwright

School of Mathematical and Geospatial Science, RMIT University, Australia

Abstract

Science and technology has been embraced by cartography as a means to ensure that what is presented is scientifically 'correct' – products are considered to 'work' if they are scientifically 'elegant', technologically 'buildable' and 'deliverable' using contemporary communications systems. But science or technology, need not always take on primary roles, and there now is a need to address the role that design needs to take to facilitate the further development of contemporary cartography, especially in the areas where new media has been applied to facilitate the building of geographical visualization tools.

This chapter addresses how, by incorporating ART elements into the design criteria of geographical visualization artefacts, 'different' visualization tools might be provided, by considering all three elements of cartography: art, science and technology.

2.1 Introduction

Understanding how technology works is important, but the partnership between art and science, and their contributions to the discipline, are as important. Art provides the 'public face' of cartography (and if we include the cartographer's passion when designing particular products, perhaps the soul as well) and science complements this by ensuring that what is presented is scientifically correct, and what could be called 'scientifically elegant' as well. Science or technology, it is argued, needs not always to take on the primary roles in cartography. However, technology is needed to ensure that the designed product can be produced and delivered and science is necessary to ensure 'correct' and rigorous products. However, the resulting artefact, designed and produced by balancing the art, science and technology attributes, as a street artist juggler might balance a chainsaw, a watermelon and a table tennis ball, has recently been biased towards science and technology, with art being relegated to the position of 'afterthought' (thinking about the art elements after the product's specifications are 'locked' within a science foundation and technology-driven production and delivery 'envelope'. Cartography is different from other contemporary disciplines insofar as it can design, develop and deliver products with an art or a technology or a science 'flavour'. But we need to address how to make art-biased cartography as relevant as science or technology-biased cartography.

To illustrate this, take the simple process of recording spatial information by an expert recorder using a pencil and a piece of paper. If a very hard pencil is used a very precise product results. The definition between one feature and another must be precisely determined and depicted and accuracy usually would reign over artistic input. The resultant map, whilst showing precise positions and clearly demarked divisions between different classes of information, cannot depict the vagaries and overlaps that are characteristic of real-world geographical information. The depiction can only show clearly defined edges of natural and cultural phenomena. The map user can only interpret what is depicted by lines, points and polygons and is unable to 'read between the lines' to comfortably interpret what may lie in the zones between discrete classes of depicted data. If a very soft pencil were used, a more flowing and interpretive map, or a more artistic approach to information recording would result. The drawer of the map, not restrained by the imposed need for accuracy and precision of a sharp, hard, thin pencil point, can render the paper with a different type of portrayal of the real world. Areas of vagueness can be illustrated as such, interpretations can be made and impressionistic drawings produced instead of planimetric map-type drawings produced using the soft pencil's precision-demanding counterpart. A different portrayal of the same information would result. Different users of these maps might prefer one information depiction device to another, and some might prefer to use both maps, in different ways, to gain a better understanding of the part of the real world being depicted.

An example of this is using a number of artefacts to gain visual information about a city. It is argued; if different media are used, then a different viewpoint is provided, and perhaps a different interpretation of the city is had. For example, a painting (one medium) can provide one viewpoint, the map another. Compared to 'standard' maps, like street directories and topographic maps, artistic interpretations can provide completely different viewpoints of a city. Users are provided with different platforms from which to view 'reality'.

With the current provision of 'conventional' maps (including conventional maps delivered on mobile devices and the world wide web (web) the question arises as to whether the users of conventional artefacts for depicting geographical information only being allowed to view information using one particular type of hard/sharp 'pencil' drawing? Would a better understanding of the real world be had if many different methods for its depiction were made available? It is argued that the present devices for viewing the complex relationships that comprise the real world, even with the application of new technology, still only allow users to see that information in one manner – the hard, precise, sharp pencil manner that is a legacy of paper mapping. It is further argued that there is a need to investigate how the use of other, 'softer' presentation methods that complement today's generally 'hard map'-biased devices portray geography and the mental images of reality constructed.

2.2 Science and Technology and Restrictions Imposed on 'Art' in Cartography

Science and technology has provided cartography with the means for providing accurate maps in a timely and efficient manner. However, at a cost – the loss of control over part of the design process. In order to apply technology to map production and replication cartographers were required to amend their documents so that they accorded with the technology used to produce them or to communicate the geographical 'message'. This section of the paper looks at some of the technologies adopted by cartography and how they affected the actual design of the map.

2.2.1 Printing

Aligned to paper map production, printers became part of the map production team and in many ways they dictated the 'look' of maps for many years due to their technologically-imposed specifications on the map production process. Map design had to follow function and Cartographers had to adapt design and production to take into account the particular restrictions which printing placed on maps. Printing 'formalised' map design and the actual 'look' of maps reflected this.

However, it must be noted that many maps produced via the printing press do have a certain quality that can be termed 'elegant cartography'. Here, the precise replication of detail provides a map that, whilst still designed to conform to the demands of the printing press, were completed to a high standard which can be admired as excellent examples of the cartographers work and the engraver's skill.

2.2.2 Introduction of Computers

With the 'invention' of the computer everything changed in the scientific world, including cartography. Cartography applied computer graphics for artwork production and output. Unfortunately many early inferior map products produced with these early computer systems were readily accepted as substitutes for the precise and elegant scribed and printed alternatives only because they were produced quickly and by new computer systems. Just because the results of many calculations could be displayed using the newly-adopted computer drawing packages, users were sometimes willing to accept the initial crude outputs only because they were produced quickly and from data which resided on a massive database.

Examples of these maps were ASCII (American Standard Code for Information Interchange) maps, maps produced by the output of simple maps, with variations in greytone density achieved by overprinting characters on a black and white printer. The demands of this method of map production, made the map design accord with the replication technique. The resultant maps were 'chunky' and almost illegible when more detailed information was needed. *Figure 2.1* shows a typical example of this map production era.

As long as the products were not exposed to too much critical analysis all appeared to be well with the carto-graphics. But, still, some below standard map products that resulted from the attempts to portray spatial information using early computer graphics systems were foisted onto the user. However, once mastered, the technology can provide beautiful maps.

```
+------------------+-----------------------------------------------------------+
| ***        **    |                                                           |
|****  *********   |                      ARCTIC SEA        Nordkapp           |
|  *****ICELAND**  |                                       ////                |
|   R************  |                                 //   //////////++++++     |
|  *************   |                                /  ///////////////+++++++++++  |
|    ********      |                            L// ////////////####/+++++++++++++|
+------------------+                           //  ////..###///####++++++++++++++|
|                                                ////::::###########++++++++++++|
|                                              ///.:::::::::##########++++    ++++|
|    ATLANTIC SEA                          /////:::::::::::::########++++++      |
|                                         ////::::::::::::::#########+++++++     |
|  %                                   /////::::::::::::::  ########++++++++++++|
| F%  %                               /////:::::::::::::   ###########++++++++++|
|                                   ///////.:::::::::::    #############+++++++++|
|                               //////////::::::::::   ###############++++++++++|
|                             ///////////:::::::::     #####FINLAND####++++++++++|
| Capital cities:            ///NORWAY//:::::::::     ###############+(Russia)+|
|                            //////////::::::::::.     ###############+++++++++++|
| C = Copenhagen              ////////O:::::::::::::    ##########++++    +++++++|
| H = Helsinki                 /////// ::::SWEDEN:::: #A   ###H         ++++++++++|
| O = Oslo                     /////    ::::::::::S:           ++++++++++++++++|
| R = Reykjavik   NORTH                 ::::::::::         + ++++++++++++++++++++++|
| S = Stockholm    SEA         %         :::::::: .:    ++ +++++(Estonia)++++++++|
|                             %%%        :::::::  :G           ++++++++++++++++++|
|             DENMARK -->    %%%%%  %%% :::                ++   ++++++++++++++++++|
| Islands:                  %%%%   %%%C      BALTIC  +++++++++++++++++++++++++++|
|                            %%%          B%   SEA    ++++(Latvia)++++++++++++++|
| L = Lofoten Isl.           ++++                   ++++++++++++++++++++++++++++|
| B = Bornholm               +++++++     ++++++   +++++(Lithuania)+++++++++++++|
| F = Faroe Isl.   ++++  ++++++++++++++++++++++++++++++++++++++++++++++++++++++|
| G = Gotland    ++++++++++++++++++++++++(Poland)++++++++++++++++++++++++++++++|
| A = Åland    +++++++++++++(Germany)++++++++++++++++++++++++++++++++++++++++++++|
|             +++++++++++++++++++++++++++++++++++++++++++++++++++++++++++++++++++|
+--------------------------------------------------------------------------------+
```

Fig. 2.1. ASCII map of Scandinavia. Source: University of Utrecht, 1998

2.3 Discrete Media

The introduction of discrete media saw maps being produced on CD-ROM. Initially, the potential of the large storage capacity of CD-ROMs for the distribution of geographical information fostered interest in publishing digital maps using a new medium (Rystedt 1987). Products like the Digital Chart of the World (DCW) and the World Vector Shoreline (Lauer 1991) were some of the first products to exploit this storage medium. The initial products mimicked their paper cousins, but later products on this medium added interaction and user tools. In the street directory genre, British company, Nextbase produced a street directory of London. Whilst this product provided innovative inclusions like layers of information that could be turned on and off, enabling information to be made available when required, enhancing what could be immediately viewed – an apparent electronic version of a paper street directory (see *Figure 2.2)*. However, what was produced was still a 'standard' street map – with interactivity added. The media was not fully exploited and the maps were only electronic reproductions of paper products.

Later, the appearance of Desktop Publishing packages made every graphic artist a cartographer. A flood of 'crude' (from a design and consumption perspective) computer generated maps depicted everything about everywhere. This relatively recent transition of mapping from large electronic purpose-built systems housed

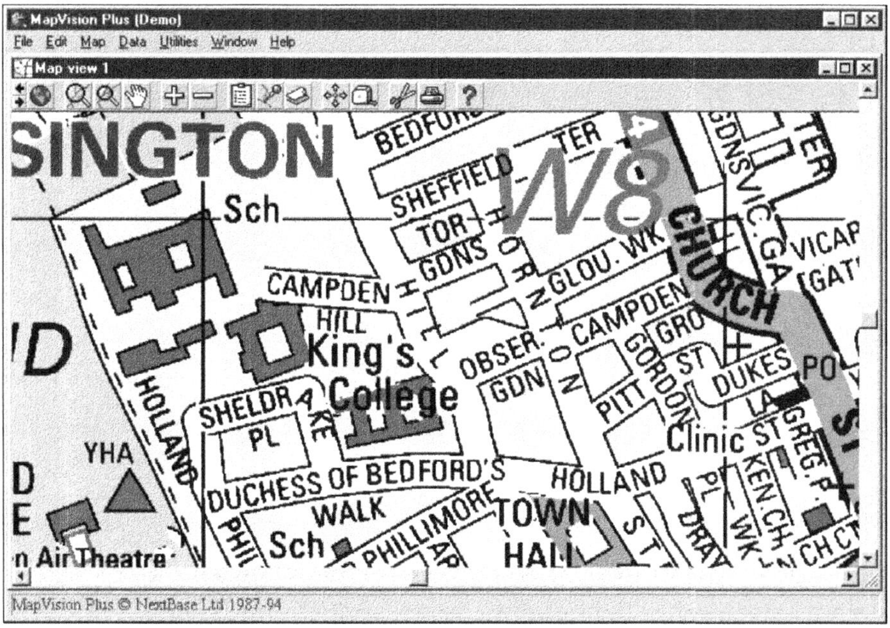

Fig. 2.2. Nextbase London Street Directory. Source: Sargent, 1994

in a map production company to individual desktop has also been a revolution for small map producers. Individual cartographers were able to produce maps as sophisticated as their corporate counterparts. Equipped with a powerful microcomputer plus a scanner, plotter/printer and modem the individual becomes part of the distributed digital electronic mapping community.

2.4 The Web

The arrival of Web publishing initially mimicked somewhat the maps produced in the early applications of CD-ROM. Paper maps were scanned and collections provided on this optical media. Maps weren't designed specifically for the medium and 'compromise' products resulted. Early Web mapping sites provided access to map collections. These were initially scanned maps, like those in the CIA World Fact Book (https://www.cia.gov/cia/publications/factbook/docs/refmaps.html) and the PCL Map Collection (http://www.lib.utexas.edu/maps/). Whilst delivered immediately, these maps provided low resolution replications of maps that were printed on paper in higher resolution than the 72 dpi computer screens on which they were displayed. A typical map availability query and resultant map from the PCL collection is shown in *Figure 2.3.*

Later, as bandwidth improved and map designers addressed the Web as a real publishing media products did improve from a design viewpoint. Maps were

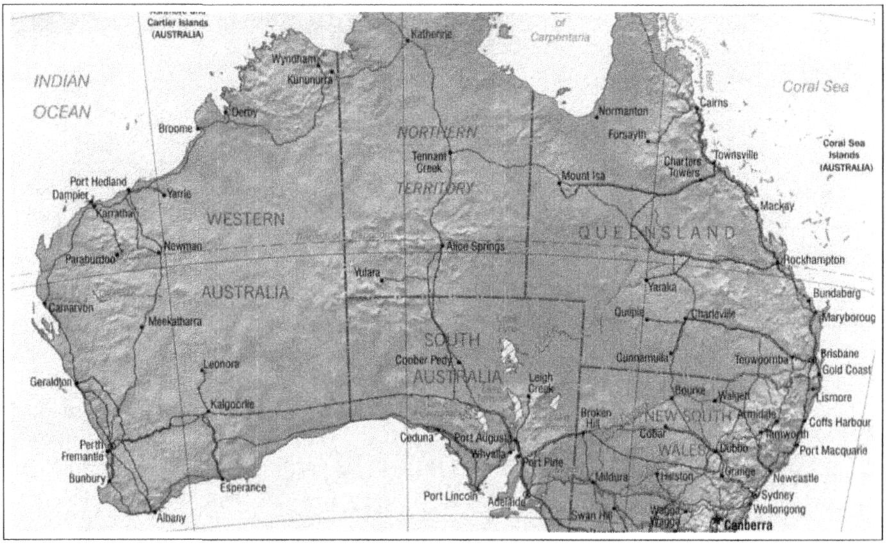

Fig. 2.3. Access interface and map of Australia from the PCL map collection.
Source: http://www.lib.utexas.edu/maps/australia/australia_rel_1999.jpg

designed to 'work' with the restrictions imposed by the communications medium. This can be seen to parallel somewhat the printer-cartographer relationship that evolved with printed maps. Map design and production methods were altered to fall into place with Web publishing. Most recently maps have been published on the Web by producer-users using a process called 'mash-ups' with Web 2.0 and Social Software. Web 2.0 is the use of the Web by individuals and groups of individuals to provide and share information, including geographical information. It provides a new model for collaborating and publishing.

But – the maps are *still* conventional mapping products, from a design perspective. In many cases the innovative use of social software and the Web has not produced innovative products, just conventional maps. And, in many cases these maps are naive design outputs. Even with the innovation of contemporary media, in most cases, conventional maps are being produced. Science and Technology has been harnessed to produce maps quickly, efficiently and collaboratively. Good design has been applied to mapping, but Art has been overlooked.

Maps show one 'view' of geography – that composed and provided by cartographers. However, according to Peters (2007), a postmodern account of geography would not consider any one representation of a place as any more valid than any other representation. So, are maps and map-related artifacts produced by cartography enough? The following section of this chapter explores where Art has been used to communicate about geography. The section is unashamedly pictorial rather than wordy, as the figures provided illustrate the richness of art produced to depict place.

2.5 Art and Representing Geography

2.5.1 Words

The power of the narrative can be used to paint 'word pictures' of the world and give the user/viewer a sense of geographical place. Authors use words to put the reader into a "time cell" of reality. For geography, stories can just provide statements of facts, where no embellishment is required and the user only wants to know 'the facts'. These facts can be stand-alone, or supported by 'on-line' experts who are able to give expert opinions on the geographical space being explored. It may be a narrative, where a documentary-type video, supported by a comprehensive, and interactive, narrative can 'walk' a user through 'unknown territory'. Users may construct their own story, or be 'talked' through an area, where they construct a story using programme support materials and aural navigation aids. Finally they may decide that they wish to experience a landscape by investigating a 'literate landscape' by being told a story.

2.5.2 Paintings

Artists have produced wonderful depictions of geography. For example, Tyrolean landscape artist HC Berann (Troyer 2007) provides beautiful paintings of alpine terrain that allowed the grandeur of such landscapes to be appreciated. He did not produce maps, but his paintings of alpine landscapes provide rich interpretations of geography.

As well, there are numerous contemporary landscape artists who, according to Peters (2007, p.1) "... represent an infinite subject in a finite space". One of the contemporary artists assessed by Peters (op cit.) is Benjamin Edwards, who is exploring the "architecture of suburbia" (op cit., p. 2), developing pieces that depict landscapes that are created by the American Interstate highway. One of his paintings, Immersion, which is his depiction of part of New York, the result of the ultimate American consumer environment. This is shown in *Figure 2.4.*

2.5.3 Installations

Installations provide 'other' interpretations of places and space. They use various means and techniques to achieve this. For example, *URBAN LEGENDS: The City in Maps* was an exhibition conducted by CITY|SPACE, "a cultural organization dedicated to exploring the built environment through events and exhibitions in a wide range of disciplines, including design, visual art, cultural landscape research,

Fig. 2.4. Benjamin Edwards – Immersion, 2004 oil on canvas. Source: http://mocoloco.com/boston/archives/edwards_immersion_dec_05.jpg

and film (www.city-space.org). This was a juried exhibition that took place over two weeks in the summer of 2004 at the Oaklandish Gallery in, Oakland, California.

From the site: http://www.city-space.org/Pages/past3_legends.htm:

"Urban Legends: The City in Maps," an upcoming exhibition of maps in all their forms and meanings. The show will explore the idiosyncratic ways in which we seek to understand the world around us, and in particular the urban environment, with a visual record of our paths and patterns, both literal and figurative.

For this exhibition, we seek submissions of finished original work in all media that address the urban landscape through maps. We encourage the broadest possible interpretation of this theme in a range of media and disciplines. Artists, scientists, geographers, planners, designers, activists, or anyone interested in the theme may apply. Works may be two- or three-dimensional. Time-based media (film, video, digital) may be submitted, but all work must be suitable for display in a gallery setting".

Another installation is Worldprocessor (http://worldprocessor.com), described as "An attempt to do justice to the term 'political' and 'geo-political' globe. ... Trying to tell the lie (of abstraction and visualization) that tells the truth." There are more than 200 different globes that can be viewed on the Web site. A selection of 36 globes was shown in 1996 as part of the exhibition "Global Change" in the Museum of Natural History in Madrid, Spain. Detail of a globe visualising international data availability can be seen at http://worldprocessor.com/index_vis.htm.

Pavement – An Installation at Cartography 2000 by John Wood and Paul Harrison

'Pavement' was installed at the Worcester City Art Gallery and Museum in May 2000. It consisted of 24 TV monitors set into a false gallery floor, each playing a different sequence of human feet. The concept of 'flood' was alluded to by showing rising water levels on the monitors, where one of the artists was engulfed by water. (http://www.fvumbrella.com/projectdetail/venue/W/27/Cartography)

The ATLAS MAPPING exhibition (1998) (http://www.kunsthaus-bregenz.at/ehtml/aus_atlasmap.htm) was a co-production with the O.K Centre for Contemporary Art, Linz. It presented an artists' view of cartography. The artists illustrate the world from a different perspective. *Figure 2.14* shows one of the illustrations from the exhibition catalogue.

2.5.4 Human Movement

Burningman is a one-week event in Black Rock Desert, held each year, and attended by up to 25,000 people. It invites artists and attendees to participate in events that include installations and dance (Steenson 200?). *Burningman* (2000) is a seven day ritual focussed on the 7 chakras of the body. Each day the Body Cartography dance company performed movements that were related to the space in which dancers performed. Also, each day was a different 'colour' day. Their 'white day' performance in 2000 is available at http://www.moonski.net/burningman/burningman/burning05.html

About their work the Body Cartography dance company (2007) notes:

"We investigate the bodies' relationship to the physical, architectural, climatic, technological and social landscapes that we inhabit in urban/wilderness and private/public contexts is the material of our work.

We activate space and challenge social and perceptual limitations of physical freedom.

We engage and provoke audiences in diverse contexts through collaborative processes with independent artists, and national and international companies and communities. Our ventures include dance performances, installation works, film/video projects and educational workshops."

As well as these visual interpretations of geography, sound, touch and smell has also been explored.

2.5.5 Sound and 'Soundscapes'

Soundscapes are described as "the overall sonic environment of an area, from a room to a region" (Porteous and Mastin 1985, p. 169). The theory of using soundscapes is generally attributed to Granö (1927). His pioneering work produced an agricultural soundscape, which illustrated cartographic representations with acoustic sensations of human activity, birdsong and grazing cattle on the Finnish island of Valosaari (Porteous and Mastin 1985).

Work was undertaken by the Spatial Information Architecture Laboratory (SIAL) at RMIT, Australia (Pettit et al. 2003). Sounds of the Central Business District of Melbourne, Australia were recorded – at each street intersection. Then a soundscape of the city was built using interactive media.

2.5.6 Touch

Tactual or tactile mapping has been produced to assist blind and visually impaired people better understand geography. The tactile relief map of the City of Basel, Switzerland *(Figure 2.5)* is a typical example.

Fig. 2.5. Tactile map of the City of Basel, Switzerland. Photograph: William Cartwright

2.5.7 Smell

In 2003 the University of Minnesota Design Institute undertook a project called the Twin Cities Odorama: A Smell Map of Minneodorous and Scent Paul. They described this as an "Olfactory experiences emanating from the intersections of people, enterprise and physical environments" (University of Minnesota Design Institute 2007a). This was undertaken under a larger 'umbrella' project – the Twin Cities Knowledge Maps, part of the Design Institute's Twin Cities Design Celebration 2003 (University of Minnesota Design Institute 2007b). The product was a map/poster illustrated smells and odours in the twin cities of Minneapolous and St. Paul. (http://design.umn.edu/go/person/SMELLMAP)

2.6 The Case for Maps

It has been argued by geographer/cartographers like Castner (1981) that the map model works best, that users have to appreciate the 'grammar' of cartography in order to fully understand the 'language' of maps and how it depicts geography (including the associated 'lies' that maps have to tell to illustrate the 'truth' about what the map reader needs to see on the map in order to have the best 'view' of reality. Using the defined geographical 'picture' that is used in the corridor provided by the tour or itinerary only a small view of what reality really is provided and therefore a true appreciation of what constitutes the 'real world', and where the user 'fits' into that world, cannot be had.

2.7 The Case for 'More than Just Maps'

Calvino's "Invisible Cities" (1979) gives some insight into what an articulated map reader/interpreter may wish to read into media that shows interpretations of geography:

"*... relationships between the measurements of its space and the events of its past: the height of a lamp-post and the distance from the ground of a hanged usurper's swaying feet; the line strung from the lamp-post to the railing opposite and the festoons that decorate the course of the queen's nuptial procession; the height of that railing and the leap of the adulterer who climbed over it at dawn; the tilt of a guttering cat's progress along it as he slips into the same window; the firing range of a gunboat which has suddenly appeared beyond the cape and the bomb that destroys the guttering; the rips in the fishnet and the three old men seated on the dock mending nets and telling each other for the hundredth time the storey of the gunboat of the usurper, who some say was the queen's illegitimate son, abandoned in his swaddling clothes there on the dock.*"

Considering that Cartography has been described in terms of science, art and technology, it is perhaps necessary to re-visit the description in the light of the application of computer technologies. Is cartography any different when delivered using computer-driven devices? Does cartography need to be re-defined because of the revolution that has taken place with both the way in which information is now communicated and the type of information that can be transferred, almost instantaneously, globally?

An argument could be put that cartographers become involved in the elements of cartography that they have both mastered (either academically or technically, or both) and that they also enjoy doing. Personal satisfaction in producing an elegant and aesthetically-pleasing design or mastering some scientific problem – both resulting in an as-near perfect a solution that is possible – can be a major part of what provides cartographers with motivational input that encourages further refinement of skills and better mastering of particular scientific problem-solving strategies that are unique to cartography. From the producer/consumer perspective, the need to produce a map that would serve a purpose, as well as mastering a technological skill, provides the motivation to complete a product that works (for them), but they could be uninterested in producing maps for anyone but themselves).

In the current situation, where maps are no longer the focus for every interpretation of geography, should cartographers view what they do differently, and do consumers of cartographic products influence the art/science/technology balance? Seeing that contemporary cartography could be seen to have as much to do with making a movie as producing a scientific document (although the scientific integrity of all cartographic products is as important now as ever (perhaps more important because of the 'casual' appearance given by the immediate facade of contemporary mapping, before the 'shell of innocent art' is broken by further exploration)) the art components need to be considered as equal partners to the scientific counterparts. Science and technology may only be something that 'gets in the way' of best depicting geography.

2.8 Conclusions

Understanding science and harnessing technology is important, but the partnership between art, science and technology, and their contributions to the discipline, are as important. In my opinion art provides the 'public face' of cartography (and the cartographer's passion when designing particular products perhaps the soul) and science complements this by ensuring that what is presented is scientifically correct, and what could be called 'scientifically elegant' as well. Science and technology, it is argued, is not the only way to address geographical depiction. It is most necessary to ensure that the designed interpretation product can be realised. However, the resulting artefact, cannot always be the result of only science or technology interpretations of geography.

References

The Body Cartography Project (2007) The BodyCartography Press Kit, http://www.bodycartography.org/news/sched_item.php4?sched_id=27. Web page accessed 11 March2007

Calvino I (1979) Invisible Cities, London: Picador

Castner HW (1981) Might there be a Suzuki method in cartographic education? Cartographica, vol. 18, no. 1, pp. 59–67

Granö JG (1929) Reine geographie, Acta Geographica, 2:1–202

Lauer BJ (1991) Mapping Information on CD-ROM, Technical papers of the 1991 ACSM-ASPRS Annual Convention. Baltimore: ACSM-ASPRS, vol. 2, pp. 187–193

Peters K (2007) The Contemporary Landscape in Art, paper presented at AAG2007, San Francisco: American Association of Geographers

Pettit C, Cartwright WE, More G, Burry M (2004) Synthesizing Visual and Acoustic City information for Understanding and Navigation, proceedings of GeoCart2003 National Cartographic Conference, Taupo, New Zealand: New Zealand Cartographic Society, pp. 97–105

Porteous JD, Mastin JF (1985) Soundscape, Journal of Architectural and Planning Research, vol. 2, no. 3, September 1985, pp. 169–186

Rystedt B (1987) Compact Disks for Distribution of Maps and Other Geographic Information, proceedings 13th ICC, Morelia, Mexico: ICA, vol. IV, pp. 479–484

Sargent W (1994) Road Warrior, Multimedia and CD-ROM Now, September, pp. 56–60

Steenson M (200?) What is Burningman?, http://www.burningman.com/whatisburningman/about_burningman/experience.html. Web page accessed 11 March 2007

Troyer M (200?) The world of H.C. Berann, http://www.berann.com/panorama/index.html#national. Web site accessed May 7, 2007

University of Minnesota Design Institute (2007) Twin Cities Odorama: A Smell Map of Minneodorous and Scent Paul, http://design.umn.edu/go/person/SMELLMAP. Web page accessed 11 March2007

University of Minnesota Design Institute (2007) Twin Cities Knowledge Maps, http://design.umn.edu/go/project/TCDC03.2.KMap. Web page accessed 11 March2007

University of Utrecht Department of Information and Computing Sciences (1998) Nordic FAQ – 1 of 7 – INTRODUCTION. http://www.cs.uu.nl/wais/html/na-dir/nordic-faq/part1_INTRODUCTION.html. Web page accessed 12 March 2007

3 Rejecting Illusionism: Transforming Space into Maps and into Art

David Fairbairn

School of Civil Engineering and Geosciences, Newcastle University
United Kingdom

Abstract

This chapter addresses the interaction and balance between two types of human activity: the general creative impulse, and current mapping activity based on a scientific framework. It is pre-supposed that there is some parallel distinction between the former, represented by art, and the latter, represented by contemporary cartography. The intention is to examine the nature of the distinction, and the boundaries or transition zones between them. An exploration is made of the rules and norms which define an object as being on one or other side of such a boundary – being either cartographic or artistic.

3.1 Introduction: Cartography and Art are Different

The link between representational art and cartography has been longstanding, since the first attempts by humans to represent space for decorative or practical purposes. Creativity and artistic impulses have informed map products for centuries. The dominant paradigm of cartographic expression from the eighteenth century onward, however, was scientific, and twentieth century technology in particular divorced the creative instincts of cartographers from their utilitarian output. The intention of this chapter is to explore some of the enduring common ground between cartography and art, yet to suggest that, today, these two activities are significantly different, and may even be mutually exclusive. The starting point for this differentiation is to examine the basic nature of each activity – cartographic and artistic. In both cases we can detect similarities in the means by which they act on 'raw material' (defined in specific ways), transforming it in different ways and for different purposes. These transformations are investigated and the dichotomous nature of cartography and art is reflected in the differences in such transformations.

The transformations inherent in artistic activity, involve the creation of an artefact resulting from human interpretation of senses, feelings, emotions, an understanding of phenomena, views of reality and other experiential elements. Cartography has also been studied from a transformational perspective, involving the compilation of spatial data and its subsequent modification into real, or sometimes virtual, maps. An attempt is made here to identify the map artefacts which result from cartographic transformations, and to distinguish them from artistic output: the result is to more clearly define the boundaries of cartography with respect to art and to conclude that they are separate human activities.

3.2 Cartographic Transformations

When photography was being developed, in the late nineteenth century, as a technique for portraying scenes, there were commentators who predicted the imminent demise of representational art (Marien 1997). Most leading and perceptive critics, however, did not share this view and, of course, it did not happen, although in many respects it did affect the manner in which art developed in the twentieth century (Harrison et al. 1998). What is notable, however, is that a similar response regarding the future of mapping was *not* evident when aerial photography developed during the first half of the twentieth century. Although aerial photography could be used for a number of purposes connected with mapping – including reconnaissance, intelligence gathering and, with the development of rigorous stereo-photographic based methods, spatial data collection – it has never (at least until now) been regarded as capable of *superseding* mapping as a means of representing either topographic or socio-economic space. There are some contemporary methods of spatial data delivery, notably web services such as Google Earth, Virtual Earth and Pictometry systems, which *do* yield aerial photographic images at the expense of map representations. However, the attraction and utility of such methods is actually in the novelty of the interface, rather than the nature of the data displayed. Google Earth's aim is to organise information geographically, not to represent geographical information. The representation of spatial data is a side-effect of the main thrust of systems such as these, and the map overlay (however much we may disapprove of its cartography) has as much value as a metaphor and tool for such organisational purposes as the aerial photography. In fact, cartography has always rejected the illusionism of photography: there has never been an aim in cartography of creating a representation of reality which tricks the eye into thinking that reality is being directly observed, or revealing to the eye a hyper-realistic image reflecting reality in each aspect of its myriad complexity.

What cartography does is transformational: the activities of observing or measuring reality, are followed by a range of general abstractions, modifications, interpretations, generalisations, symbolisations and possibly analyses (see Chrisman 1999 for a detailed examination of analytical transformations of spatial data), along with specific transformations such as changing dimensionality, applying scaling and altering geometry. Such transformations are wide ranging and impact on every cartographic process and object.

It was Waldo Tobler's early work (Tobler 1979), concentrating mainly on the transformations inherent in map projections and applying coordinate systems, that synthesized the geometric and analytical possibilities of cartographic transformations. In an attempt to develop a wider and more applicable definition of the word 'map', Moellering (2007) expanded these ideas in the context of the possible transformations in the form of map. This is based on a classification of 'real' and 'virtual' maps, the latter divided into visible, and/or hard-copy map products. The cartographic transformations examined in this paper, however, are transformations of data, rather than Tobler's transformations of geometry or map type. In a sense they acknowledge, as did Chrisman's work, the 'surface' and 'deep' structure classification of Nyerges, which suggests that there is a distinction, and possible transformation, between spatial data collected in reality (surface structure) and the meaning, semantics, relationships and interactions inherent in, applied to or extracted from that spatial data (deep structure).

3.2.1 Limitations of Cartographic Transformations

If cartography was to strive for illusionism, it would create and disseminate representations of reality which would possess the visual and semantic properties of that reality (however defined). Thus, it would attempt to incorporate a realistic viewpoint, a visual matching with real-world features or processes, a feeling of immersion in the scene through depth cues and environmental modifiers, and a sense of familiarity to the viewer's perception. This does not happen in cartography: what does is that, as a result of the transformations mentioned above, there is an adjustment of the spatial data from reality to its representation on a map product – a real map (or indeed a spatial data base, defined by Moellering as a virtual map). What is clearly evident in cartography is that the visual and semantic properties of reality are transformed: the 'viewpoint' is not realistic (it is orthographic and geometrical transformations are applied to effect it), the real-world features are generalised and symbolised, there is no attempt to give a feeling of immersion, and the viewer's perception is challenged by the symbolised representation which results from the cartographer's creativity.

To me, therefore, it is clear that cartography is primarily a set of transformations of spatial data, and the cartographic process is consequently *data-driven*. This is clearly evident, in a Toblerian sense, in the geometric transformations which occur when we capture data in the field or we geocode a real world location into a spatial data base. Rules exist to ensure that real space can be transformed into coordinate space at a reduced scale, and these rules act on data from the real space. Other transformations in cartography are similarly dependent on invariant rules acting on variable data: contemporary research into cartographic generalisation, clearly a transformation of spatial data, attempts to develop generic guidelines to achieve objective and data-specific results. Even symbolisation, a specific transformation which is also inherent in cartography, and one in which use of creativity might be expected to overcome rigid transformation rules, could be regarded as data-driven. Bertin (1981), for example, presents an opinion that particular data types possess properties and attributes, which the skilled cartographer must ascertain in order to present an optimal symbolisation of that data. "Bertin propounds a single formula for matching graphic constructions to particular information types" (MacEachren 1995, p. 271). It should be noted that this singular and rigid view held by Bertin (that it is the nature of the spatial data being handled which dictates the transformation of symbolisation with one unique solution) is not shared by all graphic designers: Tufte (1983), for example, celebrates the flexibility of alternative symbolisations and the creative possibilities of design, although he too suggests that the most important determinant of the process of symbolisation is the range of possibilities afforded by the inherent characteristics of the spatial data itself.

It is clear that the transformations inherent in cartography are driven by the location and nature of spatial data: despite this, the end product of the cartographer does not attempt to represent that data by replicating its visual appearance directly. Illusionism is not considered to be an option for cartography. Landscape mapping is not landscape photography; nor is it landscape painting.

3.3 Transformations in Art, and the Role of Emotion

Art, unlike cartography is not data-driven: here is the basis of the contrast between them. However, art clearly involves transformation – using materials and skill to render an image or object which is the result of an individual view of the world (almost all art is created by individuals, whilst it can be noted that much mapping is created by groups). A romantic theory of art would suggest art is an expression of this individual view and of individual feelings and that it is, therefore, emotion-driven. Writing in 1962, the art historian Sir Ernst Gombrich suggested that the "romantic idea that art is the language of emotions has a long and complex history … and may be more firmly entrenched to-day than in earlier periods" (Gombrich 1978, p. 56). Although he considers that such an idea is narrow and problematic, he does explore its provenance and suggests why it is still viewed positively by contemporary art historians.

The development of an appealing theory of art based on emotions has been clear throughout the development of human artistic endeavour. Firstly, there is the acknowledgement that art can reflect symptoms of an inner-state (or emotion). Such symptoms occur in everyday life – smiling, frowning, wailing – and "there is no strict dividing line between these outlets of emotion in life and in the various arts" (Gombrich 1996, p. 141). Further, we can see that emotions can also be aroused *by* art, external to the human being. This is almost the reverse of the first interaction between emotions and art (which shows that emotion drives art): here, the signals of art can drive emotion. Finally, Gombrich suggests that signs and symbols can be used in art to describe emotions which are inherent in the *subject* of the art: here the emotion is not internal, emanating from the artist, or external, triggered in the recipient, but is a descriptive property of the object d'art, e.g. the feeling of a character in a Chekhov play, or the reaction of a figure in a Munch painting.

It is the second of these 'functions' of art which has been the most significant from the viewpoint of art history. Throughout the development of art, the artist has sought to control the viewer or recipient of art: the signal of the artwork is intended to guide or lead the audience to a particular emotion – empathy, sympathy, disgust, terror, nationalism, revenge, placidity etc.

From a European perspective, it was during Renaissance times that the emphasis changed and art started to use symbols, and it was the emotions of the *subject matter* that became paramount. Religious paintings of the fifteenth century, for example, depicted the extreme emotions of biblical characters, emotions not necessarily felt as symptoms by the artist himself (although in many cases an associated intention of such imagery was also to trigger emotion in the audience through signals in the artwork).

Towards the end of the eighteenth century, according to Gombrich, the situation changed once more, and artists were expected to represent their *own* experiences, there was a "need for sincerity, for genuine emotion" … artistic expression, embodied in a work of art, was no longer a signal or a symbol, but "a symptom of emotions; for the first time, the critics wanted to know what the artist himself felt; they wanted him to bare his own heart" (Gombrich, in Woodfield 1996, p. 147).

Thus, the individual artist becomes the centre of interest: the emotions of the artist are primary. A work of art produce without feeling "is really a fake … This is the theory of artistic expression as a communication of emotions" (ibid.). The impact of emotion on art has been highlighted both by practitioners (Pissaro's dictum was "paint what you observe *and feel*") and philosophers of art ("the activity of art is based on the fact that a man receiving, through his sense of hearing or sight, another man's expression of feeling, is capable of experiencing the emotion which moved the man who expressed it" (Tolstoy 1995)). This romantic theory of art has been the major paradigm for interpreting and reflecting on artistic endeavour since the turn of the nineteenth century. It has led to a widespread, twentieth century, belief that an artist has to suffer for his art (certainly a Renaissance artist such as Michelangelo, would have regarded himself more as businessman or craftsman, rather than a traumatised genius, as someone such as Van Gogh might regard himself). The contemporary view that the artist must express the symptoms of their emotions in producing their artwork is strongly held: it is the approach which I wish to use to distinguish artistic expression from cartographic expression. I repeat the conclusion, that art is the antithesis of the data-driven transformations of cartography: it involves emotion-driven transformation.

3.4 Contrasting Technical Natures of Art and Cartography

It must be admitted that such a romantic theory of art is not universally accepted. Indeed, as has been mentioned, from the beginnings of human history perhaps the major aim of artistic expressions was to improve technical performance and achieve representations which were as realistic as possible. Such a mimetic approach attempts to be emotion-free in the sense that interpretation and feelings are not allowed to intrude on the artistic goals of reflecting nature, implied immersion in the scene, and representing familiarity.

The technical theory of art examines, in particular, the way in which art can be 'imitational'. Such a theory thus confuses art with craft (skill at producing an artefact). In fact, the ability to imitate with fidelity, and without emotion, is regarded by those ignorant of art as being the supreme aim of artistic expression. For the amateur critic, a 'painting has to look like it is in real-life', and value judgements are made on whether a painting is 'good' or 'bad' by a quantification of its verisimilitude. This is clearly an inappropriate approach to either receiving or criticising contemporary art.

But could it be used in cartography? It can be and is – we can measure the accuracy (of both positions and attributes) of a map, we can test the user's response to a map, we can determine how well the map imitates reality. Criticism of cartography addresses the skill with which spatial data has been transformed into a map representation. If the measure of success of a map is the skill with which it has been produced, then one must argue that cartography is a craft, and cannot be equated to art.

A further distinction made between craft and art (Collingwood 1937) is that the former relies on a planning stage, followed by an execution stage, to reveal the artefact produced; whilst in art the planning (inspiration, emotion) and the execution are often simultaneous. An examination of the cartographic process reveals that it, too, shares the former, 'flowline' approach: traditional cartography clearly falls into the 'craft' category. (An interesting side issue is whether some contemporary forms of cartography, which can produce dynamic, user-controlled, adaptive, temporary displays still include a planning stage).

3.5 Considering Representation in Art and Cartography

So what is the role of representation? Almost all maps attempt to be representations, but according to our view of art as a conduit for the emotions, there is a limit to how much a work of art can be a representation. "A representation may be a work of art; but what makes it a representation is one thing, what makes it a work of art is another" (Collingwood 1937, p. 43). In fact, what makes art a representation is not the literal imitation of an original scene or likeness of a portrait subject, but whether the "feeling evoked by the artefact resembles the feeling evoked by the original (I call this *emotional representation*)" (op.cit., p. 53, [my emphasis]). Here, Collingwood confirms the emotional properties of art: representations can (and perhaps should, to qualify as artistic) reveal emotion. The pursuit of imitation, which may reach its zenith in some forms of photography, is different to the rendering of a representation (no matter how realistic the latter ends up being).

What can we make, therefore of attempts at illusionism in art, attempts which aim to persuade the viewer that what is portrayed is actually real, or that what is depicted is seen through a window or from within the scene itself? *Trompe l'oeil* paintings are hyper-realistic depictions of what (to the eye) appear to be real objects, whilst *sotto in su* ceiling frescoes reveal false architectural constructs or spaces open to the sky. A natural development of such images is the eighteenth century *vedute* style, initiated earlier by Dutch painters such as Vermeer, but further enhanced by Venetians such as Canaletto. The paintings of such artists attempted to achieve the highest standards of topographical accuracy, and they are appealing in their attempts to be faithful to a view of reality. In many cases, mechanical means have been used to achieve the effect of reality. There is speculation that Vermeer used a camera obscura, with a lens, to create his epic View of Delft (Steadman 2001), whilst Canaletto certainly employed perspective aids (although he did also apply some 'artistic licence' in his paintings).

We have established that art requires emotion for its full manifestation: what is clear about illusionist painting, however, is that there appears to be little emotional content in such artworks. The geometric perspective, the usual paucity of human beings, the fine brushwork, the inclusion of weather and environmental fidelity: all these lead to the view that such paintings are, in fact, closer to architectural drawings and, indeed, to photography, than they are to art. Their aim of achieving the highest levels of realism show that *vedute,* whilst representing information of the landscape or the cityscape, are not transforming emotion (they are, in fact, transforming spatial data). They are often undeniably beautiful, but it could be argued that they are illusions, not art.

3.6 Contrasting the Utility of Art and Cartography

One further distinction between cartography and art can be identified at a prosaic level. A major purpose of cartography is to create an artefact which can be used. In addition to being data-driven, cartography could therefore be regarded as task-driven. Art has some confusion as to its 'task', and it seems impossible to develop a coherent vision of what art is 'for'. Three very different, randomly selected, views of the purpose of art highlight this uncertainty: "the supreme task of art in our epoch is to take part actively and consciously in the preparation of the revolution" (Trotsky et al. 1937); "to try to recreate what so effortlessly, so gracefully, occurs in nature is the Sisyphean task of art" (Chase 2007); "Is the task of art the creation of easily identifiable objects, or does art aim at expressing the ineffable and that which is difficult to grasp, that which brings to light the deepest mystery of being?" (Jaroszynski 2002). By contrast, cartography has a specific and identifiable task: "The basic task of cartography lies in efficient and impeccable transfer of spatial information via cartographic models of geospace" (Lechthaler 2006, p. 24). The transformations which effect such transfer form the accepted core of cartographic activity; the transformations of art are much less clearly defined.

In addition to the contrasting nature of the purposes of cartography and art, there is contrast in the tasks to which the artefacts can be applied. Because we have concentrated on the difference between emotion-driven art and data-driven cartography we can conclude that a major purpose of works of art is to arouse our emotion, whilst the products of cartography are intended to represent and communicate spatial data, and allow for human activities based on such data to be undertaken more easily.

3.7 Can Art and Cartography Converge?

The implication of this chapter is that there are significant and clearly defined differences between cartography and art – differences that may transcend any commonality in the medium of delivery (such as graphical marks on a tangible medium); differences which can be reflected in their tasks and purpose, but which may be best reflected in the varying transformations which occur during cartographic and artistic endeavour.

But is it yet possible to determine whether there are any opportunities for cartography and art to mutually assist in each other's transformations?

Performance art is particularly related to space and to spatial data. Activity which can be defined as 'artistic' can and does take place with the assistance of cartographic products, and artefacts which have some cartographic resonance can be described as 'artistic'. The British artist Richard Long has produced works of a topographical nature (they present space and locations as their subject matter), sourced mainly from his emotions revealed during his interaction with the natural world. In most cases, his 'land art' is photographic or textual in presentation *(Figure 3.1)*: it is the transformation of the emotions of one artist. His early work used Ordnance Survey mapping, but the intention is not to convey any transformation of space: in Long's work, space is a given and his artistic activity takes place within its constraints. What is evident is almost a 'cut-and-paste' approach *(Figure 3.2)* when cartographic artefacts are incorporated into his work. This is shared by another, extreme, example of 'cut-and-paste' in the guise of art – the work of Layla Curtis *(Figure 3.3)*.

WIND STONES

LONG POINTED STONES

SCATTERED ALONG A 15 DAY WALK IN LAPPLAND
207 STONES TURNED TO POINT INTO THE WIND

1985

Fig. 3.1. Richard Long, *Wind Stones*, 1985

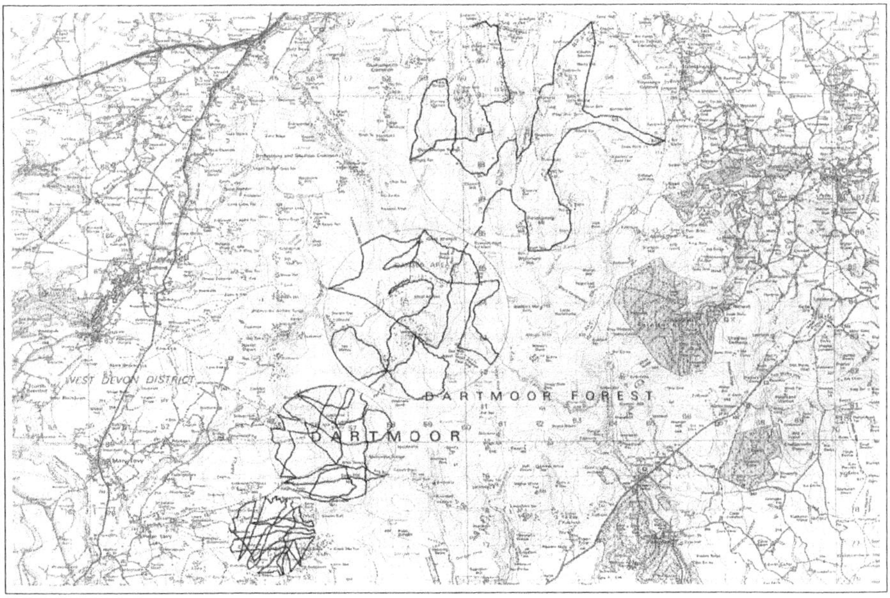

Fig. 3.2. Richard Long, *Dartmoor Walks*, 1972

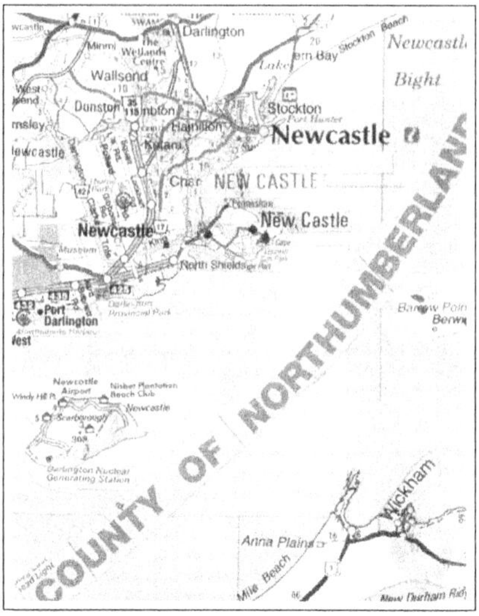

Fig. 3.3. Layla Curtis, *NewcastleGateshead,* 2005

Although created using cartographic representations, such artefacts are 'not really cartographic' – they are not using cartographic transformations. Similarly we can look at examples of paintings which are 'not really art'. It has already been suggested that the type of painting exemplified by *vedute* falls into this category. Such representations employ transformations which are almost cartographic, and because of the precision of their geometric fidelity, they could assist in the same way as photography, in creating maps. Under the valid assumption that certain constraints have been rendered effectively by the artist (for example, that building edges are rectilinear, and that vanishing points are visible using recreated perspective), it would be possible to take paintings, such as those by the Venetian artist Bellotto, a nephew of Canaletto, and use them to recreate the dimensions and appearance of buildings, based on quantifiable measurements. In such a way, his paintings of the Baroque era were used to assist in the reconstruction of the centre of Warsaw in the 1940s, and his views of Dresden reveal similar dimensional fidelity which can be used to contribute to a spatial database of the historic cityscape *(Figure 3.4).* Certainly, the utility of such representations for this purpose is considerably higher than 'true', emotion-driven art, such as that of the Expressionist school *(Figure 3.5).*

These examples show that there is some common ground which can be shared: by examining products that are 'not really art' and those that are 'not really cartography', we can see that there is some flavour of each in the other. My feeling is that they show the diverse and separate nature of art and cartography. Whilst not discounting the possible creativity which cartography can incorporate and express, an acceptance of varying transformational approaches will deny the linkage between these two elemental human activities.

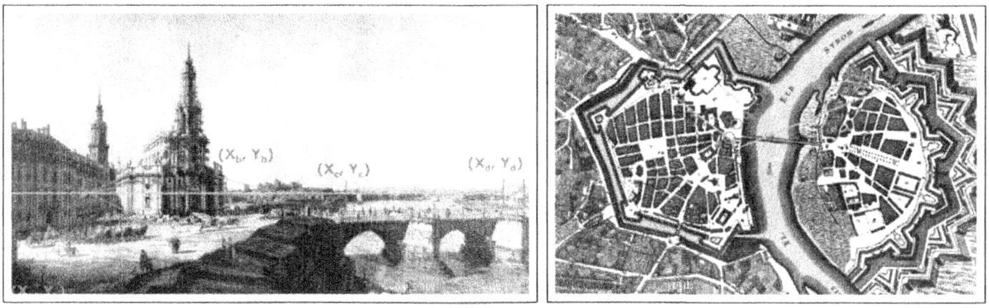

Fig. 3.4. Bernardo Bellotto, *The Hofkirche in Dresden with the Augustian palace and bridge*, 1748; Matthaus Seutter, *Stadtplan von Dresden*, 1755

Fig. 3.5. Oskar Kokoschka, *Dresden: Augustus Bridge with Steamer*, 1923

3.8 Conclusions

Just as the geometrically rigorous *vedute* and the architectural pictures mentioned earlier in this chapter can (along with art works such as botanical drawings) exhibit considerable beauty, so maps can be designed to reflect the data which they portray using a transformation which is beautiful. Such beauty can be achieved through the application of creativity during the transformation processes. However, although the transformations involved in cartography may possibly share some of these creative impulses with artistic transformations, it is argued here that these respective transformations do differ substantially in many other important respects.

Although creativity can be involved in cartography, this does not automatically require the input of emotion. So, does the potential polarisation explored in this chapter (cartography is data driven, art is emotion driven) mean that there is no room at all for emotion in cartography? It would be unfortunate if this were the only conclusion to draw.

Some contemporary discussion of cartographic transformations has drawn attention to the possible input of feelings and subjectivity. It has been argued by critical cartographers such as Harley and Wood that cartographic transformations cannot be completely mechanistic: there will always be a conscious or sub-conscious distortion of reality inherent in any map, due to the prejudices and mind-set of the cartographer, the social milieu within which the map is created, even the client-led agenda which commissions the map. Such distortions could well be sourced from the incorporation of the emotions of the mapmaker, the customer, or society in approaching the tasks of cartography. Such 'distortions' would usually be regarded as a negative factor, in their contribution to the mapping process.

The conclusion of this chapter is not definite. However, the simplistic polarisation of art and cartography, each with its different transformations, purposes, and 'raw material', may be an accurate reflection of contemporary practice and intention in both fields. Whilst creativity is a vital part of cartography, it can be difficult to find occasions by which emotion can be readily incorporated in a positive and considered manner.

References

Bertin J (1981) Graphics and graphic information processing, de Gruyter

Chase A (2007) The Sisyphean task of art, accessed at http://www.thefreelibrary.com/The+Sisyphean+task+of+art.-a0170456714

Chrisman N (1999) A transformational approach to GIS operations, Int. Jnl. of Geographical Information Science, 1999, 13 (7), 617–637

Collingwood R (1937) The Principles of Art, Clarendon Press

Gombrich E (1978) Meditations on a hobby horse: and other essays on the theory of art, Phaidon

Harrison C, Wood P, Gaiger J (eds) (1998) Art in Theory, 1815–1900: An anthology of changing ideas, Blackwell

Jaroszynski P (2002) Metaphysics and Art, Peter Lang Publishers

Lechthaler M (2006) Cartography (and Geoinformation) in Education in Austria, Cartography and Geoinformation, 5 (5) 24–39

MacEachren A (1995) How Maps Work, Guilford Press

Marien M (1997) Photography and its Critics: a cultural history 1839–1900, Cambridge University Press

Moellering H (2007) Expanding the ICA conceptual definition of a map, Proceedings of the 23rd International Cartographic Conference, Moscow

Steadman P (2001) Vermeer's Camera: uncovering the truth behind the masterpieces, Oxford University Press

Tobler W (1979) A transformational view of cartography, The American Cartographer, 6 (2), 101–106

Tolstoy L (1995) What is Art?, Penguin

Trotsky L, Breton A, Rivera D (1937) Manifesto for an Independent Revolutionary Art, accessed at http://www.marxists.org/subject/art/lit_crit/works/rivera/manifesto.htm

Tufte E (1983) The Visual Display of Quantitative Information, Graphics Press

Woodfield R (ed) (1996) Gombrich on Art and Illusion, Manchester University Press

4 From a Cartographic Glance to Synchronistic Experiences

Felicitas Thun-Hohenstein

Academy of Fine Arts Vienna, Austria

Abstract

My contribution is based on the hypothesis that the "crisis of space", which has been circulating in discussions for some time now, is not necessarily a crisis of perception theory insofar as current theories on the fine arts are still conceptualized in a largely static manner. Art, as opposed to discourse, is still far from having attained the level Merleau-Ponty had hoped for, a level on which the recipient shifts from "onlooker" to "experiencer" and, contrary to analytical animistic thought, thus becomes able to adopt a different, more direct, mature and holistic view of the "manifest world". A critical reading of *L'oeil cartographique de l'art,* a book produced by the French philosopher and cultural theorist Christine Buci-Glucksmann in 1996, provides the basis for verifying this hypothesis. I will take a particularly close look at her text on Robert Smithson's entropic gaze. The theory of a "cartographic gaze" that Buci-Glucksmann develops here is undoubtedly a sophisticated approach that has extensive integrative dynamics and provides an interesting approach also to interpreting developments in art during the second half of the 20th century. This approach also affords an extensive overview of historical developments in modern European visual perception from the Renaissance until today. Through designating the "cartographic gaze" as the central concept in her theory, Buci-Glucksmann introduces a structural method of interpretation, which situates the perception of art at the interface between fixation and motion, standstill and travel. She distinguishes between cartographies in art from a general cartography of art. Hence, questions arise regarding the critique of terms and categories as introduced by Buci-Glucksmann and in relation to their re-evaluation or re-definition under the consideration of a more comprehensive dynamics of perception.

'Cartographic Eye'

In *L'oeil cartographique de l'art* (1996) the French philosopher and cultural theorist Christine Buci-Glucksmann refers to the term 'cartography' as employed by American artist Robert Smithson.[1] The dynamics of Buci-Glucksmann's "cartographic eye" developed in this work is extensively integrative, rendering quite an interesting approach to interpreting developments in art during the second half of the 20th century. At the same time, her method lends itself to an extensive overview of historical developments in modern European visual perception from the Renaissance until today. The cartographic eye, Buci-Glucksmann's key concept, provides an interpretive structure that places the perception of art at the interface of fixation and motion, standstill and travel. She distinguishes between cartographies in art, such as Robert Smithson's, and a general cartography of art.

A way of formulating a response to the question – whether recent theories of Art dealing with perspective/view are still too static or not, would be to conduct a *critical reading* of Buci-Glucksmann, particularly her text on Robert Smithson's entropic view. Such a close reading, however, also brings up the issue of the concepts and categories Buci-Glucksmann introduces, including how she redefines or reassesses them with regard to a more comprehensive dynamics of perception.

There seem to be two decisive factors here. First, the definition of an artwork inherent in the term "cartography" points to its functionality as a tool and brings out the artwork's intellectual and conceptual properties. This inevitably leads to a reworking of the meaning of the art object in terms of releasing it from its characteristic of being a self-contained entity. The autonomy of the art object thus recedes making way for a more holistic notion of the artwork that encompasses its intellectual gesture as well. As the "key" to the work itself, the individual art object functions more as a representation of the fully formed gesture behind it, its intellectual and atmospheric conceptualizations. These conceptualizations do not necessarily have to correspond with the concept of the work itself, yet together they constitute the entirety of the work's presence. The avant-garde strategy at work here is one that no longer assumes a comprehensive form of representation

[1] I gather that Buci-Glucksmann's "cartographic eye" refers to Robert Smithson's use of the term "cartography". In a 1972 interview, Smithson repeatedly refers to this term: "... Well, gradually I recognized an area of abstraction that was really rooted in crystal structure. In fact, I guess the first piece of this sort that I did was in 1964. It's called *the Enantiomorphic Chambers*. And I think that it was the piece that really freed me from all these preoccupations with history; I was dealing with grids and planes and empty surfaces. The crystalline forms suggested mapping. [...] In other words, if we think of an abstract painting, for instance, like Agnes Martin's, there's a certain kind of grid there that looks like a map without any countries on it. So I began to see the grid as a kind of mental construct of physical matter, and my concern for the physical started to grow ..." (Interview with Robert Simthson for the Archives of American Art/Smithsonian Institution/Paul Cummings, 1972. http://www.robertsmithson.com/essays/interviews.htm, accessed 25 December 2007)

for the individual object. Moreover, it is the combination, the work's movement, its change of perspective and the accumulation of diversified views that work in unison to convey the overall message. In fact, as a response to the enigmatization or "encryption" of the artwork brought forth by its respective objects, art developed an avant-garde stance of openness toward the growing need to regulate access to power apparatuses governing the whole of society, part of which was also the 19th century concept of the museum. Modern art always revolved around gestures of freedom and opening spaces.

This brings us to the second tendency laid out in Buci-Glucksmann's cartography concept: the comprehensive and imperative integration of spatiality in thinking about art. This refers to the idea that we can grasp the textuality and meaning of an artwork or the concept of art today only if we are able to identify its key elements within a spatial matrix. Buci-Glucksmann introduces the notion of an Icarian or cartographic eye, a theoretical model fixed to the map and globe – an appropriate metaphor for this analytical process. Today, the only way of determining a location in open space, for example in the ocean that is now beset with a flood of images, objects and virtual conceptual gestures, is to negotiate it according to its spatial coordinates. In order to grasp the locational meaning of an artwork under such circumstances, art recipients must act as navigators in discerning the work's discursive latitude and longitude. The role of art theory and particularly that of art history is to provide the necessary coordinates, the methodological latitudes and longitudes.

From this perspective, the art object is inevitably viewed in a different manner, for it is no longer possible to grasp its content by simply focusing on the artwork itself. The common notion of the artwork as a tool in art theory discourse is not quite extensive enough to come to grips with the current and largely processual, conceptual, holistic concepts of artwork. As Elisabeth von Samsonow asserts, in this day and age humans and artists are still HOMO FABERI"[2] i.e. smiths or toolmakers. Their TOOLS, however, are no longer pencils or drawings, paintbrushes or paintings, chisels or sculptures, because the priority has shifted toward a desire for holism as described in Peter Sloterdijk's vision of "spheres" *(Sphären),* "bubbles" *(Blasen)* and "foam" *(Schäumen),* in which he opens up ways for developing different, less static and more performative *tools*.

But what are these tools? What do they look like? Expanding the horizon of art history, Buci-Glucksmann's cartography provides an alternative to the "artisan" conception of art as a tool. Whereas describing an art object with the term "tool" underscores its static and functional properties, I find that the term "cartography" possesses greater analytical qualities per se due to its insightfulness regarding modes of movement and particularly its expansive spatial component. Maps and globes are indeed tools, too. However, rather than simply indicating a direct impact or a

[2] Elisabeth von Samsonow: *In situ oder in motu?* In: Umzug ins Offene loc. Cit., p. 252

(sculptural) modeling, they delineate the functionality of orientation, travel, motion, velocity, migration and emigration and situation in local spatial structures.

We now move toward a conceptual and discursive alternative that follows suit with critiques of dualistic thought, as established in the work of Florenskij, Schmitz and von Förster in modernity. The holistic position represented in the art of modernity no longer corresponds with the distanced analytical gaze, which is the basis for theories of perspective. More closely aligned with the analytical system of holistic synthesis of the avant-garde, Buci-Glucksmann's cartographic eye allows for a more fundamental critique through incorporating notions of travel and motion. The continually changing view when moving from one place to another, i.e. when in motion through space(s), is not merely a visual phenomenon; for, to an even greater extent it constitutes a real movement of the body perceived holistically both on a subjective and a collective level.

Buci-Glucksmann's theoretical structure is a place in which reality is conditionally assembled and constructively synthesized. Buci-Glucksmann's analysis of radical change in perspective, the dawn of which she identifies as the "world-eye" in the Renaissance, particularly as employed in Dutch painting, provides a methodological construct, a quasi-chronological sequence of allegorical, tautological, entropic and finally virtual approaches. For Buci-Glucksmann, this construction is a field that is not yet finalized, which lends her theory a kind of fluidity and a tendency to self-referential dissolution. Her system is in fact quite flexible with a broad range of interpretive approaches for exploring modern art. In terms of its capacity for addressing recent performative developments in the fine arts, Buci-Glucksmann's theory remains bound to a notion of the cartographic object, either as an mnemotechnical trace or as a representative tool. It therefore lacks the categorization and structuring that provide the spatial and environmental conditions necessary for filling the object with content. In this way, her theory suspends the artwork in its purely object or image state.

Yet, in another vein, her theory relies on the notion that there have been historical phases in which revolutionary changes in perspective have taken place. She identifies such a paradigmatic leap in early modernity, a time when the perspective in Dutch painting radically changed. The view through the window was no longer the basis for defining the perspective, thus setting it apart from Renaissance ideas of perception. The new concept came closer to the early avant-gardist deconstruction of pictorial illusionism that, like the view through the window, also provided another view of reality. The Renaissance notion of the rigid and static view through the window is replaced in Flemish painting with an aerial view that Buci-Glucksmann calls a "world-eye" (*Weltblick*). Buci-Glucksmann's proposes the "Icarian eye" as a teleology of seeing, a panoramic view or illusion of the infinite, which takes the place of the central perspective's landscape that recedes into the distance, the fixed and indeed distanced scientific view that sweeps out the window across the landscape.

The sweeping view implied by the flying motion envisioned disorients the perspectival space, its center and horizon. The static central perspective is thus replaced by motion. What is intriguing here is this model's applicability to interpreting early modern as well as early 20th century avant-gardist strategies and objectives in art.

Indeed, the performance revolution triggered by avant-gardist deconstructionism raised issues that exceeded the scope of radical theories that deal with the perspective, including Buci-Glucksmann's. The salient point is when the meaning of an art object becomes subordinate (or even ceases to play any role), when the art object is assigned to a secondary field of meaning, the artwork's constitutive gestures, its intellectual conception and the discourse available, move to the fore.

Buci-Glucksmann's theory nonetheless entails a holistic strand of thought: she establishes that we are in the midst of a perspectival revolution today, that is similar to the shift in the early modern world from the central perspective to a panoramic or "cartographic" view. Another compelling point in her theory is that the introduction of the dynamics of motion in the "world-eye" (Weltblick) or panoramic view of Flemish painting laid the groundwork for what later became an obsession with thematizing virtual space in art in the 20th century avant-garde. To be able to understand artwork originating in this era, it is necessary to surpass the limitations of "perspective" and to conceive new qualities of perception. Buci-Glucksmann does so by ascribing two ontological regimes to the map or globe as a metaphor for an art object: "presence" and "absence". A map or globe possesses something without claiming to possess it. Buci-Glucksmann appeals to the perceivers to use their prefigured active access to imagine or hallucinate reality. By looking at a map, one can hallucinate the so-called reality or world without possessing it, as these hallucinations are based on simple coordinates and representative patters on the map. This implies that the map entails a sort of mimetic feature; it does not claim mimetic illusionism, but instead it animates viewers to plunge into the depths of their psyche in order to attain a potentially more comprehensive perception. Here, Buci-Glucksmann utilizes Deleuze/Guattari's principle of differentiating between cartography and copy. According to Deleuze/Guattari, the map is the exact opposite of a copy, because in its innermost being a copy always represents a mimetic act whereas a map only represents a cause, not a depiction of itself.

Regarding a way to address the predominant role of performativity in art today, a decisive fundamental factor comes into view when taking a closer look at the discourse on the properties of the visible opposite poles, of the map and the copy (in contrast to discussions on the "original" in avant-garde deconstructivism). Different from the copy, the map stands for or triggers a performative gesture in terms of imagination and visionary phantasmagory.

In contrast to the copy, the map is open and rhizomatic in character. The accumulation of maps and cartographic panoramas is an atlas or archive, a collection of various representations of information. This, in turn, brings up two fundamental parameters in art following the performative turn: ABSENCE and ACCUMULATION.

These parameters can be applied to crucial positions in art during the second half of the 20th century, including performative positions in contemporary fine art. An example for the latter is the meaning-conferring relation between the art object (the image) and the accumulative apparatus or "atlas" in the work of Gerhard Richter. Similar accumulative patters can be found in the work of Smithson, Roth, Beuys and Nitsch.

Robert Smithson's perception experiments greatly influenced Buci-Glucksmann's considerations. Smithson's definition of site and non-site as parameters of absence, which are so essential to performativity today are formative elements for her theory. When taking a look at works that utilize performative principles, such as Bruce Nauman's *Object and Nonobject* or Rachel Whiteread's castings of negative space, the performative gesture can only be understood if read in relation to the reciprocal coordinate system of presence (accumulation) and absence or site and non-site.

Buci-Glucksmann's closing thoughts are formulated in a euphoric chapter, that affirmatively positions modernity as having been a preparation for future demands on the human psyche and the dawning age of digital communication, biogenetics and robotics. Smithson's ideas and experiments on the entropic view thus provided a training program in phenomenology, for him to see his own way of seeing, which led to the conception of a new awareness of space and motion. Nauman links *nonobjective spaces* to a state of levitation, a feeling of weightless flotation and the potential dissolution of the body as in Guattari/Deleuze's concept of the body machine without organs, thus without pain, and in terms of the absence of the subject.

Buci-Glucksmann advocates an aesthetics of affirmation that contrasts the fascination with progress in current social and global crises, affording art with potentially uncritical esoteric properties of lightness, levitation, universe and immaterial cyberspace. The problematic of the subject, social suffering and political references in art, however, do not remain fully contained within the realm of the poetic and cosmic levitation of forces and elements. Although Buci-Glucksmann speaks of energetic elements that connect heaven and earth, the Icarian and the cartographic, she does not lack regard for the potential for the corporeality of the subject(s) to remain trapped in repression, suffering or material fixedness.

Buci-Glucksmann's theory and perspective on art hold a balance between the unsound problematic of materiality and poetic abstract relatedness, a balance that is reminiscent of the atmosphere in the work of the German performance and installation artist Rebecca Horn. Buci-Glucksmann writes that her work maps the circulation of positive and negative energies, the instability of conditions, routes and passages between life and death, the feminine and the masculine, order and chance. She employs the conceptual tool of "mapping" in terms of reproduction, representation, specification or mediation. Buci-Glucksmann refers to Rebecca Horn's installations, which hang upside down with wings that move in bizarre and jarring ways seemingly suspending the law of gravity, and attributes them with a poetic energy. This brings her to Bruce Nauman's negative, tragic spaces that provocatively and shockingly spur the viewer on to think behind the things. The element Buci-Glucksmann attributes to both Nauman's and Horn's gesture could very well be the motivation behind all art, "a violence, something insupportable, a frustration before the human condition".[3] What rings true for Nauman, namely that is a critique of violence is ultimately a more or less pronounced trait in all performative art – an art form that develops and fills spaces with gestures. In the end, the core of performance art is really to create spaces for experiencing things – not experiences immanent to the system of the avant-garde tradition, but instead experiences that are based on reflection, analysis, critique and recognition.

---

[3] Cf. Buci-Glucksmann. "Icarus Today: The Ephemeral Eye" Trans. Lang Baker, Ger Zelinski, Suzan Lord. In: Lord, Susan and Janine Marchessault (eds.) *Public. Art Culture Ideas.* Issue 18. [English translation of chapter 6 "Icare aujourd'hui: L'oeil éphémère", in Christine Buci-Glucksmann's *L'oeil cartographique de l'art* (Paris: Éditions Gallilée, 1996), 145–171.] http://www.yorku.ca/public/public/backissu/v18_5.html, accessed 25 Dec 2007

5 Marriage and Divorce: Is the Evolution of Landscape Paintings Ending in the Fields of Topographic Cartography and Graphic Design?

Markus Jobst

Institute of Geoinformation and Cartography
Vienna University of Technology, Austria

Abstract

The transmission of geospatial facts is a crucial necessity since creatures move in space and need to communicate. We can observe the collective communication behaviour of insects, which has been brought to perfection during evolution and is highly effective for these creatures. Beside gesture, verbal or dance-based communication, traditional human artefacts for a visual description of space go back 8000 years.

Whereas oldest preserved maps show very large scales, topological relations or religious influenced descriptions in Europe, the late middle ages come up with grandiose techniques of 3D landscape paintings trying to transmit a real as well as fantastic world. Refinements of presentation methods for the third dimension showed up the high potential for military purposes soon. From that point onwards the happy couple of art and graphical description of space seemed to have filed for divorce. Military purposes aimed on developing high information-density and appropriate scales. In the end information extraction with these maps asks for enormous cognitive load. On the other hand landscape painting became more artistic, fantastic and as far as degenerated art in order to attract, unsettle or criticize.

Nowadays a wide range of topographic maps uses highly developed semiotics for specific media with the aim to document, visualize and archive geospatial information. As long as a map is created and methods for archiving exist, the task of cartography seems to be fulfilled. The question for and work on impact, individual understanding and progressive linguistics barely occur. Thus the evolution of pictorial landscapes seems to have reached its end.

This contribution faces the question if there is still a perspective for landscape painting in cartographic science – or have we really reached the end of pictorial landscape evolution in cartography. How open is the science community to expand the traditional understanding of a map? If the community is open-minded, are there further perspectives for the pictorial landscape evolution especially with latest technologies? Where are potentials for pictorial landscapes, or should it be called virtual cartographic environments? May cartography come to a marriage with graphic design again? Answers, thoughts as well as remarks on these questions will be given with strong relation to history, present scientific work and outstanding examples in graphic design.

5.1 Introduction

Movement, experience and communication in space are crucial necessities for coordinated interaction as well as overall survival. As soon as these activities become disturbed, the existence of any creature is in danger. For example finding something to eat requires movement in space. If food has to be found, some experience in what can be eaten and where to find it is important. Additionally some spatial communication can shorten the search or even enable various hunting methods to be employed, if communication is done in an efficient way. Communication by these means does not necessarily use graphics or words, but some kind of linguistic, an understandable language, to establish a bi-directional information flow. For instance bees communicate by a form of dancing, which describes the way to fertile blossom grounds. Other creatures, like trees, use aromas of their blossoms, if they become damaged, to warn others. A human documentation or visual communication of interesting spatially-related activities will often use graphics and thus a graphical interpretation of space or activities in space. This step of interpretation is one of the strongest relationships between map and artefact, which allows the description of individual episodes in a very intensive, emotional and abstract way (Kroeber-Riel 1993).

In general communication, even basic level communication like gestures or mimicing, identifies some needs, which make bi-directional transmission of information possible. With these needs the importance of graphical coding can be shown and leads to the abstraction process in the brain, which is relevant for the interpretation and virtual reproduction of space. Linguistic dimensions, by means of technology, biology and culture, enable understanding and further-on interaction between communication partners. These important values lead to some overlapping of the knowledge islands between communicating partners and thus a correct decoding of information (Merten 1999).

Landscape paintings are one result of graphic communication that describe and interpret space in a sustainable way. Whereas one would think of geometrically perfect copies of the landscape with the use of various recording techniques nowadays, landscape paintings can be enormously influenced spiritually and politically.

For example representations of the world followed the strong spiritual depiction of Trinity in the Middle Ages. This adoption to philosophy brings up the strong relation to artistic interpretation of content/information. On the other hand we can observe examples in history, especially in reconnaissance, where the quest for geometric precision led to maps that were considered to be military secrets, because these maps contained information that might determine the outcomes of wars. Nowadays we are used to using satellite and aerial images of the highest quality, providing information about any part of the globe. They provide virtual reproductions of space, which again are artificially deformed by means of data enhancement, generalisation and interpretation. Depending on the needed expression and the planned effect, a virtual environment becomes more or less artificially deformed or highlighted (Barber 2005).

This contribution 'marriage and divorce' investigates the strong relationship between artistic depiction and geometric precision, which results from the communication of spatially-related topics. On one hand art at a high quality level ensures that people are aware of included information and that it is well documented. On the other hand geometric precision supports a clear, but sometimes misunderstood (due to things like map projections), description of spatial conditions. Throughout history this relationship was relevant to the transmission of spatially-related knowledge, with representations dependent on communication systems of the time. Thus the earliest historic remains of wall-paintings show hunting scenes in an impressive way and roughly explain movement in space for a specific purpose. Additionally these paintings form some part of decoration that was protected for generations. Nowadays, from a cartographic point of view, spatially-related topics are recorded in an efficient way and can be visualised by the merging of both the artistic elements and the scientific foundations that comprise contemporary mapping. But one element that seems to be absent today is emotional storytelling (or geographical storytelling (Cartwright 2004)). Focusing on the emotions of recipients will in better understanding the content of the map, whereby the storytelling makes the content more understandable in a way that the user best understands. In most cases art makes use of 'emotional storytelling', which is created by the interpretation by some artists using various artistic modes (painting, songs, ...) and often leads to discussions at the recipient side of map use (politicians, other artists or the public). Landscape painting is one kind of storytelling that makes use of the natural environment. The questions are: Is there still a relationship between landscape painting and cartography? Has there been a relationship in the past? Or will there be a relationship in the future? Cartography as a communication science cannot produce useful results without considering the effects of emotional storytelling, which can be responsible for motivation, emotional arousal and interaction. Additionally emotional storytelling seems to be heavily related to the field of aesthetics, which has impact on the overall graphical communication process (Kroeber-Riel 2000), but still relies on basic communication needs.

5.2 Spatial Communication Needs

When speaking of communication, we assume that a message/information is sent via a medium to a recipient. The person´s understanding of this message and a resultant action completes a simple communication process. Even if the message is not understood, but results in some action, the communication process is fulfilled. Yet this simple abstraction, without looking at the content, the type of media employed or the result of the transmission, can show that basic relationships between communication, recipient, content and media for communication exist (Burkart 2002). This direct communication, when a message becomes directly processed by a recipient, must be completed with indirect communication processes that effect behaviour, actions and further decisions in a co-oporative way. The co-oporative way specifies mass-media transmission, where one message is delivered to a huge number of recipients, whose manipulation of knowledge forms the impact of the message. Cartography, as well as art, mostly uses the same methodologies as mass-communication to transmit information: one artefact to many recipients (Merten 1999). One thing that stays the same for direct and indirect communication are the main dimensions of communication, which are also true for involving emotional arousal and basic level transmission.

5.3 Basic Level Communication for Spatial Content

The notion communication is used for various meanings in the fields of technique, psychology, sociology or neurology. For all fields the transmission of some information is the key action, but does not necessarily focus on graphics or sound. The process of transmission uses any kind of linguistics, which enables decoding of transmitted information. Thus gesture and mimic form some kind of basic level communication in inter-human communication. One can see the mood of a person by the expression in the face (angry, happy). Furthermore situations can express their influence on someone, if they can interpret their surrounding. For example a dangerous situation in the desert-outback is realised by environmental indicators. The person´s ability to read these indicators based either on their experience or some pre-transmission of this information by documentation like an 'outback survival atlas'. Similar to general communications, this basic level communication also follows three main dimensions of linguistics that ensure reception, decoding and understanding of the message by the recipient. Failure in any of these elements may cause either misinterpretation of indicators/message (signal received, but not necessarily understood) or interference of the signal (signal cannot be received clearly). The dimensions can be classified in technical, biological and social aspects. The technical dimension focuses on the characteristics of transmitting media and its channels. What kind of information can be transmitted with provided specifications or modes? The aim is efficient transmission, which uses all possibilities of the media in order to create an immersive presentation. Using media characteristics as tools for the communication process enables manipulation of the information as well, like pictorial illusions of MC Escher do with geometry (Schattschneider 2003).

Physiological and psychological characteristics of humans are defined within the biological dimension. These parameters are important for transmission and intuitive perception of information by means of meeting the demands of the human processing system. The understanding of information is based on the social dimension, which illustrates the influence of culture, society and mass-media (Chandler 2002). Problems in understanding information in the form of symbols/linguistic coding can be easily observed in map reading. For example one will not understand the symbol of ice-fishing when only having knowledge about the desert.

For spatial content these dimensions become much more important because human movement through space results with a very close relationship between the cognitive system, space and interaction. Artefacts and hand-made objects often follow some design which can be interpreted as some kind of symbolisation. Then the understanding of design mainly relates to a social dimension and the term 'aesthetics', which seems to be a loose notion for efficient transmission of 'useful information' about an object. For example a red big apple signifies maturation and sweet taste. This example of a simple code using the colour of a fruit lies within the human instinct, which has been genetically developed, enables decoding without a deep cognitive analysis and supports survival. It can be named as a basic level code or the result of basic level processing.

In most cases cartography uses a much more abstract presentation method by means of useful graphic values, media and geometric descriptions that relate to a primitive language, as we can observe with the oldest European pictorial writings. For efficiency, communication dimensions need to be considered along with initiating basic level processing, which then can result in a more intuitive transmission of information. Hunting scenes are graphical examples of documenting human spatial communication, movement in space and form one part of basic level communication with the help of graphics.

5.4 Early Graphical Remains of Spatial Communication

Cave and rock paintings around the world document parts of stone-age life. These pictorial artefacts describe hunting scenes, animals and even culture and religion of the stone age population, and therefore show their relationship to and use of space in a very sustainable way. The pictures can still be used/read and decoded/understood as communication artefacts. Furthermore, hunting and manufacturing tools as well as hunting strategies can be identified. Generally 'useful' animals, like horses and wild cattle (wisent, ur), form the main content of the pictures. Predators, as well as fishes and birds are rare in these pictures, with one exception – the Chauvet cave has more than 70 depictions of lions. Although the animals depicted existed at the time of creation, the number of animals shown does not mirror the main source of nutrition, but it indicates the special importance that these animals had in relationship to culture and religion. The size and form of caves indicates the use of many historical wall paintings as demonstration pictures used for storytelling and education (Hainzl 2004). The users well as the creation of these spatially-related pictures make high demands on the abstraction of space, objects, skills for creation and interpretation. Processes in the brain enable the mixture of abstract interpretation and creation, which lead to the artifical, but cognizable depiction of the world or spatially-related objects.

5.5 Abstraction of Space – Mental Processes

From a psychological and medical point of view, graphic design is a psychological learning process that is used to find and exercise the most efficient exposure to real world situations in a playful way. The process chain extends from external recording to deliberate activity (Schmidbauer 2004). External recording uses functions and various sensual modes to inform about the 'external' world and procedures of the body. Then, the cognitive system processes this recorded information, keeping this input active for further processing with tools of memory, connect and input information categorisation with existing experiences, developing interpretations and planning new activities related to movement in the external world. Responding activity then starts with activating receptors and moving in a coordinated way.

Painting and design are not art *per se* from the beginning of this process, but this pictorial form of communication has a very good chance to be understood by others. This kind of 'language' is much more direct and affective than spoken/written language ever could be (Kroeber-Riel 2000). From a basic point-of-view, graphic design is an expression of human living that reflects experiences, emotions and the functional development of learning and communication. The detailed components of this procedure are:

- With a current state of alertness the external world is recorded by all senses, especially by the visual channel;
- During the perception of space, object characteristics are collected and categorised;
- Objects are recognised;
- Recorded information is co-ordinated with internal (individual) experiences and memories by directing attention to internal values;

- Emotional objects are 'weighed' as to their relative importance in order to motivate and direct external attention, as well as reckoning;
- Semantic content is preserved within memory;
- All saved information is flexibly manipulated in order to create an operational concept for graphical design; and
- The operational concept is translated to coordinated movements (of the hand and fingers for drawing some image).

The complexity of this sequence, with associated procedures in the brain, demonstrates the sensitivity of the system to making 'mistakes' and subsequent damage. Mistakes during the sequence mostly deliver non-existing interpretations of the real world, when objects are not perceived in a correct way. For example an object that is nearer to the observer seems to be larger if indications for linear perspective are missing. Additionally, damages in the brain can be investigated by activating creativity and graphical design (Shepard 1994, Schmidbauer 2004). From this point-of-view creativity is positively motivated learning when the brain is dealing with the (real) external world. The sequence and processes when recording/creating graphical art or maps are almost the same. Thus art and cartography are closely related by cerebral procedures, but it may be that the concepts of language differ. In fact one difference lies in the use of technical tools for recording space as well as spatial information transmission. Therefore an understanding of the evolution of cartographic communication is needed.

5.6 Evolution of Cartographic Communication Using Linguistics

In general communication, especially spatially-related communication, is subject to technical and linguistic evolution. This evolution of communication is heavily related to technological changes and its use as information transmitting media, which has particularly expanded to mass-communication in the past centuries (Merten 1999). By reason that mass-communication is part of the human communication system, which uses technical tools and generally forms a uni-directional transmission from the application/transmission interface to the recipient, it represents one step in the evolution of communication and thus in the development of communication media. One of the earliest stages of communication is 'face-to-face' communication. Its impact on spatial communication is low due to the fact that geospatial descriptions are, in most cases, too complex for verbalisation, like a comprehensive description of a spatial situation. Exceptions are direct instructions for navigation that, in most cases, have to be kept short. The development of communication is then led by technical implementations, differentiation concerning strength of propagation and the number of participants reachable. Cartographic communication is on one hand definitely part of mass communication, but also part of a human-technical communication (human-computer interfaces – HCI) on the other. The stage of development increases with the implementation of applications, diversity of presentation forms and the involvement of the message recipient. This involvement enables context-dependent map content and the user directly linked to the mass-communication process (a direct input into the system/application by the recipient with some system update function).

Communication dimensions play an important role in every step of communication evolution, in particular when signs and visual codes are used. Communication dimensions then represent syntax, semantics and pragmatics of signs, which name the relational influence of signs, their compositions and especially the semantic context between content, reality and the user´s mental model that may create some gap in communication flow, which then results in loss of information and a useless application. From a theoretical viewpoint in linguistics a functional classification within semantics according to cognitive, informational and pragmatic significance has to be fulfilled in every step of the communication evolution (Chandler 2002). Cognitive significance focuses on the relationship of signs and thinking, which means that signs will only be understood correctly, if mental structures of the originator become obvious (the meaning of real-world object and abstraction is clearly the same). Informational significance concentrates on the relationship of sign and natural object, which expresses for example the importance of a city that is shown in a small scale map. Pragmatic significance changes the importance of sign according to a situational context, if for example ski-slopes appear in a tourist winter map or hiking trails on a summer map (Reichenbacher 2003, Nivala 2005).

Artistic depictions do not necessarily follow the functional classification, if artefacts intend to evoke some discussion, rethinking or use the aspects of misunderstanding instead of following some task of 'simple' documentation. In many outstanding examples of spatially-related contemporary art the artist´s expression and interpretation overlays reality and thus skips the use as a documentation artefact.

5.7 The Relationship Between Art and Space

The basic relationship of art and space seems to lay in the needs of sustainable communication. Therefore education and documentation is technically, socially, culturally or religiously motivated. Fantastic pictures that explain spatial relations or try to adapt the world to religious pictures, stories and interpretations are always an important tool in terms of communication, especially for direct manipulation (Monmonier 1996). Examples throughout history prove pictures to be tools that can manipule, which range from pictures of the world that follow the religious idea of the Holy Trinity, map projections that highlight and enlarge important parts of the world in order to show their power, to virtual reconstructions of landscapes that illustrate what landscapes might be.

5.7.1 The Step Towards Art

Art has always been used to express political structures, romantic ideas or cultural associations in various ways. Landscape illustrations are one tool for expression. Before we can define any landscape painting as art, we should have some understanding of what the notion 'art' stands for in this context. The notion of art and its understanding have changed during history, and still does. Generally art is some kind of product that results from the human creative processes which ends up as an artefact. Since modern times the creative processes itself can be the result and reflect an artifact as well. Hence the interpretation, abstraction, highlighting and deformation of landscapes in order to receive a depiction can be called art or artefact resulting from the creative processes. According to this point of view the abstraction of geographical data with the aim of specific expression make suse of the creative processes – the result of which we may call a map, a landscape visualisation, geovisualisation, scientific visualization and so on. Cartography then not only just uses techniques and formulas, but produces art as soon as a person adds some creativity within the process chain, for example by adapting informal generalisation steps.

5.7.2 Religious Pictures of the World

The influence of artefacts seems so strong that geometric distortions of geography were adapted according to cultural ideas and stories. As all of cartography is based on the collective knowledge of the Romans, Greeks and Babylonians before 600 A.D., the 'new' religion of the early Middle Ages tried to convert this knowledge, along with depictions, in order to fit into the Christian view of the world. Nowadays research can show that the geographic accuracy of these antique geo-images was not only due to the misunderstanding of this old knowledge, but were transformed in a systematic way to reflect the divine order of the Universe from the point of theology (Englisch 2002). For this reason the world with its known continents Europe, Africa and Asia was depicted as diagram, so called 'T-in-O' maps or diagrams. Generally these diagrams were oriented to the East so that the lines for rivers dividing Europe, Africa and Asia create a T, inside the earth – the 'O'. Starting with the 8th century copies of these simple diagrams were supplemented with pictures, which were irrelevant for a geographical picture, but elementary for contemporary associations. The main elements consisted of biblical names and stories like the sons of Noah Sem,

Ham and Japhet as synonym for Asia, Africa and Europe. The art of mapmaking at that time included the skill of supporting holy stories/writings with geographically distorted pictures and pictorial storytelling, mainly because the wider public was illiterate. General maps were used to support the stories from the Old Testament, the basis of Christianity. It took several hundred years until European culture needed general/global maps as navigation tools for exploration that the inaccuracies of the pictorial geo-artefact of the religious world was deemed to be no longer sufficiently accurate for scientific undertakings.

5.8 Precision and Military Secrets

Extended seafaring and explorations, like those of Christopher Columbus or Amérigo Vespucci, called for more precise maps in order to determine a ship's position and to deliver extended information of new continents. The main problem was longitude, which could not be measured by the stars but required precise clocks. New information about continents completed the view of the world. The implementation of maps was still influenced by mythological creatures that relate to biblical stories and marked dangerous areas along trade routes (Barber 2005). For example the view of the world as a plane resulted in the story of 'antipodes', fantastic creatures who lived on the other side of the world. Explorations helped to get a more realistic view of the world (e.g. the world as a sphere) and to leave mythology out of geographic depictions.

Changing the view of the world to a more realistic one influenced the use of maps as decorative art and governmental tools. In the late 16th century land surveys began in order to have detailed information about the structure, population, economy and size of counties. These surveys resulted in glamorous views of countries supplemented with emblems and territorial dominion descriptions as decorative art. However, they were also used for military planning, for example rescue planning in case of invasion (England – Barber 2005, p. 116). Changes in printing (especially the move to copperplate printing) and new surveying techniques (triangulation) led to more precise surveys and realistic pictures of Earth´s surface. Interpretation, 'religious' storytelling and mythology were avoided for official, government maps. Instead, cartographic language was developed so as to be more effective in providing a means to store comprehensive data, in a way that heights, nature, population and main movement axes could easily be read or decoded from the symbols. For example a detailed recording of Upper Austria with its heights was kept highly confidential after the Austrian empress Maria Theresia lost the seven year war against Prussia. Public maps had to be converted from height-lines to the less accurate depiction method of hill shading or hatchuring in order not to provide any information of military importance at that time. The military importance and confidential status of maps existed up until the 20th century. Furthermore maps were used as political propaganda so as to distortreality in order to support martial activities and political decisions.

5.9 Are Spatial Representations Becoming 'Artificial' Again?

Technological development for recording and dissemination (printing) techniques were introduced to provide high precision and more widely availabile spatial representations (as maps). In order to guarantee high precision standards in surveying and as maps, governmental organisations followed these tasks. Nowadays these organisations assure the actual data-sets and map series across countries and as cross-border projects at regional and international levels (EuroGeographics 2008). The use of maps is no longer restricted to a limited group of people, but is available for everyone. The main premise of topographic maps to provide the means to comprehensively record topography, whilst the information content is adapted to various scales due to perceptibility. Each element within a map is clearly identified and thus provides pertinent information. Here, 'art' has little influence on these kinds of spatial representations, with their graphical values. More artistic influences and creative margins can be observed when the rules for mapmaking are not that strict, like the case in artificial environments or virtual worlds (3D-Berlin 2008, 3D-Stadtmodelle 2008). The cartographic 'corset', which ensures readability and understanding of the map content, seems to be too restrictive for real creativity. As access to geographic data is not as restricted as it once was, the cartographic corset may be considered by mapmakers or not. Creativity, interpretations and distortions can now influence spatial representation again and leave the cartographic corset behind, even if the resultant information content may well be of no use for the recipient in some cases and situations.

Huge amounts of data and information, spatial and non-spatial, call for visualization with the aim of expressive understanding or altered access (VRVis 2008). From a technical point of view the mixture of information is enabled with various kinds of services in terms of data connection, operating system independency, data carriers, data formats, data semantics or security issues. From an artistic point of view this technical framework and new media, which can be subsumed in multimedia applications, enable new transmission forms and provide areas of "creative visualization".

One main intent of computer-graphics is the virtual reconstruction of real world environments (Döllner 2007). Hence lighting, camera characteristic and modelling should imitate real-world parameters and mostly result in virtual 'natural' sceneries, although the use of various scales, perspective distortion and massive occlusions leads to undifferentiated pixel areas with which no information can be transmitted. Alternative ways consider non-photorealistic rendering of virtual environments, which provide graphical values to ensure clear information transmission. For the purpose of clear spatial communication, non-photorealistic rendering in consideration of transmitting media characteristics and geometric projections can provide a framework to support geo-communication. Compared with the cartographic corset in traditional mapmaking, non-photorealistic rendering offers a powerful toolbox within virtual environments to add creativity in a focused way (Nienhaus 2005). Rebuilding reality in virtual space is a first step, an artistic distortion/implementation with non-photorealistic tools the second, if the environment should follow some specific communication needs. A cartographic use and the building of geomedia corsets in term of effective communication are subject of investigation in future fields of cartography. This point of view tries to expand cartographic tasks to a third and fourth dimension. From another viewpoint maps and geo-spatial data become increasingly important for graphical design.

Graphical design plays an important part in advertising industry. Should the occasion arise that geo-spatial data/information plays an important role in this economic sector, then the simple transmission of spatial content may not be sufficient (Steffenhagen 2000). It is rather the question of reaching a specific goal in advertising, which evokes needs at the recipient side. Therefore maps can play their role as metaphors, for example when emotions are drawn as a landscape in a fantasy map. Additionally, maps may be used as an add-on in an advertising campaign, e.g. using maps or location-based services to reach a client. As well, geospatial information may be directly used as main carrier of information, like geo-political maps. This economic sector gives a nice example, where the corset of cartography is not enough for communication to be effective and expressive. Rather the cartographic corset and creative input is needed to reach a maximum desired impact.

5.10 Conclusions

The relationship of cartography and the cartographic toolset for data modification in order to receive perceivable information, to art, the creative processes of expression, has historically followed a metaphor of 'marriage and divorce'. The oldest preserved spatial pictures, maps and even spatial depictions of the Middle Ages, show how their creation has been influenced by the creative processes in context with mythology, refinements of presentation methods for the third dimension and the accuracy of military mapping. From that point onwards the happy coupling of art and the need to graphically depict space seemed to have filed for divorce. Military purposes aimed at developing high information-density and appropriate scales. In the end information extraction with these abstract maps demands an enormous cognitive load. On the other hand landscape painting became more artistic, fantastic and even distorted geography (as art) in order to attract, unsettle or criticize.

Now, a wide range of topographic maps uses highly developed semiotics for specific media with the aim to document, visualise and archive geospatial information. New techniques in computer-graphics support the production of not only virtual copies of the real world, but also tools for information highlighting, selection or abstraction. A cartographic framework for these tools barely exists. That is one reason that creative processes can make use of these tools without restriction. Furthermore, economic dedications use the potential of geo-spatial communication with all possibilities of creativity. These economic intentions are supported by the characteristics of a basic level of spatial communication, an evolution of communication in terms of linguistics and the creative processes required to understand representations of distorted space. With the intention of maximising information impact, emotional influence and behavioral evocation (the recipient wants to buy something) a cartographic framework for spatial data modification is not enough. The cartographic framework has to be supported by further creative processes, namely art, that help to reach these intentions. Thus it can be stated that again cartography and art address their relationship more intensively.

References

3D-Berlin: http://www.3d-stadtmodell-berlin.de/online/3d/seite0.jsp,accessed: 10/1/2008

3D-Stadtmodelle: http://www.stadtmodelle.com/, accessed: 10/1/2008

Barber P (ed) (2005) The Map Book, London: Weidenfeld & Nicolson

Burkart R (2002) Kommunikationswissenschaft – Grundlagen und Problemfelder, 4th ed, Wien-Köln-Weimar: Böhlau Verlag

Cartwright WE (2004) Using the web for focussed geographical storytelling via gameplay, Proceedings of UPIMap 2004, The First International Joint Workshop on Ubiquitous, Pervasive and Internet Mapping, Tokyo, Japan: ICA Commission on Ubiquitous Cartography, pp. 89–109

Chandler D (2002) Semiotics – The basics; New York: Routledge Taylor & Francis Group

Döllner J (2007) Real-Time Virtual Landscapes, in Cartwright W, Peterson MP, Gartner G (eds), Multimedia Cartography, 2nd ed, Berlin, Heidelberg: Springer

Englisch B (2002) Ordo orbis terrae. Die Weltsicht in den Mappae mundi des frühen und hohen Mittelalters. Berlin: Akademie

Englisch B (2003) Imago mundi. Der virtuelle und der reale Raum in den mittelalterlichen Weltkarten, in: E Vavra (ed) Virtualiät und Realität (Akten des 10. Symposiums des Mediävistenverbandes, Krems a. d. Donau, 23.–26.3.2003, Wien: Österreichische Akademie der Wissenschaften

EuroGeographics: http://www.eurogeographics.org/eng/00_home.asp, accessed: 10/1/2008

Hainzl M (2004) Zeichen an der Wand, Höhlenbilder – Felsbilder – Graffiti, Band zur Sonderaustellung, Lebensspuren Museum Wels, Wels:trod.ART Verlag

Kroeber-Riel W (1993) Bildkommunikation, Imagerystrategien für die Werbung, 5nd ed, Stuttgart: Kohlhammer Verlag

Kroeber-Riel W, Esch FR (2000) Strategie und Technik der Werbung, 5nd ed, Stuttgart: Kohlhammer Verlag

Merten K (1999) Einführung in die Kommunikationswissenschaft; in Aktuelle Medien- und Kommunikationsforschung, Bd.: Grundlagen der Kommunikationswissenschaft 1999, Münster: Literatur-Verlag

Monmonier M (1996) How to lie with maps, 2nd ed.; Chicago, London: The University Chicago Press

Nienhaus M (2005) Real-Time Non-Photorealistic Rendering Techniques for Illustrating 3D Scenes in their Dynamics, PhD, Potsdam:University of Potsdam

Nivala AM (2005) User-centred design in the development of a mobile map application. Licentiate Thesis, Helsinki: Helsinki University of Technology

Reichenbacher T (2003) Mobile Cartography – Adaptive Visualisation of Geographic Information on Mobile Devices; PhD, Munich: Tech. University

Schattschneider D, Emmer M (eds) (2003) MC Escher's legacy: a centennial celebration, collection of articles coming from the MC Escher Centennial Conference in Rome 1998. Berlin, London: Springer

Schmidbauer M (2004) Das kreative Netzwerk – wie unser Gehirn in Bildern spricht, Wien, New York: Springer

Shepard GM (1994) Neurobiology. 3rd ed, Oxford University Press

Steffenhagen H (2000) Wirkungen der Werbung: Konzepte-Erklärungen-Befunde, 2nd ed., Aachen: Wissenschaftsverlag Mainz

VRVis: http://www.vrvis.at/home.html, accessed: 10/1/2008

URL

Supplemental figures can be found at the following locations:

1. http://en.wikipedia.org/wiki/T_and_O_map
2. http://en.wikipedia.org/wiki/Image:Escher_Waterfall.jpg
3. http://en.wikipedia.org/wiki/Cave_painting
4. http://www.practicalpainting.com/Articles/Surrealism/Salvador_Dali_Geopolitical_Child.htm
5. http://www.hpi.uni-potsdam.de/doellner/projects/illustrative_city_models.html

Section II: Aesthetics and Cartographic Design

6 Are we Living in a Cartographic Illiterate Society?

Karel Kriz

Department of Geography and Regional Research
University of Vienna, Austria

Abstract

This contribution focuses on the question whether geo-communication in general and cartography in particular is or can be of use to our society. What role has graphicacy got in a society that primarily focuses on other forms of representation and communication utilising verbal and numerical information?

Cartographic examples are presented to assess on the proposition that every cartographic representation has a structure and should consider basic cognitive rules for communicating information. It is therefore important to understand the architecture of a map in order to communicate information efficiently. As documented in many (carto)graphic products, it seems to be unfortunately evident that our society has difficulty dealing with multimedia representations for means of geo-communication – at least from a cartographic perspective. Whether it is a simple one dimensional printed map or a multidimensional, interactive installation in many cases the benefit for communication is not always conclusive.

This contribution should initiate a discussion that questions the fundamentals of cartography and geo-communication as well as their legitimisation for our society.

6.1 Introduction

The meaning of literacy is to posses the ability to read and write as well as to use language proficiently. Numeracy comprises the capability to work with numbers and to deal with mathematical concepts. Articulacy embraces the competence of being able to express oneself fluently and coherently in a distinct, clear and definite manner, having the power of speech. These skills are essential to our society and form the bases of our educational system as well as the foundation of communication. Without the expertise of literacy, numeracy and articulacy dissemination of information in our society would be very tedious. It implies that indispensable structures of our modern communication would fail.

Though how do we deal and cope with the competence of communicating with graphics? Do we learn to utilise this capability at all efficiently as the above mentioned skills literacy, numeracy and articulacy? In everyday life we are constantly confronted with manifold information. This then gets filtered, processed and utilised leading finally to a decision that in many cases can or should trigger an action. Much of this information is stored in graphic forms. Symbolisation in road traffic with all its vast (heterogeneous) variety of road signs as one example underpins the presence of graphic communication in our daily life. These are not always beneficial as can be seen in the following examples.

Fig. 6.1. Road signs Melbourne (left), Vienna (right). Photograph: Karel Kriz

The proverb "A picture is worth a thousand words" refers to the concept that complex information can be transported efficiently in just a single image. This has been and still is the case in cartography. However, it is important to be aware of the fundamentals and their functionality. Not all pictures, graphics or maps communicate efficiently. The following example in *Figure 6.2* shows a (maybe) correct geo-information depiction; however a hopelessly overloaded cartographic representation that is insufficient of communicating geo-spatial information efficiently.

The distinguished Swiss Cartographer Eduard Imhof once stated that cartography is like music (Imhof 1972). You have single notes with specific attributes connected to them. These resemble the fundamentals of cartographic depiction – the graphic variables. Each note is then aligned and combined in such a way that the arrangement can harmonise to produce pleasing music or disharmony.

Graphicacy could be described as the ability to understand and use symbols, graphics, maps, plans, in other words to communicate visually.

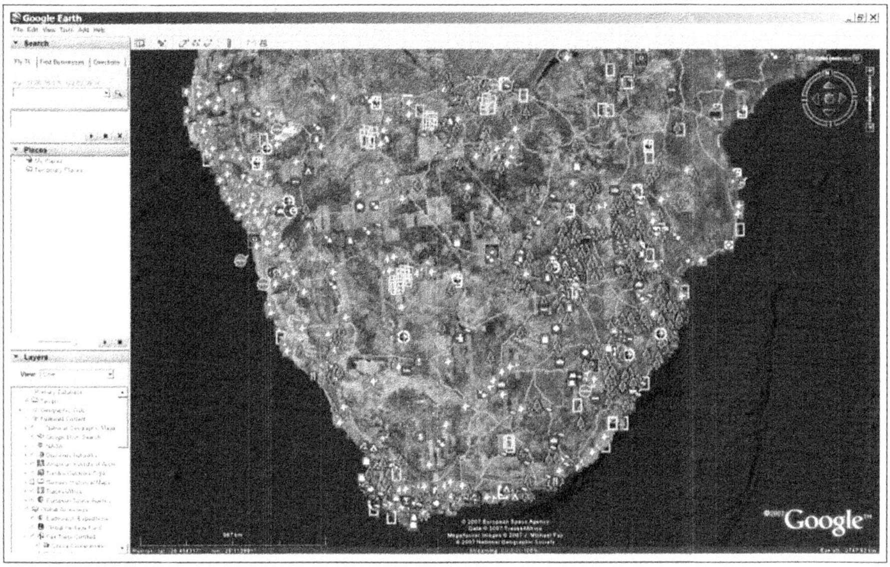

Fig. 6.2. *Google Earth* – close-up of South Africa

Communication is a process of exchanging information. It requires the use of our senses. Cartography can be seen as a communication process that has been mainly utilising the visual component for dissemination in forms of paper maps. However, in the course of computerisation and modern technology it has been recently contemplating with other varieties of distribution methods embracing all fashions of multimedia. Virtual reality, 3D depiction, interactivity, internet mapping are just some of the fields that dominate today's cartographic vocabulary. Nevertheless, they all still deal with space and have to solve the same cartographic problems of communicating spatial information in a digestible and easy understandable style. Thereby, generalisation and design issues are inevitable to the process of information distribution.

The art of generalising or in other words simplifying real or virtual three-dimensional space as well as thematic information into a spatial model without omitting essential features still remains an important task for cartographers. The outcome of this process is the basis for a map. It is a structured model of spatial relationships that has the goal to transport information in order to gain extensive insight.

Design and esthetics should therefore not only be seen as the domain of art. They play an important role in cartography, if they are utilised correctly to communicate spatial information. Thus, a map as the outcome of this process can be considered as a model of reality not only bound to paper.

Interactivity and virtual realities are the new areas of interest for cartography. The user may interact with the new artificial environment by analysing, stimulating and interpreting processes in detail. Unrestricted positions and viewpoints enable an immersive holistic three-dimensional insight. Thematic and dynamic-temporal contents may be freely combined and manipulated. Opportunities not previously possible for cartography in general are unlocked by this outstanding new technology.

Furthermore, topics such as the integration of new ideas and possibilities, moving beyond the traditional boundaries of so called classic cartography, adopting methods from neighboring sciences, interdisciplinary cooperation and the need to redefine the cartographic axioms and values become incorporated in the general discussion. It is therefore important to understand the holistic approach in this field without neglecting the meaning of esthetics and design in context of cartography.

6.2 Theory and Methodology

A map can be seen as a metaphor for all products that are produced within cartography. It can therefore be characterised as a structured model of spatial relationships that has the goal to transport information to the user in order to gain extensive insight. Based on the assumption of disseminating these spatial relationships effectively, cartography can be described as a communication process between producer and user. Kolacny (1969) defined this process in his triangular representation of cartography, defining cartographers and users as communicative partners. Adapting Kolancy's model this paradigm can be still applied to current cartography. On the one side the cartographer interprets and defines the spatial problem with his specific cartographic knowledge in order to translate the information into a common understandable representation – a map that does not necessarily have to be only on paper. This representation can then be seen as the interface to the user to extract information. On the other side it is essential that the user has basic cartographic knowledge to translate the expert model into usable code for further processing. This is the crucial point. If the information is not transported correctly or misunderstood the communication cycle is interrupted and no insight is gained. This is often the case due to certain circumstances. On the side of the cartographer the 'map' can for example be too complex, such as utilising too many symbols or colors, or even incorrect using wrong styling and coloring that can lead to misinterpretation. On the side of the user the most common problem is in lack of cartographic knowledge in order to translate the cartographer's intensions. Not understanding the basics of (carto)graphics is a crucial problem that leads to misconception, sometimes even not realised by the participant.

In order to alleviate this issue it is important to understand the three conceptual cornerstones of cartographic communication. Starting with the cartographic initialisation the spatial related problem is identified, specified and finally formalised. This is in most cases the initial phase within the communication cycle where cartographer and user – who can be in theory identical – meet and decide how to communicate the problem. Thereafter the design phase is launched. This incorporates the implementation that is based on the initialisation decisions, utilising specific cartographic knowledge, solutions, conceptions as well as tools and methods. The outcome is a product that results in the process of knowledge acquisition. The user then must interact, interpret and perceive in order to gain insight. If the result is satisfactory an action is the result otherwise modification can be applied and the product can be adapted.

In most cases the design phase is the crucial part. Although there are many suggestions and principles in and around cartography on how to design a 'map' effectively, they are all very vague due to the fact that there are so many ways to do the same thing. Principles of cartographic design as defined by the British Cartographic Society Design Group at their meeting in Glasgow and posted to the Society of Cartographer's listserv, 26 November 1999 (British Cartographic Society Design Group 1999) or explained by the Swiss cartographer Eduard Imhof in his book "Thematische Kartographie" (Imhof 1972) are just two examples of recommendations on how to design a cartographic representation from a visual perspective. They both give a basic overview on the principles of cartographic design without however pinpointing the user to concrete situations on what to do and what not. One could imply that the only definite rule in cartographic design is that there is no rule. Many principles and concepts are vague but give the user freedom to interpret in a very individual way. Concept before compilation, hierarchy with harmony, keep it smart and simple or engage the emotion to engage the understanding are just some examples on how broad these principles can be defined. They do however give a very good impression on how cartography can be utilised – if the cartographer is able to understand and then take advantage of them.

In order to utilise design issues in cartography it is also important to be aware of the limits on our capacity for processing information. It is known in psychology that human working memory is generally considered to have limited capacity. One of the earliest quantification of such limits was introduced by the cognitive psychologist G. A. Miller (1956). He introduced the magical number seven, plus or minus two to be significant on how information is stored and utilised. Once this threshold is reached strategies must be developed to solve the problem. This is the cognitive load that refers to the load on working memory during problem solving, thinking and reasoning. Either information is grouped to chunks and further processed or it is lost. Retention of discrete units of information before information loss occurs is related to as chunking. This is a strategy that refers to structures that can be used as units of perception and meaning. It is a strategy for efficient use of blocks of information or chunks and the capability to understand more complex information. In everyday life we all automatically apply these strategies in some way or other. If for example a telephone number is to be memorised, a specific grouping method is applied to remember the number.

[country code] [area code] [institution code] [depart. code] [room number]

This magical number seven, plus or minus two is to a certain degree also applicable in cartography. This does not mean that only seven items, levels, colors, symbols must be used. It merely implies that strategies must be developed to "chunk" information into digestible portions of (carto)graphical forms of representations. It is therefore important to understand the architecture of a map in order to communicate the information and desired message efficiently.

6.3 Examples

Cartographic products are manifold and with current technology many different kinds of representations are possible to communicate spatial information. However, all cartographic artifacts do have a similar if not common structure.

Every map has at least one layer that contains information for orientation. This topographic-orientation layer serves as a base for locating the thematic information in a spatial context. It can vary from a very simple depiction all the way to a multi-dimensional complex base map. Superimposed on this layer thematic information is then included. This can be either single or multi-dimensional depending on the information density and depth.

Keeping in mind the cognitive limits mentioned before every map can be analysed according to its cognitive overload. The simple rule of thumb not to exceed seven, plus or minus two objects is helpful when evaluating the effectiveness of a map. This does not mean that only seven, plus or minus two objects are allowed in a map. Objects can be grouped together to create 'chunks' that help structure and organise the overall cartographic representation.

The following examples show three different types of complexities based on the assumption that no cognitive overload is produced. They all utilise basic cartographic methods based on the conventional graphic variables. Furthermore, different ways of depiction are used to illustrate the diverse methods of communicating spatial information.

6.3.1 Map "Population in Austria"

This map shows the population density of Austria in 1991 based on administrative boundaries using choropleth mapping techniques. The topographic base information for this map is kept very simple. It resembles the administrative boundaries of

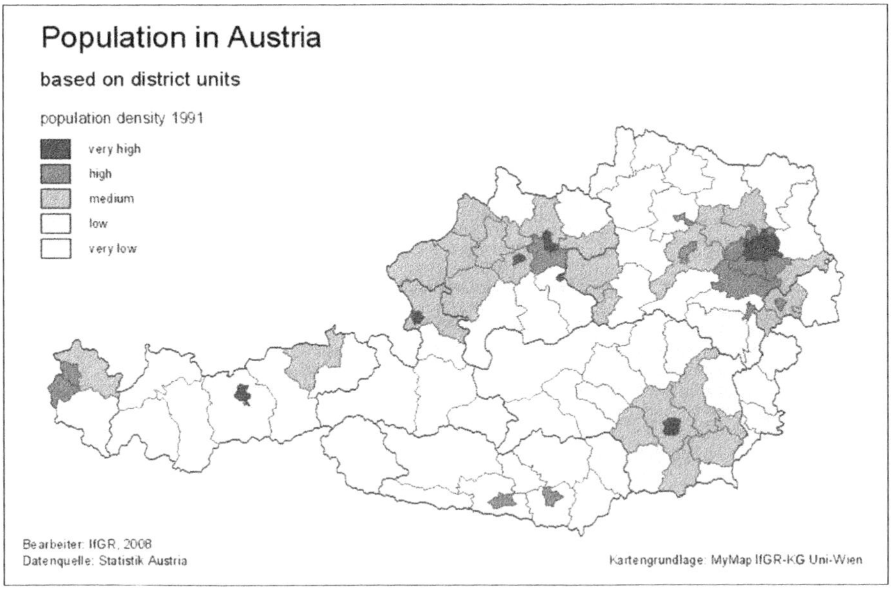

Fig. 6.3. Map and Legend "Population in Austria". Source: ÖROK Atlas

Austria differentiating between districts and province units using distinct line width for discrimination. The thematic information is single dimensional and classified in five classes utilising a monochromatic scheme. There are 99 units subdivided in nine provinces that hold the thematic information. The overall impression is clear and very transparent. The viewer is not distracted by any unnecessary extra information and can focus on the essential message.

It is clear that not all cartographic products can be reduced to such simplicity; however it is also obvious that this unembellished depiction does transport the message very clearly and directly.

6.3.2 Map "Mean Annual Air Temperature 1961–1990"

The portion of the map "Mean annual air temperature 1961–1990" from the Hydrological Atlas of Austria shows a typical example of a thematic map using associative coloring with a bi-polar color scheme for thematic depiction. The topographic base information for this map is more substantial than the previous described map and accommodates five levels of hierarchy. The goal of the topographic base map was to support the user in localising the thematic information in space without stressing too much on topography or even distracting the user from the principal thematic information. The first level includes a monochromatic grayscale hill shading to emphasise the topography and to communicate the specific morphological situation of the area. Superimposed on this level is a dense river network to accentuate the hydrology that is a key topic of the atlas. This contains the rivers and lakes that are delineated in five line widths depending on their priority as well as selected river names aligned to the river. They are placed in such a way in order to be correctly identified and if possible should not intersect with any rivers. The third level encompasses the administrative boundaries that are depicted with a grey linear symbol as well as being accentuated with a dashed line. Two line widths were used to distinguish national as well as international boarders. In order not to be in conflict with rivers that may also coincide with a boundary the dashed lines were omitted when ever this occurred. The next level includes settlements with place names. This was realised using point symbols with text information that were placed in such a way not to interfere with other objects. The final level incorporates the geographical reference in degrees that is depicted through a solid grey line. The line color is prominent enough to standout against the grey hill shading however not too dominant, as black would be, in order to blend in the overall subtle composition.

The thematic information is depicted using a bi-polar color scheme and grouped in 10 classes. This classification would normally produce a cognitive overload, if reproduced in a monochromatic scheme. However, utilising a bi-polar (two) color scheme from red/yellow to green/blue the representation is split (chunked) into two groups. In this case, forced by the coloring, the observer perceives two levels (red/yellow above 4°C and green/blue below 4°C) or even three, if yellow is taken as a single chunk. This is more obvious in the map then in the legend. Even though the colors change slightly in the map due to the diffusion of the colors with the hill shading, the overall impression is still evident, because the color weight between the classes is large enough to keep the unique identification indisputable.

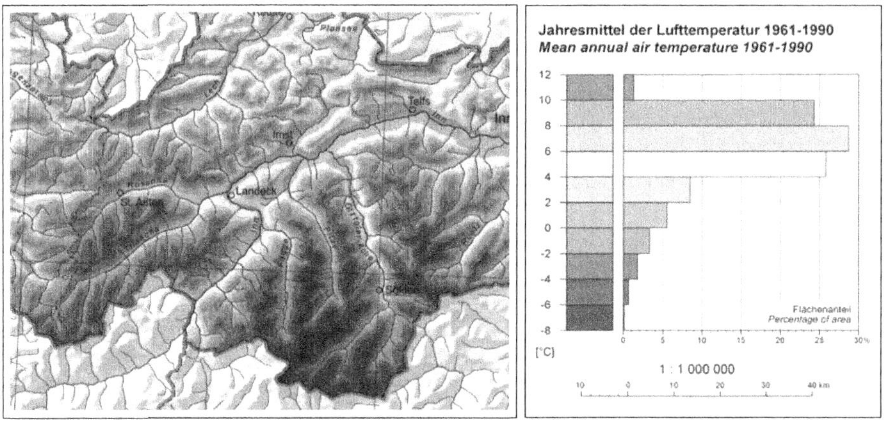

Fig. 6.4. Map (left) and Legend (right) "Mean annual air temperature 1961–1990". Source: Hydrological Atlas of Austria

6.3.3 Map "Mean Depth to Groundwater Table"

The map "Mean depth to groundwater table" from the Hydrological Atlas of Austria described in this section has the same topographic base map as the map "Mean annual air temperature 1961–1990" however incorporates multiple thematic information. It can therefore be classified as a multi-dimensional complex thematic map. At first glance the representation seems to be pushing the limits of being readable and appears to be overloaded. It becomes though apparent that all layers are comprehensible after examining the structure of the map. Altogether there are seven levels of thematic information depicted in this map. Five of these levels are linked to area features and two levels to point features. The first level shows the interpolation from observed data onto an area depicted using an associative bi-polar color scheme from blue (wet) over green to yellow (dry). The red color scheme was deliberately omitted in order to be available for the singular forth level "no regional estimation possible" that must outstand visually and be unique in the map. In order to visualise similarities, such as estimations in the second level "Estimation of the typical region's average", a coarse visual raster was introduced with the same color scheme of level one. This gives the observer the possibility on the one side to associate the information with the corresponding level by color. On the other side the raster enables a clear distinction even though the color weighting may coincide with other classes. The third level "Mean values at individual stations" is also related to the first level however refers to a point feature. This level only occurs in areas where no regional estimations were possible. The symbol in which the color is incorporated is large enough to transport the same information as in level one and has an outline boarder color to emphasise the color within the symbol. The fifth level "Groundwater stations – used for interpolation" refers to point features and distinguishes two classes. This is realised using different dot sizes. In this case qualitative information is communicated. The sixth level "Mean groundwater table of selected regions" depicts quantitative information based on the same areas as

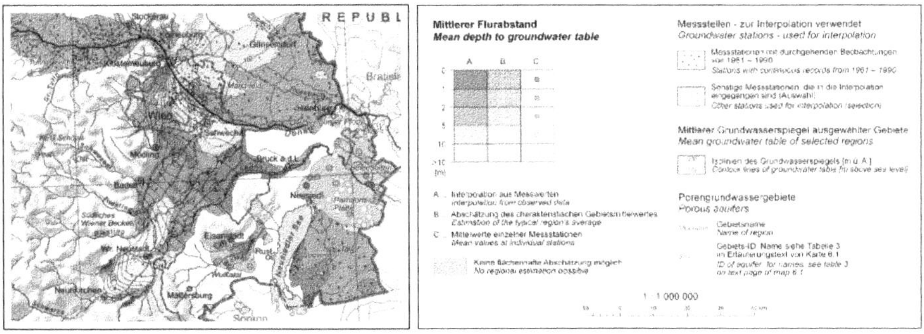

Fig. 6.5. Map (left) and Legend (right) "Mean depth to groundwater table". Source: Hydrological Atlas of Austria

level one. Therefore a different representation method apart from area coloring had to be realised. Contour lines with 10 m (solid line) and in some cases 5 m (dashed line) were installed. This linear depiction method of area information allows two levels to be visualised simultaneously without interference. The final seventh level shows textual information based on area units that have specific thematic information linked to them.

6.4 Conclusions

It is apparent that cartography has the potential to promote spatial communication in our society efficiently – if utilised correctly. It depends on the one side how cartography is able to communicate its products based on cartographic knowledge acquisition as well as cognitive design rules. On the other side the user must be capable of understanding and interpreting the symbolised cartographic reality that is being transported. In order for this communication process to function it is essential that graphicacy becomes an integrative part of our societal system. Graphicacy is an important asset that, if applied and understood correctly, can have a strong impact on modern communication within our society.

References

BMLFUW (ed) (2007) Hydrologischer Atlas Österreichs. Wien: Bundesministerium für Land- und Forstwirtschaft, Umwelt und Wasserwirtschaft

British Cartographic Society Design Group (1999) http://www.mckinleyville.com/cart/cabinet/cab_cartprinc.html

Imhof E (1972) Thematische Kartographie. de Gruyter, Berlin

Kolacny A (1969) Cartographic Information – a fundamental term in modern cartography. Cartographic Journal 6: 47–49

Miller GA (1956) The Magical Number Seven, Plus or Minus Two: Some Limits on our Capacity for Processing Information. Psychological Review, 63, 81–97

7 Aesthetics of the CORINE Land Cover Maps

Ján Feranec and Ján Pravda

Institute of Geography, Slovak Academy of Sciences, Bratislava, Slovak Republic

Abstract

After 1985, when works on the CORINE land cover started, maps that represented land cover at scales 1:100 000, 1:250 000, 1:500 000 and smaller in several European countries were produced. The subject of this paper is the comparison of the appearance and partially functionality of these maps i.e. mediation of information to user, for instance about size and land cover class frequency. Aesthetics of land cover maps issued in several West European countries (such as Belgium, France, Ireland, Luxembourg or Spain) is based on the principal mean of expression (the graphic variable) that is to say colour and its hues, number of which is rather elevated (in some cases more than 40). On the other side, aesthetics of maps issued in Slovakia is based on a smaller number of colours and a higher number of colour patterns. Both groups of maps though, boast similar artistic/aesthetic qualities, as the comparison of their fragments with those of Pierre Auguste Renoire's painting suggests. Authors believe that the artistic and technical elements and aspects of a map should be in harmony in order to sup-port and enrich its practical functions.

7.1 Introduction

Diverse Earth surface objects and their properties are represented by different maps. Contents and ways of representation applied to cadastral, topographic or thematic maps are distinctly different. Some maps are simple, others are more complicated in terms of graphics. But the uniting elements are functions and aesthetics (Keates 1996, pp. 207, 221) of represented objects and their properties. Keates considers important the following two aesthetic aspects in relation to maps:

- On the one hand is the fact that as the map is a graphic creation that inevitably affects the map user.

- On the other hand, it is possible to regard maps as of interest in themselves, apart from their ostensive functions.

The functional aspect is normally linked to the aim that the map should fulfil and simultaneously it determines the way of its use.

This paper is involved with the cartographic expression, the aesthetic (colour) and functional (purpose-bound and pragmatic) aspects of the map produced by the CORINE land cover (CLC) methodology (Feranec et al. 2007).

7.2 Corine Land Cover Data

Collecting of data concerning land cover of Europe based on satellite images started in 1985, as part of the CLC Project (Heymann et al. 1994). Data obtained by this Project characterise the state of land cover (representing the material manifestation of natural and socio-economic processes that take place on the Earth's surface) of almost the entire of Europe from the 1990s and from 2000 (+/– one year), as well as its changes during approximately a decade (Feranec et al. 2007). In January 2007, the follow-up activities under this project started with the aim to characterise the state of land cover in Europe in 2006 (+/– one year) and its changes in 2000–2006. Data about land cover of Europe from three time horizons and its changes in two periods (1990–2000 and 2000–2006) are compatible *inter se* as they derive of one methodology at scale 1:100 000. The size of the smallest identified land cover area is 25 ha, the minimum width of the identified area is 100 m and the minimum area of identified changes is 5 ha (Büttner et al. 2004, Feranec et al. 2007). The unified nomenclature, which characterises land cover, consists of 44 classes *(Table 7.1,* Heymann et al. 1994, Bossard et al. 2000). It should be emphasised that they manifest in satellite images by means of physiognomic features – colour, shape, texture and pattern (see *Figure 7.1).*

Manifestation of characteristic patterns on satellite images should guide the choice of graphic means, provided the land cover map is created on the basis of data obtained by interpretation of satellite images. The brief overview of so far published CLC maps (see *Figures 7.2–7.10)* is meant to demonstrate to what extent the manifestation of land cover classes on satellite images was respected by the choice of the map representation means.

Table 7.1. CLC nomenclature (Heymann et al. 1994, Bossard et al. 2000)

1	**Artificial surfaces**
11	*Urban fabric*
111	Continuous urban fabric
112	Discontinuous urban fabric
12	*Industrial, commercial and transport units*
121	Industrial or commercial units
122	Road and rail networks and associated land
123	Port areas
124	Airports
13	*Mine, dump and constructions sites*
131	Mineral extraction sites
132	Dump sites
133	Construction sites
14	*Artificial, non-agricultural vegetated areas*
141	Green urban areas
142	Sport and leisure facilities
2	**Agricultural areas**
21	*Arable land*
211	Non-irrigated arable land
212	Permanently irrigated land
213	Rice fields
22	*Permanent crops*
221	Vineyards
222	Fruit trees and berry plantations
223	Olive groves
23	*Pastures*
231	Pastures
24	*Heterogeneous agricultural areas*
241	Annual crops associated with permanent crops
242	Complex cultivation patterns
243	Land principally occupied by agriculture, with significant areas of natural vegetation
244	Agro-forestry areas
3	**Forest and semi-natural areas**
31	*Forests*
311	Broad-leaved forests
312	Coniferous forests
313	Mixed forests
32	*Scrub and/or herbaceous vegetation associations*
321	Natural grasslands
322	Moors and heathland
323	Sclerophyllous vegetation
324	Transitional woodland-scrub
33	*Open spaces with little or no vegetation*
331	Beaches, dunes, sands
332	Bare rocks
333	Sparsely vegetated areas
334	Burnt areas
335	Glaciers and perpetual snow
4	**Wetlands**
41	*Inland wetlands*
411	Inland marshes
412	Peat bogs
42	*Maritime wetlands*
421	Salt marshes
422	Salines
423	Intertidal flats
5	**Water bodies**
51	*Inland waters*
511	Water courses
512	Water bodies
52	*Marine waters*
521	Coastal lagoons
522	Estuaries
523	Sea and ocean

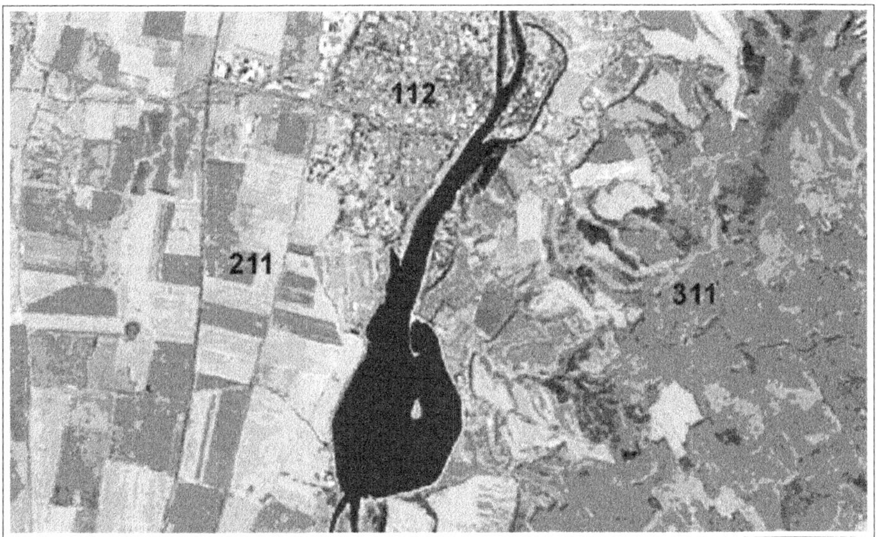

Fig. 7.1. Characteristic patterns of classes 112 (discontinuous urban fabric), 211 (non-irrigated arable land) and 311 (broad-leaved forest) on satellite images.

7.3 Corine Land Cover Maps of Some West European Countries

Authors of land cover maps in several western European countries (Belgium, France, Ireland, Spain and others) chose to represent land cover classes prevailingly by colour scale. The impression of such maps is fairly active (more than 40 colour

Fig. 7.2. Fragment of land cover map of Belgium (Reproduced with permission of the National Geographical Institute – Belgium)

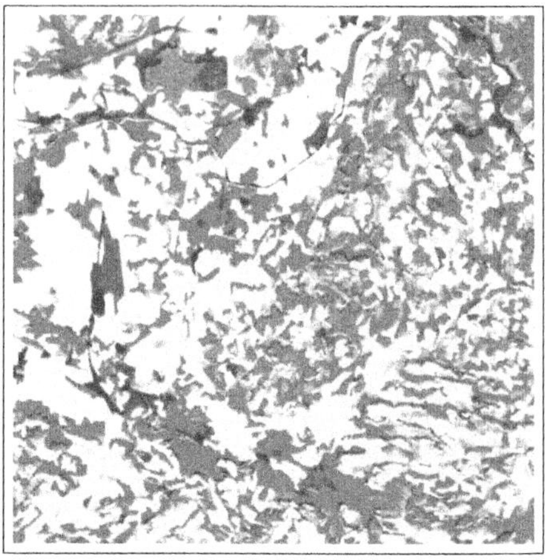

Fig. 7.3. Fragment of land cover map of France (Reproduced with permission of the Institut Français de l'Environment, Orleans, France)

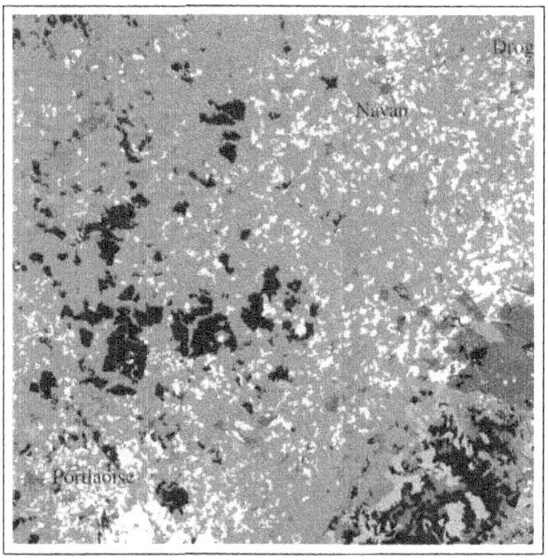

Fig. 7.4. Fragment of land cover map of LC Ireland (© Ordnance Survey Ireland/ Government of Ireland, Copyright Permit No. MP 008907)

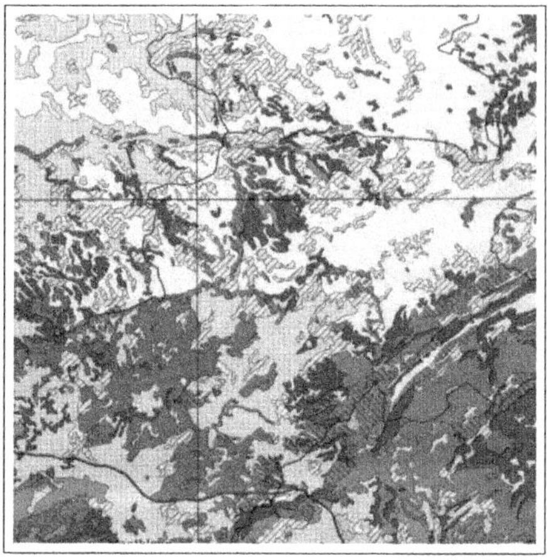

Fig. 7.5. Fragment of land cover map of Spain (© Institut Cartogràphic de Catalunya)

hues). However, in terms of functionality, the reader finds it difficult to discern and perceive the objects of visualised database (land cover classes) represented by such an amount of hues.

Reduced map fragments are examples: *(Figures 7.2–7.5)* Belgium (CORINE Land Cover, Map sheet Liège-Namur-Leuven 1:100 000, 1995), France (CORINE Land Cover France, The map 1:1 000 000, 1997), Ireland (CORINE Land Cover Map Sheet C7-Dublin 1:500 000, 1995), and Spain (Mapa CORINE Land Cover de Catalunya 1:250 000, 1993).

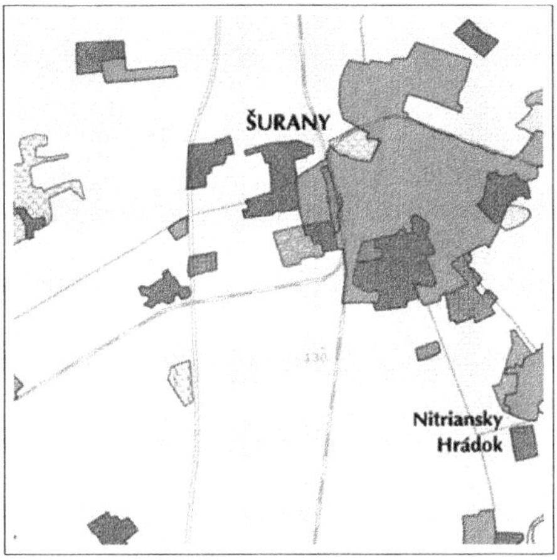

Fig. 7.6. Diminished land cover map fragment of Slovakia's territory at scale 1:50 000

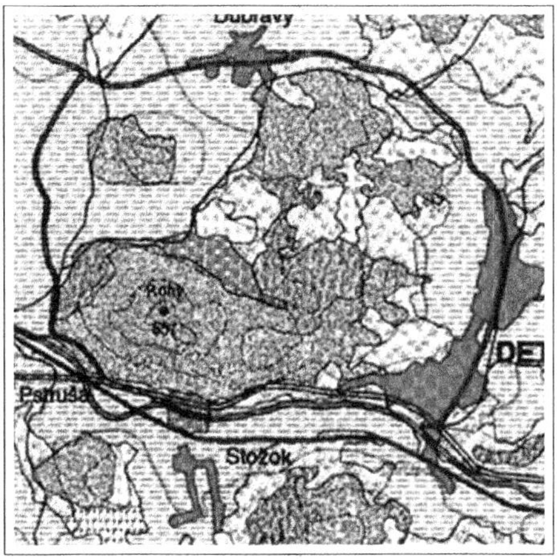

Fig. 7.7. Diminished land cover map fragment of Slovakia's territory at scale 1:100 000

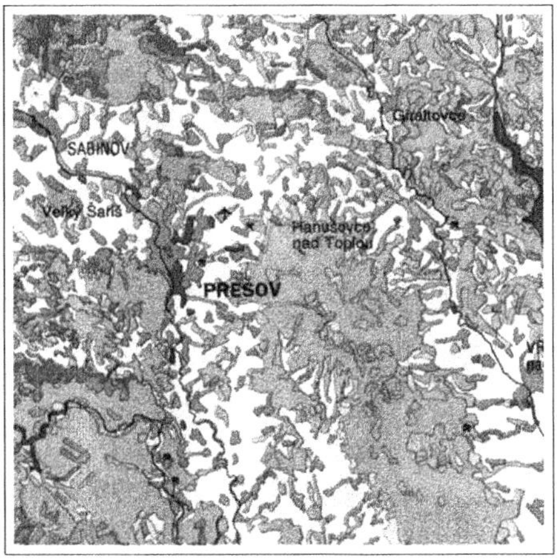

Fig. 7.8. Diminished land cover map fragment of Slovakia's territory at scale 1:500 000

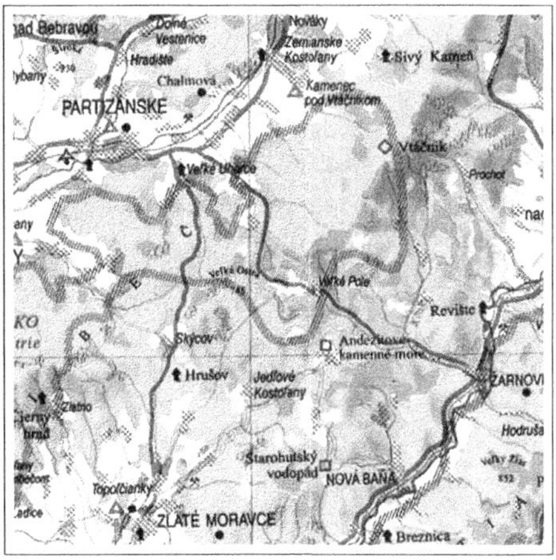

Fig. 7.9. Diminished part of land cover tourist map of Slovakia at scale 1:500 000

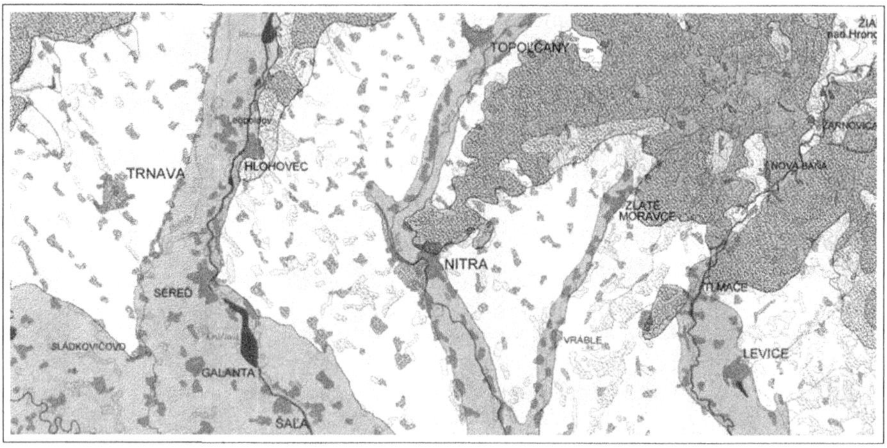

Fig. 7.10. Diminished part of natural landscape and land cover map of Slovakia at scale 1:500 000. (Reproduction of all these fragments with permission of the Institute of Geography SAS, Bratislava, Slovakia)

7.4 Maps Produced at the Institute of Geography of the Slovak Academy of Sciences

The Institute of Geography SAS applied the CLC methodology to map production that did not interfere with perception and comprehension of the map content. First of all, two map sheets at scales 1:50 000 (Feranec and Oťaheľ 1994) and 1:100 000 (Feranec and Oťaheľ 1998) were produced followed by two maps of the whole territory of Slovakia at scale 1:500 000 (Feranec and Oťaheľ 1996 a, b), where the third level of CLC classes was represented by associative patterns and at the superior second level maps the colour hues were applied. The third map at a scale of 1:500 000 (Feranec and Oťaheľ 2000) was more complicated: additional colour layer that confronted the division of the present land cover with that which originally existed (natural landscape) in the territory of Slovakia before human impact was laid over the second and third layers of CLC classes. The figures that follow illustrate these maps.

7.5 Experiment

The painting of Pierre Auguste Renoire (1841–1919) *"Dans le jardin aux arbres"* was chosen at random (see *Figure 7.11)*. Several fragments of the painting were selected (*Figure 7.12)*.

Fig. 7.11. Pierre Auguste Renoire (1841–1919) *"Dans le jardin aux arbres"*

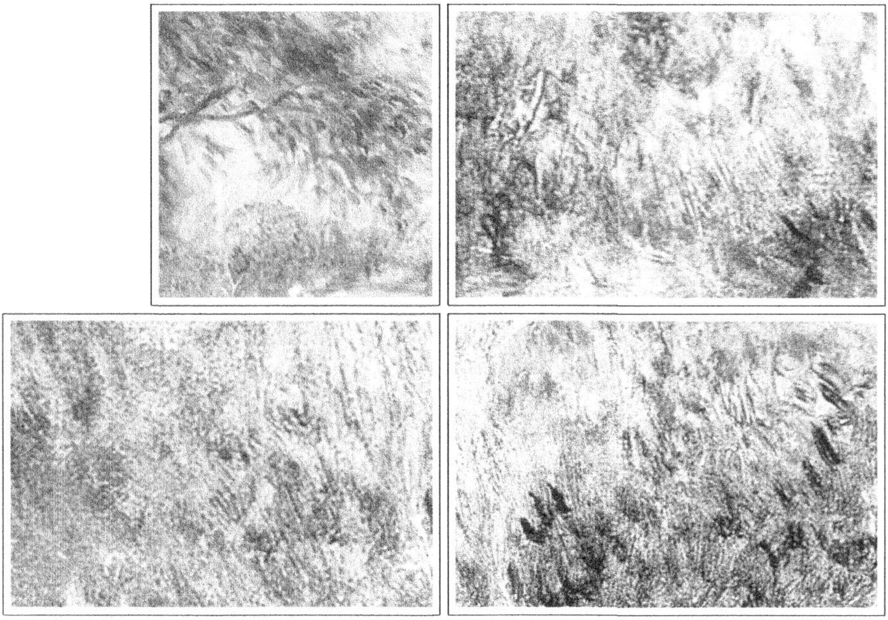

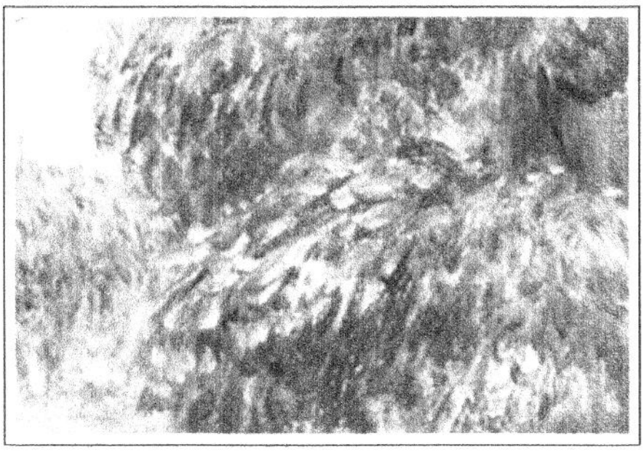

Fig. 7.12. Fragments of Renoir's painting

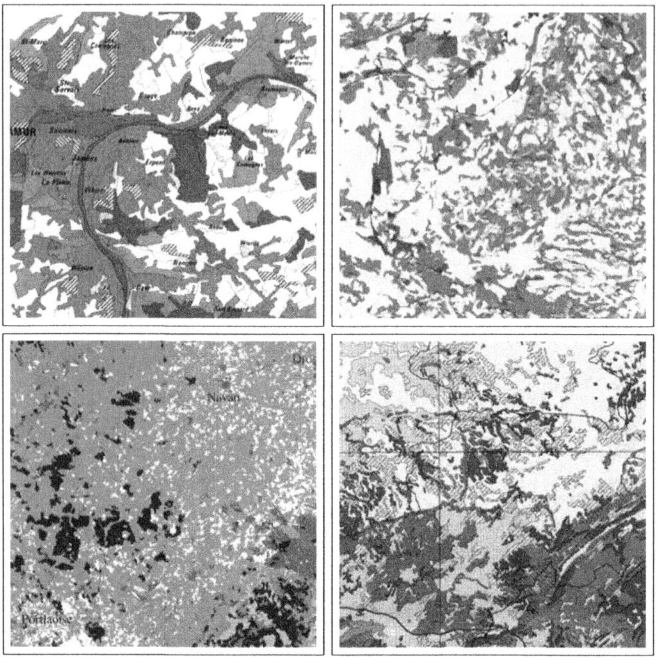

Fig. 7.13. Fragments of western European land cover maps

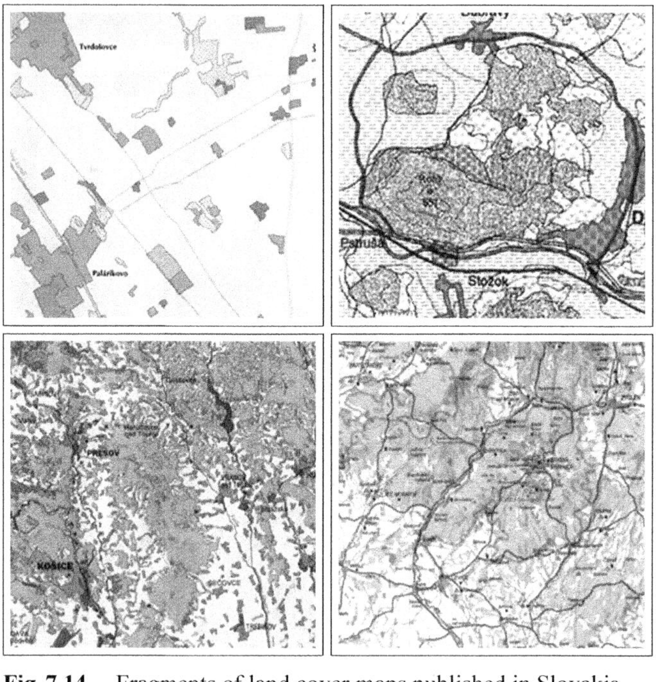

Fig. 7.14. Fragments of land cover maps published in Slovakia

Fragments of Renoir's painting were compared with diminished fragments of western European LC maps *(Figure 7.13)*, also with diminished land cover map fragments of Slovakia *(Figure 7.14)*.

By arrangement of fragments either of the west European or Slovak land cover maps a more artistic (aesthetic) effect can be reached above all if the principle of random choice of fragments their size, orientation or other characteristics is be applied. It should require further experiments.

Pursuing the classification of map styles, the land cover maps, published both in the western European countries and Slovakia, maintain their proper publishing style. As far as functionality is concerned, work is easier with the Slovak maps because of the means of representation used – i.e. associative patterns supported by associative colour hues of five classes of the first level facilitate land cover classes identification (artificial surfaces, agricultural areas, forest and semi-natural areas, water bodies). In terms of aesthetics, the final impression is more or less equal because the artistic effect is reached by conspicuous artistic means – colours. However, there is a slight difference. Land cover maps published in Slovakia make use of colours in order to discern layers of land cover classes (red – artificial surfaces, light brown – arable land, green – forests, vineyards, fruit trees and other areas with vegetation, yellow – natural grassland, turquoise – inland marshes and blue – inland waters), the proper principle applied to painting. In Slovak maps, this principle allows one to discern typical features of the landscape by means of land cover as is the basin character of the territory *(Figure 7.15)*.

Some artistic principles have been available to cartography in the past. They concerned representation of one of the most interesting phenomena on Earth – georelief (using mostly orthogonal maps). The efforts to reach an impressing representation of the elevation layers using different colours were only possible if artistic methods were applied. One example is the principle of colour succession of hypsometric scales: the scale based on the principle "the higher the darker" (Franz Edler von Hauslab), the scale with regional colours (Theodor Emil von Sydow), or scale "the higher, the lighter" (Carl von Sonklar), which are easily and comfortably

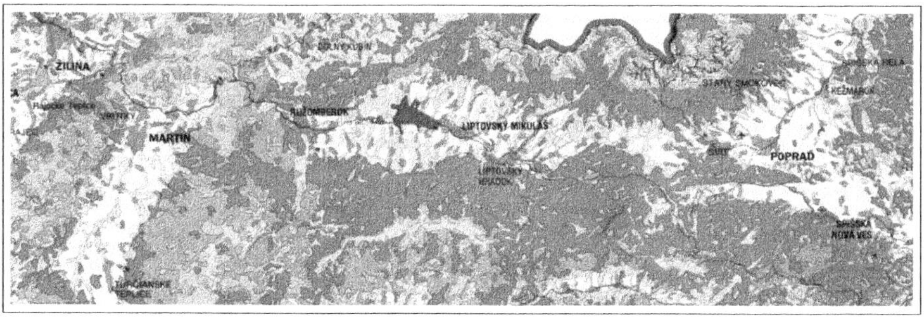

Fig. 7.15. Basin character of Slovakia's territory on land cover map

perceived. The variant "from dark, over the light back to dark colours" also appeared – the interesting thing is that all scales (above all with shading added) render good to excellent 3D results in term of perception. This can only be explained by artistic effect. We should not forget the contribution of Karl Peucker and Wilhelm Ostwald in the area of applying colours to planar (flattened) map, because this type of map also confirms the artistic role of colours in cartography. Artistic inputs of Eduard Imhof (1965) and the whole Swiss school in the area of shading applied to representation of georelief which applied the principle of 'air perspective' in representation of mountain relief on orthogonal map are also doubtless.

7.6 Conclusions

The fact that a map can provoke kind of reader's reaction similar to that induced by an artistic works is certain. The map fulfils its cognitional functions by means of signs that are visualised thanks to graphic variables (Bertin 1967) – size, shape, colour, texture, value and orientation. If additional tools of artistic painting such as purity or luminance of colour, warm versus cold colour, contrast, fore- and background including resolution of layers are applied the map perception is enriched.

Ignorance of such graphic and artistic tools would mean giving up the important factor, which completes, strengthens and enriches the cognitive function of the map and discerns the map with artistic elements from that which the result of mere technological procedures.

Acknowledgements

The paper is one of the outputs of the *Structure of the rural landscape: analysis of the development, changes and spatial organization by application of the CORINE land cover databases and the geographical information systems* Project No. 2/7021/7 pursued at the Institute of Geography of the Slovak Academy of Sciences in 2007, supported by the VEGA Grant Agency. Authors are grateful to the Institut Cartogràphic de Catalunya, the Institut Français de l'Environment, the National Geographical Institute – Belgium and the Ordnance Survey Ireland for the consent to publication of the quoted CORINE land cover map fragments. The authors are also grateful to Hana Contrerasová for translation of this paper into English.

References

Bertin J (1967) Sémiologie graphique. Les diagrammes, les cartes. Gauthier-Villars, Paris

Bossard M, Feranec J, Oťaheľ J (2000) CORINE land cover 2000 – Addendum 2000. Technical report, 40. European Environment Agency, Copenhagen

Büttner B, Feranec J, Jaffrain G, Mari L, Maucha G, Soukup T (2004) The CORINE land cover 2000 project. In: Reuter R (ed) EARSeL eProceedings, 3. Paris, EARSeL

CORINE Land Cover France, The map, 1:1 000 000 (1997) Institut Français de l'Environnement, Orléans

CORINE Land Cover, Map Sheet C7-Dublin 1:500 000 (1995) Ordnance Survey of Ireland, Dublin

CORINE Land Cover, Map sheet Liège-Namur-Leuven 1:100 000 (1995) L'Institut Géographique National, Bruxelles

Feranec J, Otahel J (1994) The forms of landscape, Map sheet M-34-133-C (Šurany) 1:50 000. Geografický ústav SAV, Bratislava

Feranec J, Otahel J (1996a) Slovakia – CORINE – land cover map 1:500 000. Geografický ústav SAV, Bratislava

Feranec J, Otahel J (1996b) Slovakia – CORINE – land cover tourist map 1:500 000. Geografický ústav SAV, Bratislava

Feranec J, Otahel J (1998) Land cover map. Territory: Zvolen – Detva – Krupina 1:100 000. Geografický ústav SAV, Bratislava

Feranec J, Otahel J (2000) Natural landscape and land cover of Slovakia. Geografický ústav SAV, Bratislava

Feranec J, Hazeu G, Christensen S, Jaffrain G (2007) CORINE land cover change detection in Europe (case studies of the Netherlands and Slovakia). Land Use Policy 24:234–247

Heymann Y, Steenmans Ch, Croisille G, Bossard M (1994) CORINE land cover, Technical guide. Office for Official Publication of the European Communites, Luxembourg

Imhof E (1965) Kartographische Geländedarstellung. Walter de Gruyter, Berlin

Keates SJ (1996) Understanding maps, 2nd ed. Longman, Essex

Mapa CORINE Land Cover de Catalunya 1:250 000 (1993) Ampliació de la llegenda europea. Institut Cartogràphic de Catalunya, Barcelona

8 Making Legends By Means of Painters' Palettes

Sidonie Christophe

National Geographic Institute (IGN), COGIT Laboratory
Saint Mandé Cedex, France

Abstract

Current on-demand map design tools on the Web do not support cartographic and semiologic rules. Moreover they rarely offer to users to customise their styles. We propose a dialogue application in order to take the user's needs into account and to bring the cartographic expertise he may lack as well. In this paper, we argue that on-demand map design should benefit altogether from user artistic capacities and from cartographic theory. Our approach consists in analysing the exploration of colours and how colours support the cartographic rules. Then we propose a process to conceive legends by means of painters' palettes illustrated with an example of a painting by Matisse.

8.1 Introduction

In recent years, on-demand map design tools on the Web have been greatly expanded. They propose to load remote data, to design new objects, to apply predefined styles in order to conceive on-demand maps. Actually on-demand map design tools do not support cartographic rules. Moreover they do not offer to users to customise their styles. So we define an on-demand map as a map that is:

- *efficient,* i.e. cartographically correct
- *satisfying,* i.e. that answers to the conceiver needs.

The aim of our research is thus to assist users to conceive real on-demand maps and particularly to design efficient and satisfying legends on the Web.

The specificity of a map compared to a work of graphic art is to transmit an unambiguous message related to a geographic space. Graphical signs on the map must altogether convey a visual message that should be well understood by the users of the map. Robinson (1952) argued that maps designed based only on an artistic approach are more often not functional. He proposed to analyse characteristics of perception so that symbolisation and design decisions can be based on objective rules. A map should tell the truth in a certain way, through lying about certain aspects (Monmonier 1991). Graphic semiotic has been defined by some cartographers to correctly use graphical variables (Robinson 1952; Bertin 1967; MacEachren 1995). It consists in semiotics and conventional rules about visual variables. The graphic semiology yields constraints on the map design process but does not explain how to create the styles that satisfy all constraints. Thus, each cartographer has his own way to design a map according to his own expertise. This is a creative process that takes a significant part in the map design process (Krygier 1995). We have to find a way to formalise the cartographic expertise required for the styles definition.

This chapter outlines work being undertaken in an on going PhD Thesis in the COGIT laboratory. We are conceiving a dialogue system to pilot the design of a map by a user through cartographic rules (Christophe et al. 2007).

In this paper we handle explicitly the map design as a creative process guided by rules. First we present our approach of the process based on the exploration of colours and how they support the cartographic rules. Then we analyse a new strategy to conceive legends from painters' palettes.

8.2 Approach

In this part we analyse some steps of the legend conception and especially the choice of visual variables and the specification of a symbol. More specifically we focus on the choice of the variable 'colour'. When a user conceives a legend, one of the main tasks is to choose the symbol colour. We propose here to help the user in such a task, while measuring the respect of the cartographic rules.

8.2.1 The Exploration of Colours

The exploration of colours is an important step in the process of legend design. We distinguish two main tasks: to choice/change a colour and to brighten/darken a colour. We look for the best tool to realise these tasks. The current colour choosers in GIS applications propose many colours. Their organisations make them difficult to explore. Moreover the user cannot know if two close colours are effectively distinguishable between one and another. *Figure 8.1* presents an example of the colour chooser of the QGIS application.

Chesneau (2006), then Buard and Ruas (2007) established the colour reference system LUCIL from of Itten's chromatic circle in twelve colours (1967). The circle is organised according to the visual perception of colours and allows a user to explore more easily a wide range of colours:

- by hue: the hue helps to differentiate the colours; and
- by value: the value helps to brighten/darken a colour.

The hue and the value of a colour are very intuitive, even for an inexperienced user of colorimetry. *Figure 8.2* shows on the left a contrast between two hues, e.g. the bigger the angle between hues the more different the colours are. On the right in *Figure 8.2* is shows a contrast between two values for one hue on the circle, e.g. the colours are ordered. We see also that a user can easily perceive the distance between two colours: the bigger the angle between hues the bigger the distance.

In the next section we focus on the cartographic rules integrated with colours in the colour reference system LUCIL.

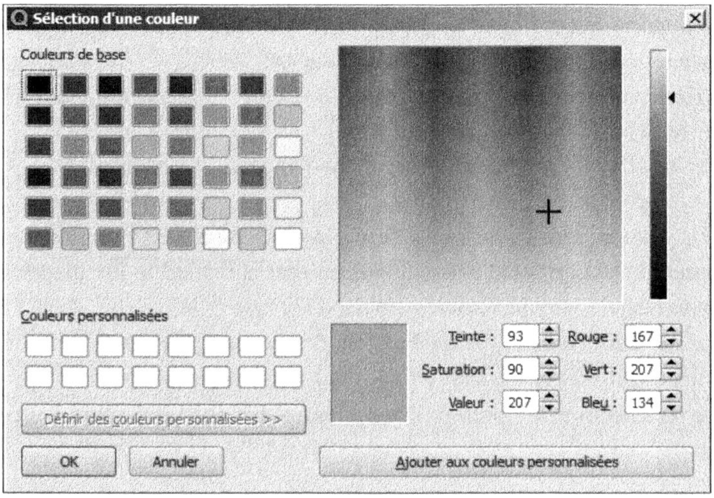

Fig. 8.1. Colour chooser of QIS application

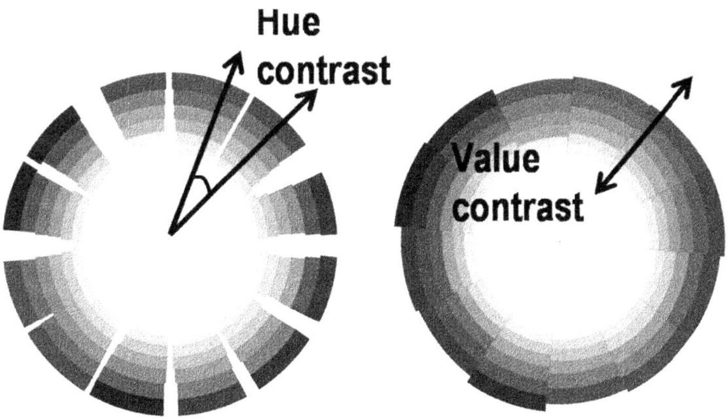

Fig. 8.2. The colour reference system LUCIL (Buard and Ruas 2007)

8.2.2 The Cartographic Rules Integrated in the Chromatic Circle

A legend is composed of objects organised according to semantics relationships *(differential, association, order)* that they share (Bertin 1967, Weger 1996). We propose to structure the information into *themes* in order to show the differential relationships. Then inside a theme, the objects are grouped in association and/or sorted according to their order. The various relationships must be clearly represented through the choices of the graphical signs and more especially through the choices of the colours. So the reader of a map can interpret its intended message from the meaning of the graphical signs used by the cartographer.

In the previous section we presented two types of contrasts in the colour reference system LUCIL. They facilitate the rendering of the relationships between themes and into themes. According to the position of hues, the hue contrast allows to render *difference* or *association* between graphical signs:

- A differential relationship between two themes is rendered by a maximised distance between hues *(Figure 8.2);* and
- An association relationship between two themes is rendered by a minimised distance between hues *(Figure 8.2).*

Next, the value contrast allows rendering an *order* between graphical signs: it is thus rendered by a distance between values that is related to the distance of order between the data – whichever hues are used *(Figure 8.2).*

In our context of a dialogue application, the system relies on the chromatic circle to help the user to make the best choices in terms of colours.

In the next section, we present the strategies that embed the tasks of conception detailed.

8.3 Strategies to Conceive a Legend

We try to capture strategies of the map design process in order to integrate them in our dialogue system. To do so, we imagine different map design processes that a cartographer may have. Some prefer to begin with a blank page and to design and render all themes one by one; others prefer to use existing maps and put great pieces of them together in order to make a new one; some could prefer to use existing combination of harmonious colours. Therefore we propose two different ways that correspond to the last two options.

The first strategy 'Select and Refine a map sample' (presented in Christophe 2007) consists of putting together some elements of a map in order to conceive a new one. It proposes to the user to choose the preferred map sample from a database of map samples (Domingues and Bucher 2006). In order to converge towards an appropriate legend, the user chooses samples and colours on the interface. Then the system transforms these choices in constraints and proposes samples according to these constraints. Once the user chooses his best map sample, it may be further refined by changing some colours.

The second strategy 'Palettes' works by proposing various combinations of colours, so-called Palettes, to the user in order to conceive the legend. We use the word palette here like the complete colour set used in a painting.

In the following sections, we detail the second strategy Palettes. We first explain the conception of palettes and then the methods used for applying them on the user's data.

8.3.1 The Conception of Painters' Palettes

We first have to conceive our palettes. Thus we propose two kinds of palettes: colours taken from famous painters' palettes; and colours taken from National Mapping Agencies. We focus here on the palettes of painters, actually, and only famous ones. In this section we present the conception of a painter's palette in order to propose it to a user of our application of dialogue. We only use characteristic paintings whose colours could be recognisable. We first choose a characteristic painting of the art of *Matisse, La tristesse du roi* (1952).

Then we choose the most characteristic colours in the painting. We record 14 colours: three greens; four blues; two yellows; a pink, a black; a white; and two greys. They all contribute to what we call our 'Matisse Palette', shown in *Figure 8.3*, which is a selection of colours in the painting.

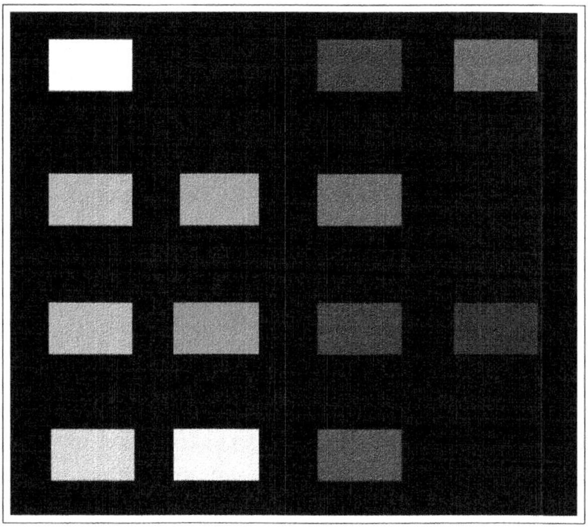

Fig. 8.3. Our palette of "La tristesse du roi" Matisse (1952)

This selection of colours is still approximate because we select the average value of RGB-codes of pixels in a colour solid. Nevertheless, even if colour value may be easily improved, the result is already visually satisfying. Other painters' palettes have yet to be conceived.

8.3.2 The Use of the Palette on the User's Data

We have to define methods in order to apply the colours of our Matisse palette to the data. A first issue of this application is that the number of colours in the palette does not correspond to the number of themes in the map. We have 14 colours and we use a geographical dataset on Royan, France (BDPays IGN©), an urban zone near the sea, with 5 themes: roads (with different clasifications); sea; communal background; wooded area; buildings. We thus have to find strategies to select 5 colours from these 14.

8.3.2.1 Use of Matisse's Rules

We can first conserve the characteristic composition of the painting of Matisse. In the painting *La tristesse du roi,* we observe that the blues, the greens, the rose and the black are only used for large surfaces and also rectilinear ones. The yellows are only used for the little patches all over the painting. The white is used the two ways as well. Therefore, we can keep this arrangement of colours in our future map. These characteristics are called the painter's rules. Thus this conception is done without the use of any cartographic rules.

So, in our example of the city of Royan, the blues, the greens, the rose and the black have to be used only for large continuous surfaces and the yellows only for more punctual objects with little surfaces. *Figure 8.4* shows two examples of this use of colours.

The characteristics of the art of Matisse are recognisable here. The second example is the one which seems to best render the Matisse palette. These maps are totally fanciful because they are absolutely not conventional – the sea is in pink and the communal area is in pink or blue. These choices are not usual. Moreover the colours of Matisse here are very intense, which is not usually the case with topographic maps. In the logics of graphic semiology, these choices would disturb the intended message of the map.

In order to retain the possibility having available a wide of innovative legends, we formalise in our model of the map design process new information as well as our initial data: the total area concerned by a theme; and the type of repartition (continuous, blotchy, ...). During this step of the application, the system looks for this information so as to choose the best theme for a colour, if a user clearly wants a creative legend in the style of Matisse.

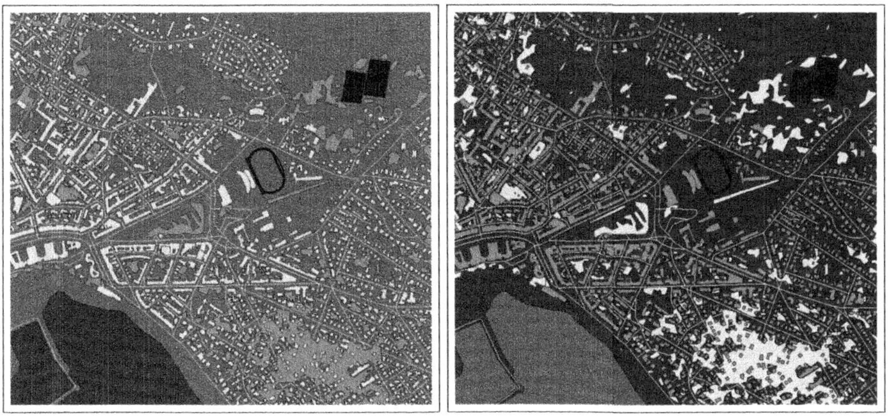

Fig. 8.4. Two artistic applications of the Matisse Legend

8.3.2.2 Use of Cartographic Rules

We now use elementary cartographic rules for the colours of our Matisse Palette – firstly conventional ones, then ones having more contrast. The conventions accord to the following rules:

- The sea is in blue,
- The wooded area is in green,
- The background is generally clear enough,
- ...

For the wooded area and the sea, we face no difficulties because we have 3 greens and 2 blues in our Matisse Palette.

For the communal background, the white in our Matisse Palette seems to be the most adapted to the communal area because the others colours should be more suitable on a white background.

For the other themes, we have to choose from the other colours, with the help of the contrast rules. The different themes must have different hues. Moreover the themes in superposition with the background have to be effectively contrasted with it. Thus we can use pink, yellows, greys and black to effectively differentiate them. *Figure 8.5* presents *conventional* application of our Matisse Palette on the dataset of Royan.

In our application of dialogue, we especially focus on rules about contrasts and conventions. Firstly the system could integrate the conventional rules. Then, when there are no more conventional indications, the system integrates the contrast rules. The system applies one blue to the sea and one green to the vegetation and for the others colours it makes it 'as good as possible' in terms of contrast. Therefore it is

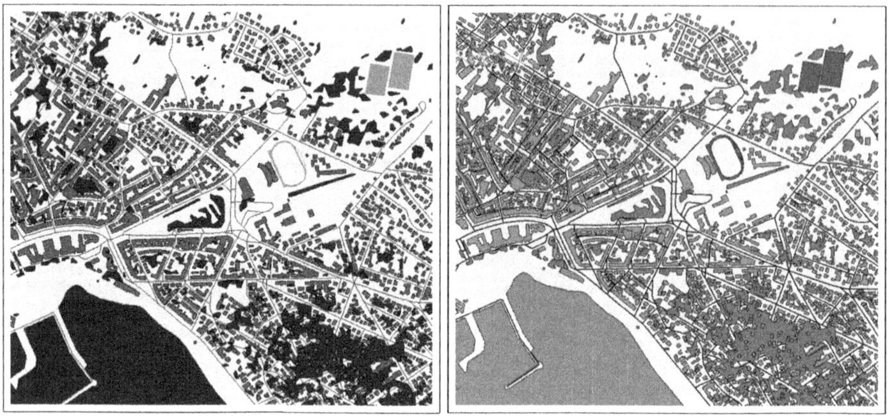

Fig. 8.5. Two cartographic applications of the Matisse Palette

obvious that for a unique palette, even with the respect to the rules employed, the system has several legends to offer users: and, for instance, they could be sorted according to their level of contrast.

8.3.2.3 Combination of Painter' Rules and Cartographic Rules

We can also combine Matisse's rules with the cartographic rules. *Figure 8.6* presents this kind of maps that result.

Finally, the legends obtained beforehand are rather traditional due to the use of conventional and contrast rules. We compare them to a traditional map from IGN France shown in *Figure 8.7.* It is an extract of the database BDTopo with its legend. Our maps prove to be close to this traditional one.

However, the use of the Matisse colours makes them really aesthetic and less common than the very traditional ones. Of course, we can propose various painters' palettes as illustrated in *Figures 8.8* and *8.9.*

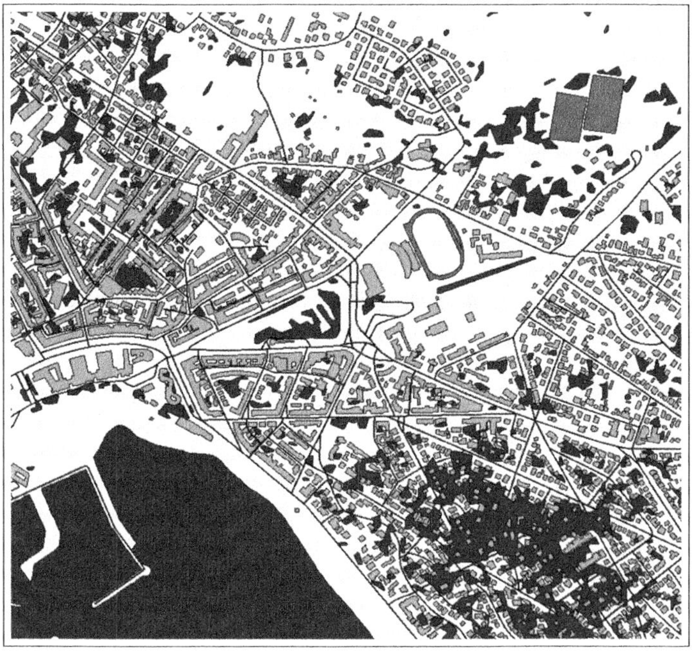

Fig. 8.6. A combined artistic and cartographic rules application of the Matisse Palette

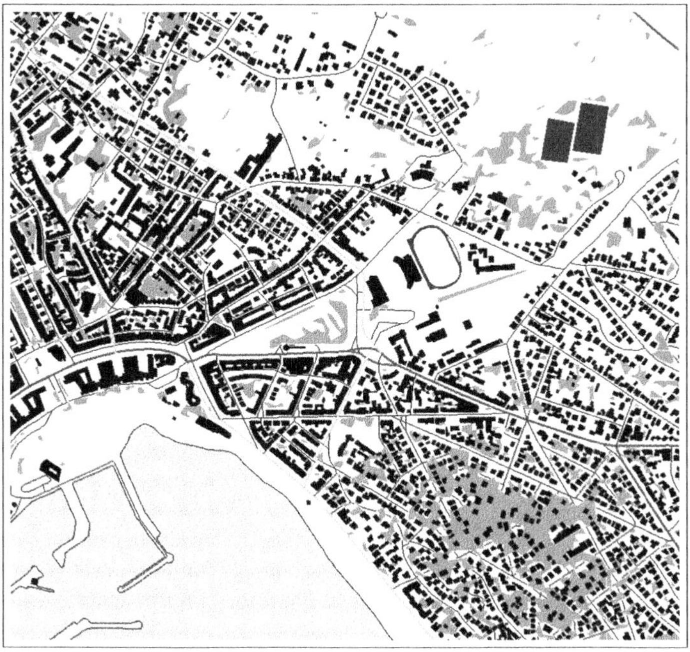

Fig. 8.7. A traditional NMA legend (BDTopo IGN France)

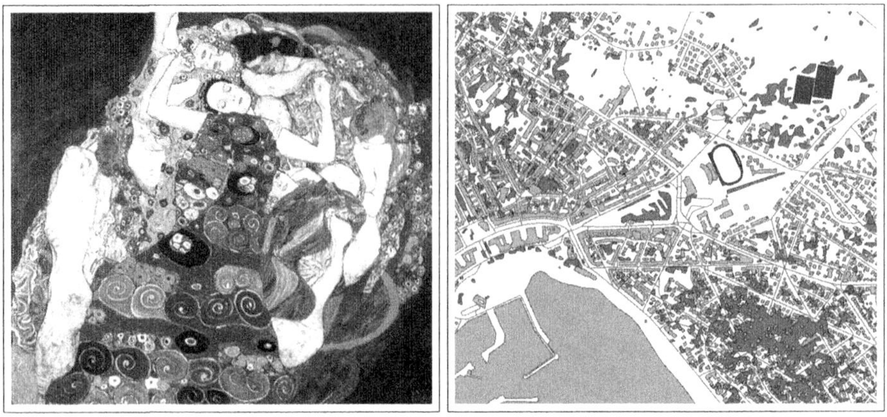

Fig. 8.8. Use of colours of a Klimt's painting

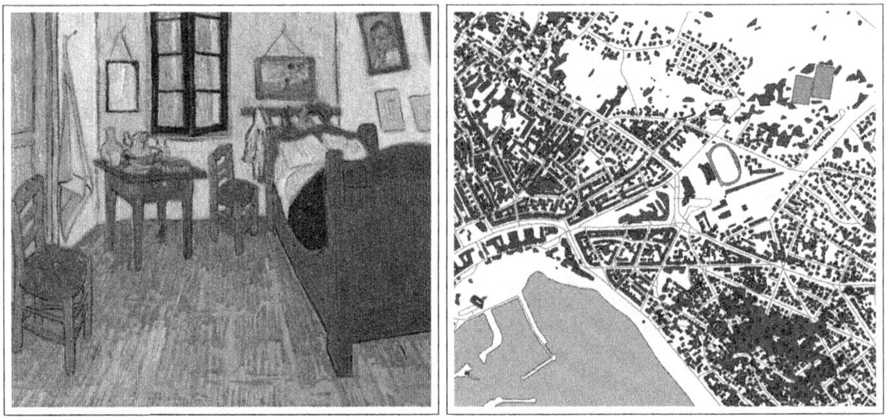

Fig. 8.9. Use of colours of a Van Gogh's painting

8.4 Conclusions and Perspectives

This chapter has presented cartographic conception as a creative process guided by the graphic semiology theory. We highlight that the map design process relies on cartographic rules, but there are no rules that explain how to best choose accurate styles. We clarify our approach by detailing the exploration of colours and the cartographic rules supported. We explain the new creation strategy – Palettes – that proposes to the user various painters palettes can be applied to their data, so creating a legend 'in the style of' a certain painter. Therefore, in this chapter, we argue that on-demand map design should benefit altogether from user's artistic capacities and from cartographic theory.

A prototype of an automatic tool of legend conception is under development. Our choice of colour as the only visual variable to study might seem to be restrictive, but we needed to limit the influence of visual variables for the test of our application of dialogue. We have also to explore several painters' palettes in order to find more aesthetic and powerful legends while respecting cartographic rules.

Moreover we need to propose various paintings according to the type of user' data to be used. In our example, our zone of interest is an urban zone near the sea and the colours of Matisse are great for this particular area. But, if we only have rural zones, we should prefer to use a painting with more greens and browns.

Finally we could characterise our legends in indicating their level of 'Matisseness'.

References

Bertin J (1967) Sémiologie Graphique. Paris, Mouton

Buard E, Ruas A (2007) Evaluation of colour contrasts by means of expert knowledge for on-demand mapping. In proceedings of the XXIIIe International Cartographic Conference, Moscow, Russia: International Cartographic Association

Chesneau E (2006) Modèle d'amélioration automatique des contrastes de couleurs en cartographie – Application aux cartes de risque. PhD Thesis University of Marne La Vallée, France

Christophe S (2007) Legend design by a novice user on the Web. In proceedings of the 5th Geographic Information Days. GIDAYS'07, Münster, Germany

Christophe S, Bucher B, Ruas A (2007) A dialogue application for creative portrayal. In proceedings of the XXIIIe International Cartographic Conference, Moscow, Russia: International Cartographic Association

Domingues C, Bucher B (2006) Legend design based on map samples. In proceedings of the 4th GIS Science conference, Extended abstracts, Raubal, Miller, Frank and Goodchild. (eds) IFGI Prints, Münster, Germany

Itten J (1967) Art de la couleur. Original edition in 1961: Kunst der Farbe. Eds Dessain and Tobra

Krygier J (1995) Cartography as an art and a science? The Cartographic Journal 32: 6. pp. 3–10

MacEachren AM (1995) How maps work. Guilford Publications, New York

Monmonier M (1991) How to lie with maps. University of Chicago Press

Robinson AH (1952) The Looks of Maps. Madison: University of Wisconsin Press

Tufte E (2006) Beautiful Evidence. Graphic Press

Weger G (1996) Cartographie: La couleur. Cours ENSG-IGN

Note: The original figures in this chapter are in colour. For colour reproductions, refer to e-book online materials.

9 What Can We Learn from the Masters? Color Schemas on Paintings as the Source for Color Ranges Applicable in Cartography

Lucie Friedmannová

Department of Geography, Masaryk University, Brno, Czech Republic

Abstract

The technology boost in the second half of the 20th century caused a partial resignation of map makers concerning the artistic value of maps. Now it is the time, that technology itself is not enough. Technology, as much as it is important for development, it does not bring beauty to our work. Nowadays it isn't the truth that beauty equals quality, as it was for example in the 16th century; however, harmony which is the main part of what we understand under the term of beauty is what brings us to the next level of comprehension. Let's look on the work of landscape painters with 'cartographic' glasses. The masterful paintings full of vibrant colors are an excellent source of color harmonies. As such, there is possibility to use them as a template for creating better maps.

9.1 Introduction

According to John Ruskin (1846) or Kenneth Clark (1949), landscape painting was one of the leading themes of 19th century art in Europe. In these days landscape painting and cartography walked hand in hand (let's not forget Vermeer's painting "Geographer" from 1668). A landscape painting was as much a document as a map was an artistic object. Impressionism brought less analytic details and more feelings to painting. On the contrary, cartography naturally followed path leading to its more technical and scientific aspects. The difference between painting and 'snapshot' photography is the same as the difference between map and aerial 'photomap'. Painting and map are all about interpretation, capture of the 'unseen'.

Almost from the beginning of cartography, maps were perceived as a graphic work on the same level as prints from artistic woodcutting, engraving or etching.

The techniques, employed by artists as Albrecht Dürer (1471–1528) were the same as the ones used by cartographers. Old maps are often swamped by additional graphic art picturing landscapes, towns and even people in local costumes. It is reasonable to believe it was the effort to brighten those pictures around the map what brought colorization into cartography. For prints the use of watercolors or colored inks was ideal, but there are also known large wall maps/paintings rendered as frescos (for example the Gallery of Maps in Vatican Museum from the end of 16th century) and let's don't forget the world of tapestries, which are especially known for use of bright and rich colors.

When 'thematic' cartography started to solve its own graphic and design problems, art and cartography have gone separate ways for a long time already. For many cartographers it is difficult to create required color ranges for thematic maps. From the simplest choroplet or chorochromatic maps to the most complex synthetic maps there is requirement on color schema to give a true picture of that part of reality which is represented on the map.

The study of a map, even without a legend, should give the user a general picture about the extend of a phenomena and also about their relative importance and impact. Colors are the primary tool for affecting map users. The art of painting has the ability to evoke ideas based on emotional experience and is therefor a well of knowledge for cartographers.

This chapter is the result of the long-term effort to define suitable color ranges usable for thematic mapping in a wide range of geography branches. The idea is to use reliable color combinations which would call on human emotions to help readability and understanding of maps. Even with color models based partially on psychology, as for example the Munsell color system, it is not easy to create sufficient color ranges. To use well known and loved paintings as base for making color schemas for maps is maybe a safer way to success than just trusting purely mathematical models. For color analysis of paintings Color Schemer Studio was used which allows to choose from two to nine random or selected color samples. The amount of color samples is fully sufficient for most of simpler maps.

9.2 Impressionism – The Story of Light and Color

The second half of the 19th century brought together several events which had influence on the art of painting in the sense of its meaning and used techniques. Many of them were, maybe ironically, of a technical origin. The most important ones included the spread of photography, boom of railways and development of oil paint in tubes (Kilian 1999).

The first one took from art of painting the necessity to document (the number of painted portraits radically decreased) and the others enabled for the first time in history to take "atelier" easily into the landscape. The portability of tube paints allowed "plein air" (outdoor painting). Until then only drawing or sketching was possible to be done outside. As mentioned earlier landscape painting was a popular theme in 19th century European art, so the possibility to go outside was liberating for many painters. Now artists could mix colors quickly and easily without having to grind pigments. E. H. Gombrich (2006) wrote in his Story of Art: *"Édouard Manet and his friends,"* nowadays widely known as 'The Impressionists' even if Manet was only loosely related to them, *"became aware, that when they were looking on landscape under the open sky, they did not see objects with one defined color, but rather as a bright mixture of colors, which are merged together into final color in observers' eye or more precisely in his mind"*. It would be naïve to think that impressionists were the first to understand this statement (let's not forget the color rich Venetian School, paintings of Peter Paul Rubens or wild scenes of Eugène Delacroix, who was convinced that colors are more important than drawing), but they were the first to fully sacrifice the art of drawing and shape the world by color itself (see *Figure 9.1)*.

Fig. 9.1. Claude Monet, *The Waterlily Pond*, 1899, oil on canvas, 88.3x93.1cm, National Gallery, London, UK. Source: http://commons.wikimedia.org, traced 10.11.2007.

The founder of the Impressionistic movement, Claude Monet (1840–1926), is famous, among others, for his brilliant landscape paintings. A new type of oil paints allowed him to move painting outside the atelier. He alone was capable to depict rainbows of color harmonies from nature and preserve them through his paintings up to the present times. However Monet and his followers were not guided only by "the beauty of the moment". They aimed to use colors scientifically, according to the color theories of the French chemist Michel E. Chevreul. This fact makes them an excellent source of inspiration for cartographers. Painters had to make deliberate decision about which colors to use and inadvertently created exquisite color ranges.

Claude Monet was one of the most devoted painters to "plein-air". Monet's goal was to catch the beauty of the moment in constantly changing nature in completely different way than even contemporary photography or film can. He wanted catch not only the reality of the landscape, but also its spirit. In his own words *"Artist has to use quick brushstrokes directly on canvas and the detail subordinate to general impression"*. As the result of this rule, impressionists cease to mix colors on a palette and started to lay colors on canvas separately as small blots and lines. Thanks to this "new technique" impressionistic paintings are full of much more brighter tints than ever before (Gombrich 2006).

Terms scale, gamut or range means ideally evened sequence of elements, be it tones, tints or numbers. The principle of a harmonic color range is based on the theory of complementary colors. By the analysis of some of Monet's paintings we can see the impossibility to choose sample of tints without harmony.

Beautiful color schemas can be found also on landscape paintings by Paul Cézanne or Vincent van Gogh. The separation of color led to an extreme style called 'pointillism' by Georges Seurat (see *Figure 9.2)*. Seurat juxtaposed small dots of pigment according to his interpretation of scientific theories of color and optics (The Minneapolis Institute of Arts 2007) and according to Pierre Auguste Renoir *"created the image-world where the continuous form is built up from the discrete"* (Harden 2007).

It is this 'new style' in laying pigments on canvas which makes impressionistic paintings such a good source for cartographers. Critics in early stages of impressionism condemned painters for a 'lack of details' and 'roughness' (Gombrich 2006). It was due habit of this time to look at painting close up as people in 19th century were used to study details on academic paintings. An impressionistic painting is necessary to study from distance – the distance allows the human eye and mind to capture its magic and merge color spots into shapes and objects. Nevertheless, it is this roughness and bold brushstrokes that allow curious cartographers to look close up and see 'into the painter's kitchen'. What we see is a wide range of color hues from brilliant to light pastels but also a scientifically based system.

Fig. 9.2. Georges Seurat, *La Parade* – detail showing pointillism technique, 1889, oil on canvas, 99.4x147.8cm, Metropolitan Museum of Art, New York, USA. Source: http://artchive.com, traced 10.11.2007.

Even if we are nowadays maybe seeing science in solely artistic work of impressionists[1], they surely came to similar results by pure observation as 19th century scientists by experimentation. Balance and harmony which they were able to achieve is lying probably purely in their talent.

9.3 Color range – From Chaos to Order or from Order to Chaos?

The important point for cartographers is, that impressionist painters, with Claude Monet leading the way, managed to create relatively easily distinguishable color

[1] *The Principles of Harmony and Contrast of Colors* by Michel E. Chevreul, which had proved influence on post-impressionists as Seurat, was first published in 1839. Impressionists established themselves in the 1870s, but according to G. Roque (1996) there is little of actual prove they had knowingly applied his theories on complementary colors and simultaneous contrast.

ranges on their paintings. These ranges have everything cartographers demand. They have a more than sufficiently wide number of tints, great level of contrast and last but not least undisputable harmony between used color hues. The last is maybe the hardest for non-artist to reach. Also as impressionistic paintings follow theory of complementary colors, there is an option to represent phenomena with bi- or tri-polar characterization. In geography these are typical in almost every sub-discipline – for illustration let's name temperatures for climatology or unemployment for demography. The bold, clash colors of some of the modern art are not that easily applicable in harmonic color ranges usually used on maps as color harmonies of impressionists.

So, why is it so relatively easy to distinguish colors or tints from impressionistic paintings? As it was written before, impressionists used bold brushstrokes and that means relatively homogenous color spots left on canvas. In previous times (for example academic style), the brushstrokes were very fine which means smooth transition from one color hue or tint to another and thus very hard identification of threshold from one tint to another. The best-known representative of academism is the ingenious portraitist Jean Auguste Dominique Ingres (see *Figure 9.3)*.

There is no discussion on color harmonies on academic or any other type of paintings, however, the impeccability of tint transition is not playing into mapmaker's hand. On the other side, on impressionist's landscapes we can relatively easily recognize borderline between color tints, because there is depict color contrast created by outdoor light, which is brighter than the one in ateliers. That does not mean that there are no transitions at all, just that there are areas of colors big enough

Fig. 9.3. Jean Auguste Dominique Ingres, *Luise de Broglie, Countesse d'Haussonville*, 1845, oil on canvas, 131.8x92.1cm The Frick Collection, New York, USA, Source: http://www.artchive.com, traced 20.11.2007 – on detail of the gown is possible to see smooth transition of tints.

for us to be possible to "pick" color in one clean tint or shade and put it into our "palette" or better said color scale.

Let's have a detailed look on one of Monet's paintings of *Saint-Lazare Railway Station* (see *Figure 9.4)*. The painting itself depicts view from inside of the railway station hall with trackage, boosting with smoke from the steam locomotive. Color schema is relatively simple – hues from light yellow to deep ochre for ground and buildings, grays and blues for smoke and sky. Black engine makes pivot of the painting. This is an excellent example of the contrast of complements as it is defined for example by Itten (1976). According to this theory, complements complete each other and create the feeling of harmony and simultaneously maintain ample level of contrast for creating bipolar color scale.

Such color scale is possible to use, as it was mentioned above, in many geographic themes. Bi-polar schemas are also known as diverging (Brewer 2005) and are used when from focal point, which is relatively insignificant (it can be median, average or other, for dataset meaningful value, such as zero for temperatures or natural increase/decrease), values spread to two sides, when one is usually of positive and second of negative nature. Apart from the contrast of complements there is often the contrast of cool and warm used (Itten 1976). These two are naturally intertwined for complement colors laying on opposite sides of the color circle. To determine specific twosome of color hues for creating such a schema and assign them to a specific theme is more a problem of association (again for example temperatures: red = high, blue = low or precipitation: yellow = dry, in sense 'desert', green = wet, in sense 'a lot of vegetation' (Friedmannova 2001)) than of color harmonies. Due to a high number of themes with spatial dimension there are always more than several suitable subjects for using a created color scale even if we are being picky

Fig. 9.4. Claude Monet, *Gare Saint-Lazare* – detail, 1877, oil on canvas, 75.5 x 104cm, Musée d'Orsay, Paris, France. Source: Denis Kilian (ed.), *Your visit to Orsay*, Versailles 1999.

about which associate together. So if we want to create a map based on colorization of our favorite painting, it is not a problem. Nevertheless, a mapmaker should be aware of standards and traditions in visualization of data in specific geographical sub-disciplines.

In this case we could probably use *Saint-Lazare Railway Station* as base for choropleth map of cloudiness and sunshine (tints of blue-violet for cloudiness x tints of yellow/ochre for sunshine) or for representation of Age Structure of population (tints of yellow/ochre for young and blue/violet for old). An example of a color scale 'picked up' from Monet's *Saint-Lazare Railway Station* is shown on *Figure 9.5; Figure 9.6* is an example of a simple map where it is applied.

R	115	174	213	235	238	218	98	50	82
G	52	29	107	140	193	211	139	110	97
B	19	0	19	22	102	218	193	160	116

focal point
(point of divergence)

Fig. 9.5. Bi-polar color scale "picked up" from Monet's Saint-Lazare Railway Station with RGB values

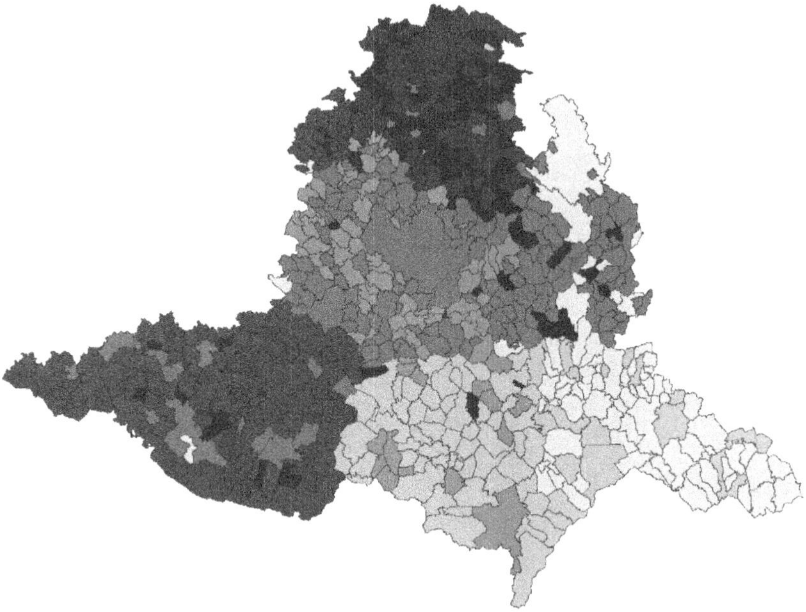

Fig. 9.6. Monet's Saint-Lazare Railway Station transformed into simple map

Complementary (see *Figure 9.7a)* is naturally not the only color schema we can choose from. There is an analogous schema *(Figure 9.7b),* monochromatic (or sequential – *Figure 9.7c)* schema and an tertiary (or triad – *Figure 9.7d)* color schema which is used in cartography or rather in visualization of data from geographic disciplines. Combination of sequential and tertiary color schemas *(Figure 9.7e)* or complementary and analogous (called harmonic (Parramón 1998) – *Figure 9.7f)* is possible to use when there is necessity to distinguish between kindred phenomena on one side and absolute opposite on the other side. These situations are typical for visualization in crisis management or in environmental studies. As an example we can use maps of forest fires. On the one side there is the information about the type of forest and its condition, which is necessary to differentiate, because of its importance for assessing its fire danger – these information are essentially of similar type. On the other side there is actual information about the existing fire, where the most important to know is its extend. This information is in its spirit absolute opposite, so the use of harmonic color schema is well called for.

Paintings are also invaluable in assembling qualitative color schemas. Knowledge of color theories does not necessarily mean ability to apply it into color schema usable in cartography or usable at all. Sometimes it is complicated to choose colors that do not clash. The art of painting has been attracting people with natural 'feel' for color harmony for many centuries. Outcomes of their work are rainbows of consonant hues and tint ready to be used.

Especially when we want to apply the extended theory of cool and warm aspects of color, also known as 'Theory of seasons' (it is widely used in design and cosmetics

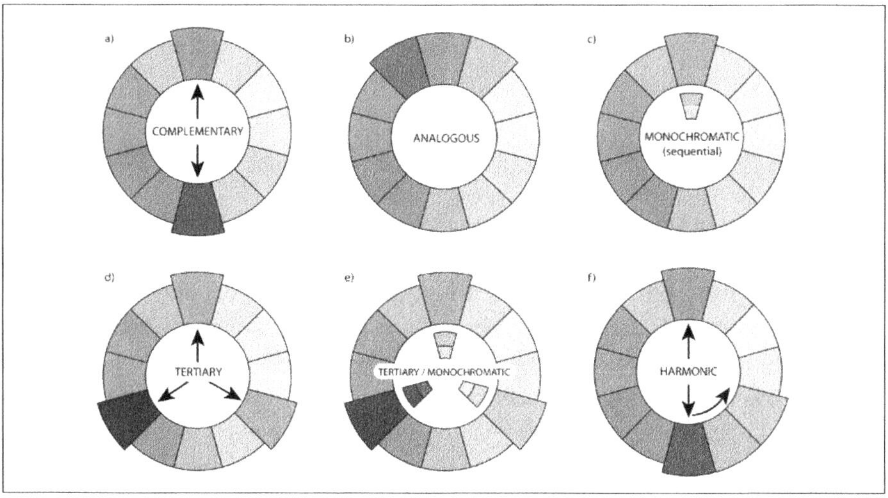

Fig. 9.7. Various color schemas on color wheel: a) complementary, b) analogous, c) monochromatic or sequential, d) tertiary or triad, e) combination of tertiary and monochromatic, f) harmonic (combination of complementary and analogous)

circa from 70s of the 20th century), there is highlighted ability of color change when blending with blue or yellow background. In short, the temperature aspect of color hue is stretched out on whole color circle. This shift is based on changes of natural daylight through day and its influence on color hues. The watercolorist Bruce MacEvoy says on this theme: *"In general, as the intensity (illuminance) of light on or through materials increases, the apparent color of the materials shifts toward yellow: blues become blue green, greens become yellow green, oranges become yellow, reds become orange; violets shift either toward red or toward blue, depending on the hue balance. The effects of reflected, filtered or shadowed light change in the opposite direction, from yellow to red or green blue to blue. (For example: layers of translucent yellow material redden as they become thicker, greenish pools darken to blue green as waters get deeper, etc.)"* (MacEvoy 2007). The second idea is the existence of specific characteristics for color tints in each particular season (in a simplified way winter colors are described as cool, rich, with blue undertones, summer colors are also with blue undertones, but fine and 'powdered', spring and autumn colors have yellow undertones, first are light, second rich). This fact is giving new possibilities not only to qualitative color schemas, but has also influence on quantitative scales running through three and more hues (spectral). Cool color schema can now include traditionally warm colors and vice versa.

Impressionistic landscapes were, also thanks to invention of colors in tubes, painted outdoors. Effects of what we call 'Theory of seasons' are embedded into the paintings themselves purely by artists' observations. The yellow or blue undertones are often hard to find for a person without trained eye. To pick up colors from paintings is a relatively safe way to avoid color swing from one 'season' to another. The tints' 'swinging' can produce misinterpretation of map. The value or interval could be interpreted as something unusual or important, albeit it is not.

9.4 Harmonies and How to Pick them

Everyone who occasionally has to colorize a map knows how hard it is to harmonize tints. And now we are only talking about simple areal maps without additional features like lines, points and names. Every areal color schema must be set against other map features and shifted in accordance with needs of the rest of map features' readability. In this situation every help with mixing color schemas is welcome. In the previous text was already established why impressionistic paintings, landscape paintings in particular, are a good source of color harmonies. But even with this knowledge it is not exactly easy to extract cartographically applicable color scales.

First – we can hardly go to Musée d'Orsay or any other place hosting impressionistic exhibition every time we need to colorize a map. Secondly – even if we could go and on top of that we could get to the selected painting close enough to use our well hidden colorimeter, the result would be in vain (brief experiment with picking colors by colorimeter directly from oil painting – albeit it was naturally not one of Monet's – did not provide satisfactory results among others because of uneven surface of painting). On the other side, there are hundreds of good reproductions, a lot of them reachable through the Internet (almost every bigger gallery has its web pages and there is lot of web galleries where it is also often possible to see paintings from private collections).

Although truth is, that even the best reproduction can not compare to original. Some paintings are more suitable than others, but the colors are never exactly the same. For example, even the best reproductions of Van Gogh's 'Sunflowers' are nothing more than a bunch of flowers in a simple vase. In reality the painting is like tied up sunshine radiating energy through the whole exhibition room in the National Gallery in London. Fortunately, cartography is much less ambitious. Color shifts caused by reproduction and digitalization of paintings are not so much troubles as it would seem. The color shift is usually uniform in the frame of painting's color range and thus do not reduce applicability as map color schema even if it somewhat reduces the artistic experience.

One way, how to separate out needed colors from selected painting is to open it in one of the many graphic programs (usually named Photo- or Image-something) and help yourself by using tools named 'Eyedropper' or 'Color Picker'. With the eyedropper tool you basically sample color from an image to create a new color swatch. You carry on as long as you need. The following process of adjustments of the created swatches is long and uncomfortable and practically erases the advantage given to us by the painting. The more you zoom in on the painting, the more colors differentiate. The goal is to effectively use aggregation of pixels with similar values.

The better way is to use one of the tools designed for making color schemas (they are often oriented on web design). In this case Color Schemer Studio was used *(Figure 9.8* – http://colorschemer.com).

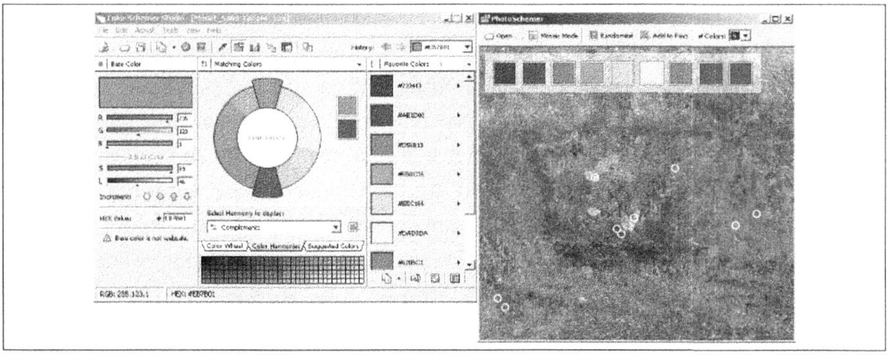

Fig. 9.8. Color Schemer Studio user face with PhotoSchemer. White circles indicate chosen colors.

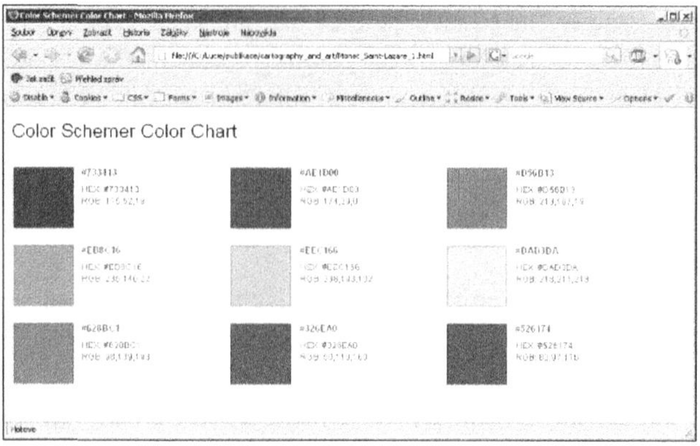

Fig. 9.9. HTML file generated from Color Schemer Studio

Part of this software is the Photo Schemer with a possibility to open any image (in gif, jpg or bmp format) and pick from 2 to 9 color samples. The picked tints are instantly seen on top of the image and they are movable even after picking. When the scale is finished we can add selected color swatches to favorites. The Color Schemer can export finished color palettes into different formats and generate RGB values of hexadecimal code *(Figure 9.9).*

The possibility to see chosen tints alongside each other is of great help when assembling color scale. When in need of generating qualitative type of color scale, it is possible to let Photo Schemer place samples randomly. Randomly chosen color schema based on Monet's paintings is always in harmony, no matter how many times the function randomize is used. This is the main fact, why cartographers should be looking outside cartography itself. For creating better colorized maps the hardest work was already done.

9.5 Conclusions

The incipient question – what we can learn from the Masters – has a simple answer: A lot! In area of color management knowledge, experience and practice from the art of painting are invaluable. Sometimes it seems as if people who call themselves scientists were afraid of things that can not be measured. This simple example of creating color schemas shows us how nowadays a basically technology oriented field can achieve satisfying results by using artistic work as an inspiration.

Professional mapmakers are probably not in great need of helping hand when creating color scale, but widespread of map producing applications brought possibility of mapmaking to many people without any or with small cartographic knowledge. Boost of people who are actually making maps in combination with not always well prepared technology has outcome in decrease of aesthetic value of maps. To look outside of cartography itself can maybe at least partially reverse this trend.

Why concentrate on impressionists and not later, even more "blotchy" artistic style from fauvism to pop art? The answer is relatively easy. Impressionism is well known for its interest in color harmonies and is worldly popular. The wide popularity means close positive resonance to the majority of the population. There is no or very little need to verify functionality of color harmonies made on base of impressionistic painting. They were already verified. As for more modern or older art – that only depends on the mapmaker's needs and taste.

This research is part of the project "Dynamic visualization in Crisis Management", supported by the Ministry of Education, Youth and Sports of the Czech Republic.

References

Brewer CA (2005) Designing better maps. Redlands, USA: Esri Press

Clark K (1949) Landscape into Art. London: John Murray

Color Schemer Studio, http://www.colorschemer.com, traced 3.12.2007

Friedmannová L (2001) The Climatologic Maps in Atlas Making – Visualization Characteristics. Kartografické listy, vol.9, pp. 67–72

Gombrich EH (2006) The Story of Art. Praha: Argo

Harden M (2007) Mark Harden's Artchive, entry Georges Seurat, http://www.artchive.com, traced 10.11.2007

Itten J (1976) The Art of Color. New York: John Wiley and Sons, inc

Kilian D (ed, 1999) Your visit to Orsay.Versailles, Fr.: Art Lys

MacEvoy B (2007) Watercolors. http://handprint.com, traced 26.11.2007

Parramón JM (1998) Teorie barev (The color theory), Praha: Svojtka a Vašut

Roque G (1996) Chevreul and Impressionism: A reapraisal, Art Bulletin, vol. 78, no. 3, March, pp. 26–39

Ruskin J (1846) Modern Painters. Vol. III. Of Ideas of Beauty. London: AN

The Minneapolis Institute of Arts, http://www.artsmia.org/, traced 10.11.2007

10 The Role of Augmented Elements to Support Aesthetic and Entertaining Aspects of Interactive Maps on the Web and Mobile Phones

L. Tiina Sarjakoski[1], Tapani Sarjakoski[1], Ilpo Koskinen[2] and Salu Ylirisku[2]

[1] Finnish Geodetic Institute, Department of Geoinformatics and Cartography Masala, Finland
[2] University of Art and Design Helsinki, School of Design, Helsinki, Finland

Abstract

Maps have become commonplace tools within mobile phones, as indicated by recent developments in commercial markets. Wayfinding and various location-based services have become one of the primary purposes for their everyday use. At the same time, multimedia phones are becoming personal, ubiquitous entertainment centres. As a consequence, not only technical and functional elements are important for a user, but also the overall user-experience, including issues such as aesthetic elements and characteristics that serve pleasure. There is very little reason to believe that maps on mobile phones are an exception in this respect. The chapter studies some aspects related to this. Such factors as graphical design, the use of screen dynamics, animation and video as well as the use of sound landscapes and icons are studied. The fundamental role of a map as a cartographic language for communicating information and its relationship with respect to aesthetic and entertaining aspects are also discussed.

10.1 Introduction

10.1.1 Background

Over the last centuries, maps have been based on surveys and scientific measurements. They have been produced by professional cartographers, who have turned space into representations through various means. Since ancient times wayfinding has been the core goal for creating maps (Akerman and Karrow 2007). Maps are the foundation of work for many occupations, and indispensable for many others. More recently, with the Web and mobile phones becoming ubiquitous, maps have become commonplace tools. Wayfinding and various location-based services are the primary purposes for their use today. At the same time, multimedia phones are becoming personal, ubiquitous entertainment centres.

As a consequence, not only technical and functional elements are important for the user, but also the overall user-experience, including issues such as aesthetic elements and characteristics that serve hedonic rather than rational faculties. There is very little reason to believe that maps on mobile phones are an exception in this respect. In fact, this is the case: they have many features that go beyond functionality, aiming primarily at augmenting user experience.

This chapter studies some of these techniques that enhance user experience. Such factors as graphical design, the use of screen dynamics, animation and video as well as the use of sound landscapes and icons are studied. The fundamental role of a map as a cartographic language for communicating information and its relationship with respect to aesthetic and entertaining aspects are also discussed.

We limit this paper to interactive maps that consumers can use either on the Web or on mobile phones. Our focus is primarily on two-dimensional maps, and we discuss three-dimensional maps only when 3D effects are used to enhance user experience. We do not deal with games that integrate sound, 3D space, and a set of characters into a story line.

10.1.2 Cartographic Communication and Previous Research

Spatial communication in cartography has a long tradition of research on how people perceive their environment. The cartographic communication model theory (Board 1967, Keates 1964, Kolacny 1969) has its roots in information and communication theory (Shannon & Weaver 1946) and semiology (Bertin 1967) *(Figure 10.1)*. According to cartographic principles, a map is an abstraction of the real world (Mackaness 2007), and throughout civilization, maps have been used as a powerful means of communication. Some of the earliest records of maps (using clay tablets) date back to 2300 BC (Turnbull 1989). Since we are by nature spatial thinkers (Kaiser 1993), our efforts to develop theories and models often revolve around visual forms and metaphors (Mackaness 2007).

The development of Geographic Information Systems (GIS), mobile interactive terminals, positioning systems and the integration of ubiquitous services enable new possibilities for exploiting location-related information. Instead of presenting geospatial information by only two-dimensional graphical map symbols, future interactive maps may be integrated into multimedia tools for communication, taking, for

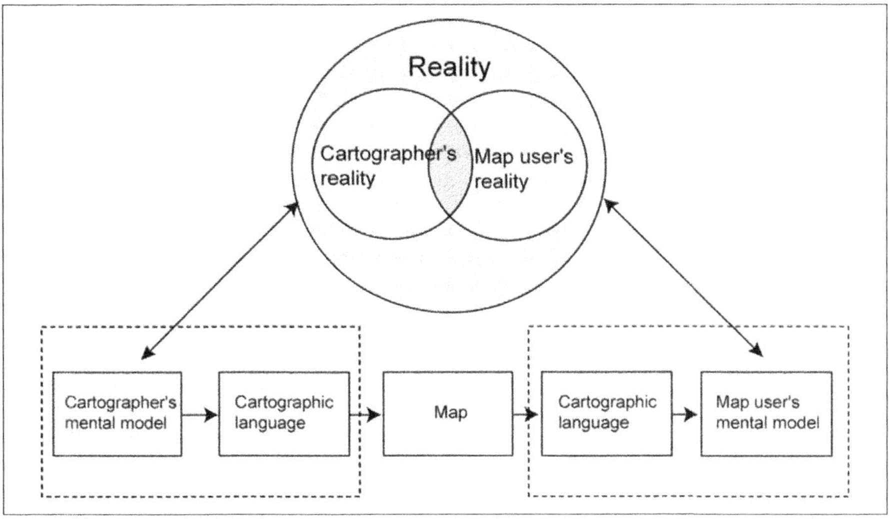

Fig. 10.1. Cartographic communication model (after Kolacny 1969)

example, advantage of embedded tags in the environment, through which the user can interactively communicate with the surrounding world (Morita 2007, Retscher 2005) and augmented elements in it (Kolbe 2004). However, the diversity of the multimodalities provided by the new technology does not necessarily imply that the user interprets the message related to the spatial context (Golledge 1996, Meng et al. 2005, Paay & Kjeldskov 2005, Weissensteiner & Winter 2004) in the best possible way. Users may also have a new role as content providers, as the spatial services also allow them to share location data and place-dependent messages. As a result, the possibilities for different use of ubiquitous maps are immense. The development of emerging mobile map applications has leaned on the traditional map metaphor in which space is regarded as flat and static (Dykes et al. 2005). As the requirements of ubiquitous maps are considerably different compared to traditional maps used, there is a need to expand the model of cartographic communication into ubiquitous interactive use, where user-centredness, personalisation and immediacy are key elements.

It is also interesting to consider what role the dualism art/science used in cartography (Krygier 1995) has in this communication model. The aesthetic and entertaining aspects of maps are related topics since both aspects aim at waking an emotional response in the user (i.e. the map reader), and thus affect the communication process of the map. This applies regardless of the media used.

Communication is the essence of cartography (Lilley 2007); cartography can be seen as the language used to communicate spatial information for the user's needs (Kilpeläinen 1989). Although location-based services are a real and current example of where technology is bringing new and exciting methods for delivering personalised data to users, according to Lilley, cartographic visualisation still does not match the innovation of the technology being employed. Krygier (1995) studied the history of the art/science dualism in cartography and information graphics in general. In 1995 he was of the opinion that the artistic component of cartography has been undervalued during the past decade, which still seems to be the case.

The focus in development has been on rational technological solutions without underpinning irrationalities such as 'user experience' or 'aesthetics'. The earlier definition of cartography – "The definition of cartography as the art, science and technology of making maps, together with their study as scientific documents and works of art" (Meynen 1973) – emphasised the artistic aspect of cartography, while the current definition focuses more on the technology and communication aspects. The current definition of cartography by the International Cartographic Association (ICA 2003) is as follows: "The art, science and technology of making and using maps. A unique facility for the creation and manipulation of visual or virtual representations of geospace – maps – to permit the exploration, analysis, understanding and communication of information about that space." Nevertheless, it needs to be realised in modern ubiquitous environments as well.

In the following we study current examples where some elements or fragments of maps have been used to enhance the user experience or the aesthetic faculties of maps. The examples are divided into three categories:

- Augmenting maps
- Community augmentation
- Sketching

10.2 User-Experience Augmenting Elements

10.2.1 Augmenting Maps

Most maps in virtual space are based on surveys and scientific measurements and, in fact, mathematics (comp. Lynch 1991). Whatever symbols are drawn in these spaces, they are also above all exact: roads, blocks, buildings, types of buildings, and height (typically measured from the sea level). Good examples of this are the maps in *Google*, *Yahoo*, *TomTom*, and *MapQuest* – or the City Transportation systems almost anywhere. If these representations are somehow augmented, the methods of augmentation typically follow the same scientific canon. Thus, in *Google* maps, one can check the traffic situation on main roads in cities like Los Angeles and New York. Some of these functions also work on mobile phones *(Figure 10.2)*. With accuracy come several desirable functionalities like the possibility to calculate distances, areas, and driving times, which, in turn, support planning and other forms of 'acting from a distance' (Latour 1991).

This baseline can be augmented in several ways to support user experience. Seeing a city like San Francisco through a satellite gives a bird's eye view of the city, but scarcely tells one what the streets look like. Experience can be supported through several means on maps. Typically, photographs have been integrated into maps. Again, this is the case in *Google* maps, which has a 'Street View' option that functions in some cities. For example, the streets of San Francisco have been photographed from cars, these photographs have been built into 360-degree panoramas, and tagged for their place. By clicking the Street View option, one can get a street view. As *Figure 10.3* shows, these pictures give not only a street overview, but can also provide details of street life – like where alcoholics sleep and where people in black suits spend their time.

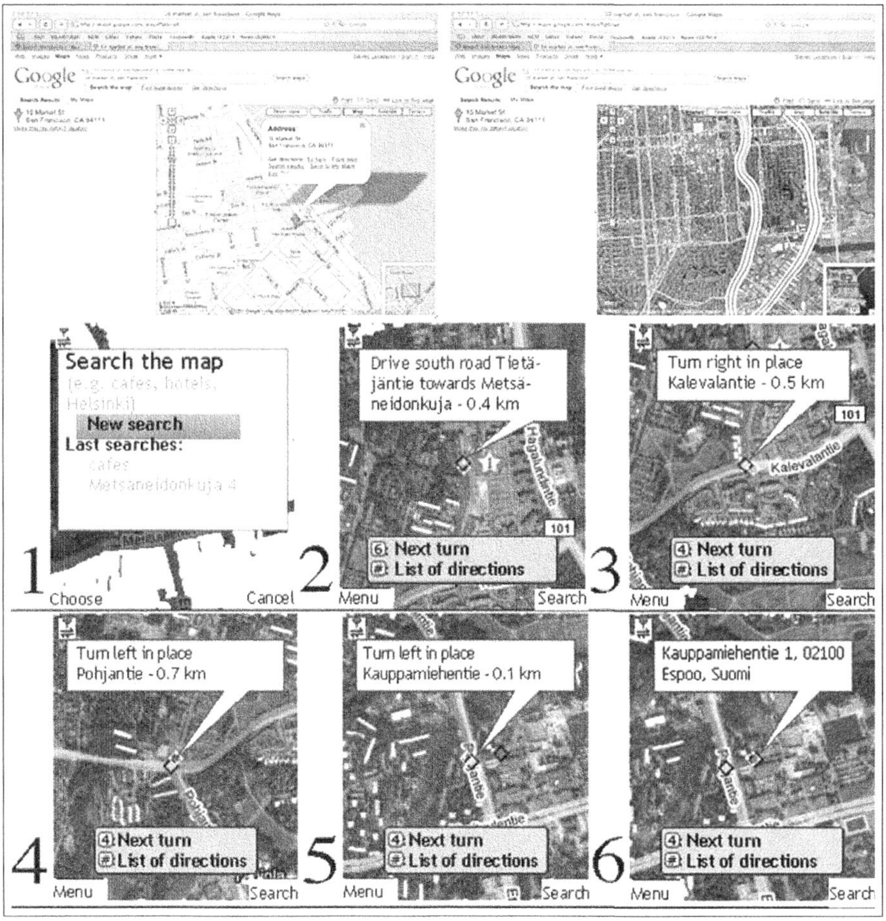

Fig. 10.2. *Google Maps.* Up, on the Web: map and traffic situation on the main freeways of San Francisco. Down, *Google Maps* on a mobile phone (left to right): searching cafes close to Metsäneidonkuja 4, Espoo, and driving instructions to a café in Tapiola, Espoo.

Fig. 10.3. Left: *Google Maps'* street view of Bay Bridge; right: street sign for Jack Kerouac Alley on San Francisco's North Beach

In an EU project, GiMoDig (2005) at the Finnish Geodetic Institute (FGI), Department of Geoinformatics and Cartography, a prototype was developed, through which spatial data were delivered to the user's mobile device by real-time harmonisation and generalisation, *Figure 10.4.* According to the map user's preferences, a vector format map delivered to the user was adapted to the current needs of the user based on season, age group and language. For example, if a tourist visits a national park during the winter, a map with skiing routes is delivered to him/her instead of a summer map showing the places where you can swim nearby.

The visualisation on mobile maps shown in *Figure 10.4* is based on the idea of enhancing the importance of the map objects by symbolising them with large comic-like icons, an idea familiar from old maps, *Figure 10.5.* This type of visualising awakes emotions, and as cited by Meng (2005, p. 11) "The design pattern's

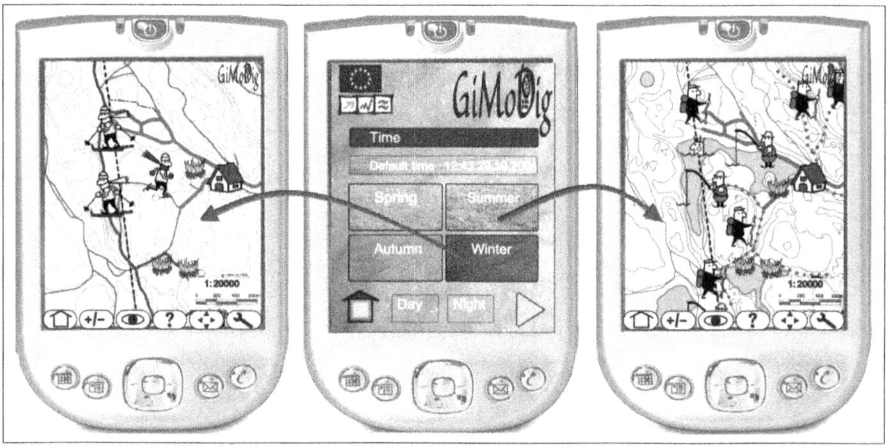

Fig. 10.4. Adaptive mobile maps delivered in real-time from the GiMoDig service. Visualisation adapted for teenagers and according to the time of the year: winter and summer maps

Fig. 10.5. The map of Udine, Italy. Source unknown

'affective emphasis' has the goal to encode the spatial information with symbols that do not necessarily carry additional information, but can possibly invoke positive emotions, and calm down in panic situations."

Another means that has been used to augment map experience is sound, which gives the user a lively feeling of life, even though this view is scarcely an accurate sample of life in that place. Recently, an American company built technology for integrating sound world into Google Earth (BBC NEWS 2007, Wild Sanctuary 2008).

One good example comes from the city of Tampere, Finland, where Aamulehti, the main local newspaper, constructed a sound landscape of Tampere for the Web (Aamulehti 2007). It is a relatively simple application in which one gets a sound sample of Tampere to empathize with by clicking hotspots on the map. *Figure 10.6* represents a collage of the site and some of its sound landscapes. By clicking round hotspots on the map, animated Shockwave files open into a new browser window with a soundtrack created in each place. By clicking the hotspots, one hears sounds from an ice-hockey arena (up, left), library (up, right), church (down, left), and a local café slightly outside the city centre (down, right). With these sounds, one gets an idea of what takes place in the city even if one does not know it (see Koskinen 2007, Ch. 5). Church bells evoke very different images of a place compared to the sounds of a political rally or a crowd watching an ice-hockey game.

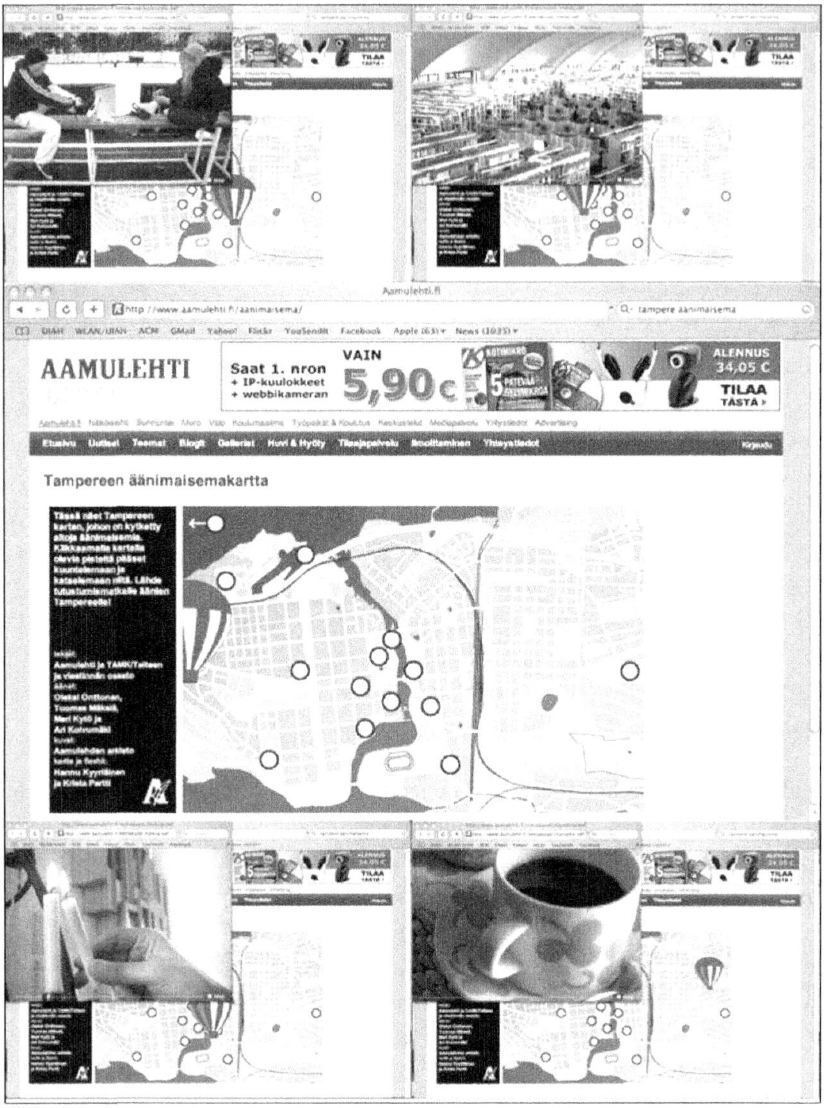

Fig. 10.6. The sounds of Tampere, Finland. In the middle is the main page of the service. Above and below are screenshots of animations one gets by clicking the hotspots (Aamulehti 2007)

A more complicated example in the mobile domain exists in Shibamata, Tokyo. Researchers at Keio University created a mobile application through which people in this township on the border of Tokyo can use mobile phones to share pictures and sound samples of the neighbourhood through a Web site. They also embedded bar codes into things such as postcards. Once people captured these bar codes with their mobile phones, they could send them to a certain number and retrieve these pictures and sound bites (for example, Kato 2006). In even more complicated examples, the sound world is integrated into a story that is dramatised. For example, soundwalk.com has produced a series of walks for places such as Lower East Side in New York (soundwalk 2008). One can experience the colourful history of this place – like the days of Meyer Lansky and Bugsy Siegel, and the heydays of the crack epidemic – by following a map and listening to dramatized histories of the neighbourhood on actual locations with a CD player.

These multimedia elements add little to the basic functionality of, say, Google maps or route maps. Their main functions seem to be experiential – giving a lively feeling of the place. Seeing palm trees and California sunshine in a picture of the Embarcadero, or seeing alcoholics and drug addicts sleeping on Columbus Avenue feed the imagination far more than seeing a satellite image of San Francisco.

Similarly, hearing sound landscapes from a less well-known town like Tampere immediately characterizes the city, transforming a map into an understandable human experience.

In another research project at the FGI, Department of Geoinformatics and Cartography, a mobile map experiment with 3D models and a sound landscape has been created, *Figure 10.7*.

The idea is that the user can experience something about the atmosphere at Nuuksio National Park on his/her mobile maps that are integrated into 3D models

Fig. 10.7. The user can experience the atmosphere of Nuuksio National Park on the mobile

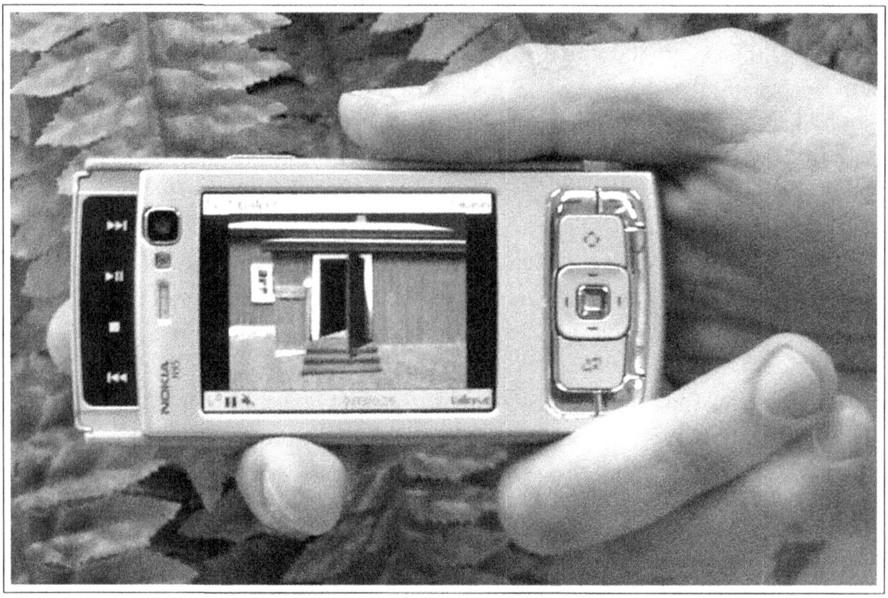

Fig. 10.8. A sound landscape is integrated into a 3D model

of the terrain and the buildings and sound landscape of the area. The user can zoom into the selected area on a 3D map, and continue zooming into a cottage in the area. While zooming nearer and nearer the user hears birds singing, a dog barking and footsteps entering the cottage and finally ending up in front of an open fire. Although this is an initial experiment, not a fully fledged application, it demonstrates the possibilities that are not used much as yet to enhance user experiences on mobile applications. It is a question of an overall experience where the user has the possibility to use his/her varying senses, and perhaps at its best, also receive comprehensive aesthetic pleasure.

10.2.2 Community Augmentation

More recently, Web 2.0 techniques have become more common on Web-based maps; maps become platforms for user community action. We can use *Google Earth* as an example to see what user communities do in these platforms. *Figure 10.9* shows again a picture of how San Francisco's landmark Transamerica Building has been rendered onto *Google Earth* (left, up). Although the movement first started in major US cities, European communities have been quick to get onto the bandwagon.

Figure 10.9 also shows how pictures and other types of user-generated content can be integrated into *Google Earth*. A picture of Helsinki Cathedral, which is rendered onto a map of Helsinki, also shows pictures that users have taken in front

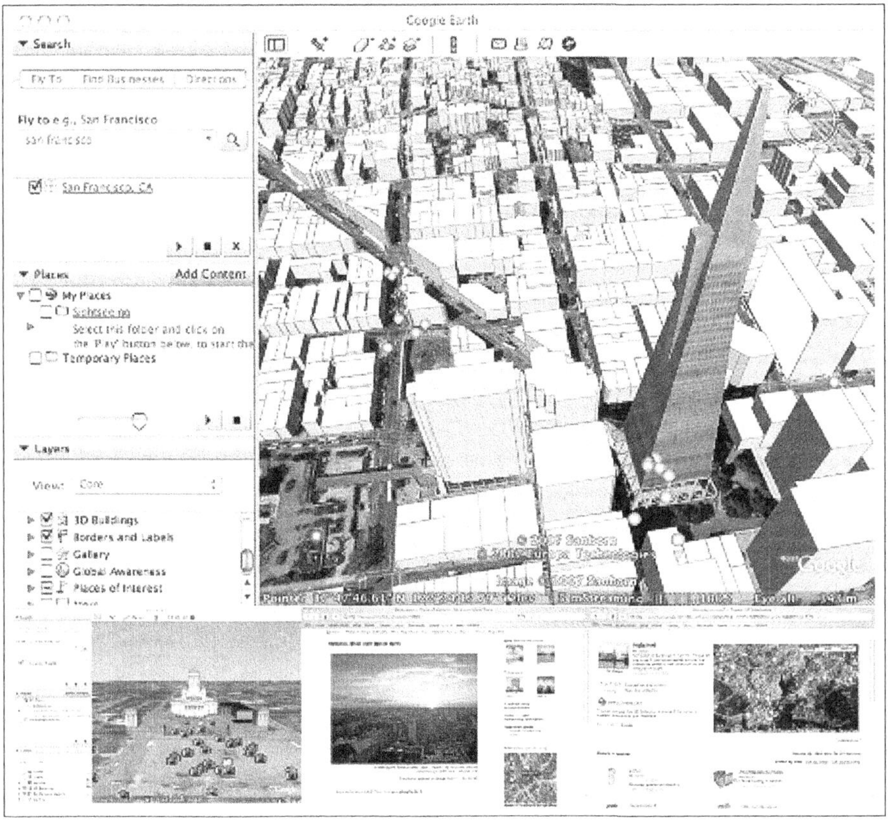

Fig. 10.9. Examples of Community Augmentation in *Google Earth*

of it and tagged to that place. By clicking the camera-shaped icons, one gets to see the picture on the Web (typically in places like *Flickr*, *Figure 10.9,* lower row).

In this usage, maps become platforms for community action rather than just business-to-consumer products. Through them communities of users can build interest communities, and probably even coordinate the actions of these communities.

Although many types of technical and artistic skills are exhibited in these platforms, their basis is still in measured representations aiming at accuracy. However, these platforms also support many types of social activities. For example, they encourage competition by encouraging communities to vote on the quality of 3D renderings and the quality of photographs on these sites. So far, community action in Google Earth has mostly been in the service of improving representations and the service rather than supporting fun and entertainment. In addition, applications have been developed to integrate thousands of pictures taken by amateurs into panoramic and 3D renderings, as can be seen by jumping into "the *Photosynth* Experience" part of a *YouTube* video (TED 2007).

10.2.3 Sketching

Some uses of maps differ radically from those that utilise maps for navigational purposes. Traditionally maps have been authored based on accurate measurements of exact dimensions, displaying these as a cartographic representation. For architects and community planners, typical cartographic presentations may not provide the best solution when they approach new tasks and try to capture the spirit of a place. In the early phases of planning, especially in projects that attempt to create a radically new perspective upon an area, traditional maps would only limit the ideas into the existing forms.

What kinds of solutions are there currently for community planning, and what kinds of developments are emerging that we could indicate? In the study of community planning between the University of Art and Design Helsinki, School of Design and the Department of Architecture of the University of Oulu, we participated in exploring new means of supporting the early phases of communal planning. Students of architecture were provided with several solutions that gave them new kinds of ways to utilise maps. One alternative was called the Web Map Media *(Figure 10.10)*, which attempted to enhance community members' participation in ideation and in bringing the place alive through pictures and comments. Another alternative was called the Ambience Wall *(Figure 10.11)*, which formed an interactive collage of pictures taken of the area. The continuously changing collage aimed to provide a location-related tool to inspire new ideas and give texture to imagination.

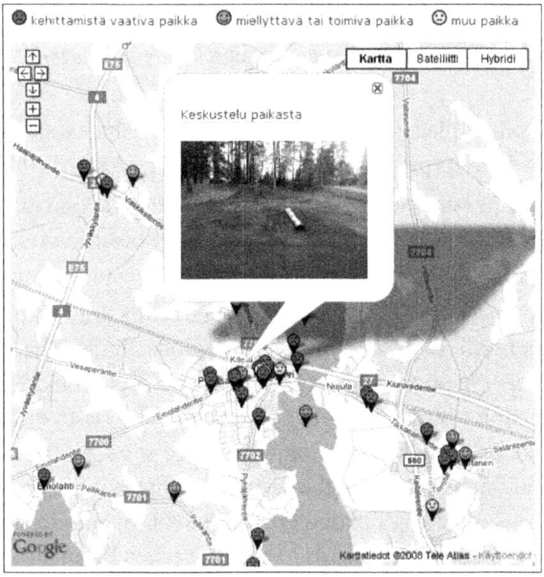

Fig. 10.10. The Web Map Media (Nuojua, University of Oulu, 2008)

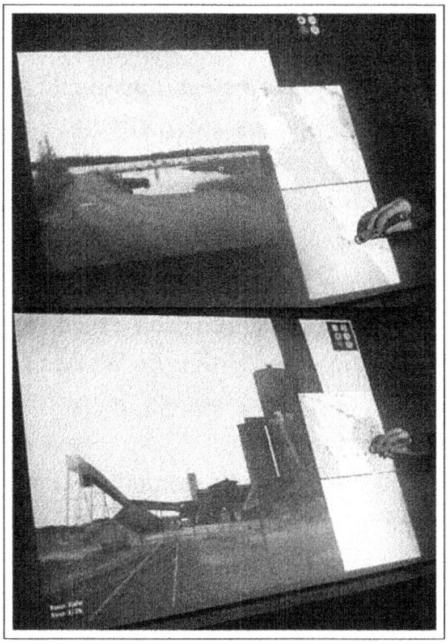

Fig. 10.11. Ambience Wall used as an interactive collage of pictures.

However, these technologies were either not yet on a sufficient level of quality to support early planning activities, or were introduced into the process in ways that gave it little support. Whatever the reason, the planners studied kept to traditional printed maps and semi-transparent sketching paper. Why do the traditional tools of sketching function so well?

One of the reasons might be that the use of sketching paper is scale-free and not tied into any particular measurement units. It frees designers and planners to draw sketches on any scale, draw perspectives, colour surfaces, and layer the papers on top of each other to create various views of the place. Another reason is roughness.

Semi-transparent sketching paper also allows making rough shapes, even draw pictures, and perspective sketches over the blurry image, which are filtered by the sketching paper. The sketching paper also supports imaginary constructs. Another reason for stepping outside accurate, 'mathematised' maps is that they make it possible to superimpose imaginary elements upon each other to try out various thoughts and hypotheses. Talking about semi-transparent sketching paper, an architect notes: "This stuff is transparent, you see the map through it. You can add several layers upon each other and compare them. There is always material for that."

Such 'geocaching' falls between the demand for accuracy and artistic use, in which accurate maps are a starting point for imagination. Artistic uses – including architecture – build on many types of demands, mathematical accuracy being just one of them. Means of representation have to support geometric imagination following cubism, ideas of society and community, ideas of history, aesthetics, and so forth.

10.3 Summary

With the Web, interactive maps have become available for the average consumer. With wireless and mobile technologies, interactive maps have become ubiquitous and, in some domains of life, a part of everyday life. These domains consist above all of maps on car navigation devices and maps for navigation on the Web and, increasingly, on mobile phones.

As maps have moved from the desks of professionals such as cartographers, geologists, architects, and geographers first into the Web and then to smaller devices, the means of expression have changed accordingly. While professional maps are characterised by accuracy based on measurements, maps aimed at consumer markets augment user experience with many types of expressive elements.

This chapter has explored some of these elements, including photographs, sounds, 3D renderings and, increasingly, interactive communities. *Table 10.1* presents some of these types and shows what kinds of functions these user-experience augmenting elements have.

Table 10.1. Types of augmentation and some of their functions

		Task assistance	Entertainment	Peer-to-peer fun	Artistic
Measurement-based	Standard applications	+			
	Augmentation	+	+		+
	Community	+	+	+	?
Non-measured	Artistic		+	+	+

Naturally, mobile devices pose many types of limitations for the kinds of elements that can be used and how they can be used. Still, technologically, most of the things studied above can already be used on mobile phones. For example, there are interactive wayfinding programs and map-based searches. Satellite images can be downloaded onto a mobile phone, street names can be imposed on them, and driving directions can be imposed on satellite images step by step.

10.4 Conclusions

This chapter has looked at some of the ways in which maps based on surveys are augmented on the Web and in mobile domains. Through our examples, we have explored some of the main categories of augmentation, including photographs, video, sound, and community interaction. As we have seen, maps have become a platform for many types of experiences rather than just something that supports such rational actions as wayfinding and planning. Simultaneously, they have also become experiences in themselves. Historically speaking, this is scarcely anything new; maps have for centuries had many types of aesthetic elements in them.

However, with new interaction techniques, these artistic elements are making a comeback. As we have seen, many types of augmentation are appearing in maps when they become interactive and as they are integrated to enhance the map-using experience. Still, what our review shows is that most of this work has been done based on technical rather than aesthetic premises. *Google Maps's* boxes and models of buildings are a beginning rather than a desirable state. As Kent (2008) suggests, the cartographers' desire to create aesthetically pleasing map may be stronger than their desire to ensure that their design meets other objectives or ideals, perhaps relating to communication. Our review shows that there are certainly new tools for creating aesthetically pleasing maps, and that an aesthetic perspective is called for to turn these tools into something useful that simultaneously engages our sense of beauty.

References

Aamulehti (2007) Tampereen äänimaisemakartta. (Eng. The sound landscape map of the city of Tampere). Online: http://www.aamulehti.fi/aanimaisema (06/05/08)

Akerman JR, Karrow Jr RW (eds) (2007) Maps – Finding our place in the world. The University Chicago Press: Chicago and London

BBC NEWS (2007) Online: http://news.bbc.co.uk/2/hi/technology/6639977.stm (06/05/08)

Bertin J (1967) Semilogie graphique, Gauthier-Villars, Paris

Board C (1967) Maps as Models. In Chorley R and Haggett P (eds), Models in Geography, London, Methuen, 671–725

Dykes J, MacEachren AM, Kraak MJ (2005) Exploring Geovisualization. In Dykes J, MacEachren AM, Kraak MJ (eds) (2005) Exploring geovisualization, Amsterdam, Elsevier, pp. 3–19

GiMoDig (2005) Geospatial info-mobility service by real-time data-integration and generalisation. Online: http://gimodig.fgi.fi/ (06/05/08)

Golledge R (1996) Human Wayfinding and Cognitive Maps. In Golledge R (ed) (1999) Wayfinding Behaviour, Baltimore: John Hopkings University Press, pp. 5–46

ICA (2003) A Strategic Plan for the International Cartographic Association 2003–2011. Online: The Association/Strategic plan (06/05/08) http://cartography.tuwien.ac.at/ica/index.php

Kaiser MK (1993) Introduction to Knowing. In Ellis SR, Kaiser M, Grunwald AJ (eds) Pictorial communication in virtual and real environments. London: Taylor and Francis, pp. 43–47

Kato F (2006) Seeing the "Seeing" of Others: Environmental Knowing through Camera Phones. In Nyíri K (ed) Mobile Understanding: The Epistemology of Ubiquitous Communication, Vienna: Passagen Verlag

Keates J (1964) Cartographic Communication. Abstracts of Papers of the 20th International Geographical Congress, London

Kent A (2008) Mapping the View: Exploring the Aesthetics of Map-Making. In Proceedings of the Symposium of Cartography and Art – Art and Cartography. January 31 – February 2, 2008, Vienna, Austria

Kilpeläinen T (1989) First Steps in Automated Map Understanding. Photogrammetric Reports (56), 1991, Reports and Collected Reprints 1988–1991, Transactions of the Royal Institute of Technology, Department of Photogrammetry, Sweden, p. 11
Kolacny A (1969) Cartographic Information – A Fundamental Concept and Term in Modern Cartography. Cartographic Journal, 6: 47–49
Kolbe TH (2004) Augmented Videos and Panoramas for Pedestrian Navigation. Proc. of the 2nd Int. Symposium Location Based Services & TeleCartography, TU Wien, 45–52
Koskinen IK (2007) Mobile Multimedia in Action. New Jersey: Transaction
Krygier JB (1995) Cartography as an art and a science? The Cartographic Journal. 32: 3–10.
Latour B (1991) Drawing Things Together. In Lynch M and Woolgar S (eds), Representation in Scientific Practice. Cambridge, MA: MIT
Lilley RJ (2007) Who Needs Cartographers? The Cartographic Journal, 44(3) 202–208
Lynch M (1991) The Externalized Retina: Selection and Mathematization in the Visual Documentation of Objects in the Life Sciences. In Lynch M and Woolgar S (eds), Representation in Scientific Practice. Cambridge, MA: MIT
Mackaness WA (2007) Understanding Geographic Space. In: Mackaness WA, Ruas A and Sarjakoski LT (eds), Generalisation of Geographic Information: Cartographic Modelling and Applications, Series of International Cartographic Association, Elsevier, pp. 1–10
Meng L (2005) Ecocentric Design of Map-Based Mobile Services. The Cartographic Journal. 42(1): 5–13
Meng L, Zipf A, Reichenbacher T (eds) (2005) Map-based Mobile Services, Theories, Methods and Implementations, Springer Berlin Heidelberg New York
Meynen E (1973) Multilingual Dictionary of Technical Terms in Cartography. Stuttgart: International Cartographic Association
Morita T (2007) Theory and development of research in ubiquitous mapping. In: Gartner G, Cartwright W, Peterson M (eds) Location Based Services and TeleCartography, Lecture Notes in Geoinformation and Cartography, Springer Berlin Heidelberg New York, pp. 89–106
Nuojua J (2008) Exploring the Possibilities of Web Mapping Technology in Participatory Urban Planning, a Master‘s thesis, University of Oulu, Department of Information Processing Science (in Finnish)
Paay J, Kjeldskov J (2005) Understanding and Modelling Built Environments for Mobile Guide Interface Design. Behaviour and Information Technology, 24(1): 21–35
Retscher G (2005) 3-D Position determination in a multi-storey building using a Pedestrian Navigation System. In Gartner G (ed) Proceedings of the Symposium "Location Based Services and TeleCartography", Vienna, Austria, November 2005, Geowissenschaftliche Mitteilungen, Technische Universität Wien, Heft Nr. 74, pp. 81–86
soundwalk (2008) Online: http://www.soundwalk.com/ (06/05/08)
Shannon CE, Weaver W (1949) The Mathematical Theory of Communication, The University of Illinois Press Urbana and Chicago
TED (2007) Talks Blaise Aguera y Arcas: Jaw-dropping Photosynth demo. Online: http://www.ted.com/index.php/talks/view/id/129 (06/05/08)
Turnbull D (1989) Maps are territories: Science is an atlas. Geelong: Deakin University
Wild Sanctuary (2008) Online: http://www.wildsanctuary.com/ (06/05/08)
Weissensteiner E, Winter S (2004) Landmarks in the Communication of Route Directions. In Egenhofer M, Miller H, Freksa C (eds), Geographic Information Science, Lecture Notes in Computer Science, Vol. 3234. Springer, Berlin, 313–326

11 Variables of Aesthetics in Maps

Felix Ortag

Institute of Geoinformation and Cartography
Vienna University of Technology, Austria

Abstract

Cartographers do not continuously work to strict rules when they design a map, some map elements are mainly designed by personal preferences, experience and habits. In this process the map creator considers aesthetics. In order to optimise this part of mapping it would be useful to know the variables of aesthetics: Which graphic elements and attributes of a map are considered when it's beauty is assessed? Colours, line styles, relief visualisation, font types, legend, map margin, overall impression, symphony of colours, composition, ...?

This chapter gives an overview of aesthetics in cartography and presents the results of a focussed interview which was done with 152 persons, where three exemplary topographic maps were compared to find out, how the beauty of maps is assessed.

11.1 Introduction

"Beauty gives blissful pleasure – which is why it is sought after." (Karvonen 2000, p. 86) Are cartographic products made under consideration of beauty? Is there a benefit in beautiful maps? Imhof (1965) says that the objectives of a map are (translated from German) "highest possible accuracy in respect to map scale, good geometric significance, good characterisation of shapes, highest possible clearness and good readability, simplicity and clearness of graphic expression and finally, as sum of all these qualities, a special kind of beauty which is inherent to a map." This could lead to the conclusion that a beautiful map is not necessarily a good map, but a good map should be beautiful. Imhof also states (translated from German), "A map is not only an image, above all it is an information medium. It has other responsibilities than a painting." But despite the fact that maps are objects fulfilling a function, having a specific purpose, globes and maps can often be found as wall and room decorations, probably because they are considered to be beautiful.

Franck (1998) thinks, that attention spent by our fellows is the new currency in economy and society and "the most irresistible of all drugs" (translated from German). Everyone lusts after attention and being attractive is a good way to get attention. In our modern times media confront us with a very high quantity of information, but only a small part gets our attention, the rest is filtered out. Therefore things which get attention can be considered as attractive. Following this definition of attractiveness and using it in cartography would mean that top selling maps are the more attractive maps, the opinion of experts is not taken into account here. But Franck also differentiates between high culture and popular culture: In the latter it is the attention of the masses which counts, whereas in high culture the attention of experts is essential.

This article does not want to find the rules of cartographic aesthetics, but it is intended to be preliminary work that can answer the question which elements of a map are important concerning aesthetics. Or which elements are mentioned above-average which could legitimate further investigations. Some may argue that maps should work, they should fulfil their function of communicating spatial information instead of being visually attractive. An answer to this could be the assumption that beauty in maps enhances the function of transferring content, but before this can be investigated one should know what makes a map beautiful.

It should be also mentioned that the question of absolute beauty is of course a permanent issue in philosophy, but as this work is driven by natural sciences philosophic approaches are not discussed in detail and the terms *beauty* and *aesthetics* are used synonymously. In an honest self-criticism one should also take in mind that the basic idea of finding cartographic rules of aesthetics will be heavily criticised by some philosophers – nevertheless the results should be useful for cartographers.

11.2 Absolute Beauty?

Trying to analyse and optimize the beauty of maps raises the question whether absolute beauty exists which is constant over space and time and independent of the beholder. The golden ratio for instance is not proven to be a god-given or natural proportion, but its use in art has a long tradition (golden ratio is approx. 1.618:1 or exactly: a is to b as (a+b) is to b). The fact that art students start their studies by learning "principles of beauty, such as symmetry and proportion, and only then eventually bend these rules to represent their overall vision of the world" inspired the Italian neuroscientists Di Dio et al. (2007) to do a survey on "objective beauty". They showed photos of ancient sculptures with canonical proportions (golden ratio) to test persons. Some photos were modified to get distorted proportions. During looking at the pictures the test person's brain activity was measured using magnetic resonance imaging technology. The authors come to the conclusion that there is a brain area for "objective beauty" that is activated by undistorted images only but when a person judges over the beauty of something, it is not only done by this area but in a combination with another area responsible for the person's emotional experience, this leads to "subjective beauty".

11.3 Cartographic Variables

Maps and the cartographic communication process can be explored through the concept of cartographic *semiotics* and its three branches *syntactics, semantics,* and *pragmatics* (Freitag 1971). A map is – as any graphic representation – a composition of points, lines, and areas. These signs and their relation to each other as well as rules concerning their generation and transformation are covered by cartographic syntactics (e.g. contour lines of differing value must not cross). The relation between used signs and their meanings (and consequently their associated real world features or facts) is discussed in cartographic semantics (e.g. contour lines and altitude). And finally cartographic pragmatics deals with the relation between signs and map users and the map using situation. At first glance aesthetics in cartography could be assigned to syntactics because it's about changing the used signs to improve readability and harmony but as the proverb says, "Beauty is in the eye of the beholder." Therefore a look into pragmatics also makes sense – the beauty of a map is not a property of the map or its used graphic language itself but a judgement given by the user. In digital interactive mapping systems it is possible to make user adaptive maps with individually adapted visualisations (see e.g. Zipf 2002). López and Balboa (2008) developed a framework to approximate cartographic products to the customer's expectations, which is also applicable for printed maps. This is not directly considering aesthetics but overall quality. Interestingly it was found out in the example implementation for a natural park map that the variables *packed size* and *weight* were the two most important ones, design variables followed with lower importance. This shows that if we consider beauty to be something the user awards to the map after observing his feelings, there is much more to improve than the graphic design of map elements to optimize map beauty. However, this is the case for any graphic design product and for a cartographer it therefore still makes sense to concentrate on the cartographic part of a map.

The aim of the interview described later was to find out variables in assessment of map's beauty. The following three chapters exemplarily give variables that are normally considered in map making. They are more or less also found in graphic design, but in difference to cartography graphic designers work with a higher degree of freedom, whereas maps are based on the immovable facts of geography and the possibilities therefore are limited.

11.3.1 Content Density

The visual appearance of a map can not be determined without knowledge about the content to be visualised. Two maps following the same design guidelines can look very different, dependent on their amount of information per space. *Figure 11.1* tries to illustrate this. For finding the right amount of displayed elements rules of graphic design can be applied, but they are based on experience rather than on facts. Imhof (1965, p. 400) provides a very true but not very precise instruction, "One shall not impose more on a piece of paper than it is able to cope with" (translated from German).

11.3.2 Method of Representation

Maps visualise geographic facts, in order to do this a code is used. Different codes can be used on the same information, which changes the look of the map. For the example of terrain visualisation this is shown in *Figure 11.2.*

11.3.3 Visual Variables

Bertin (1967, in Carpendale 2003) introduced seven visual variables: position, size, shape, value, colour, orientation, and texture, see *Figure 11.3.* These variables are

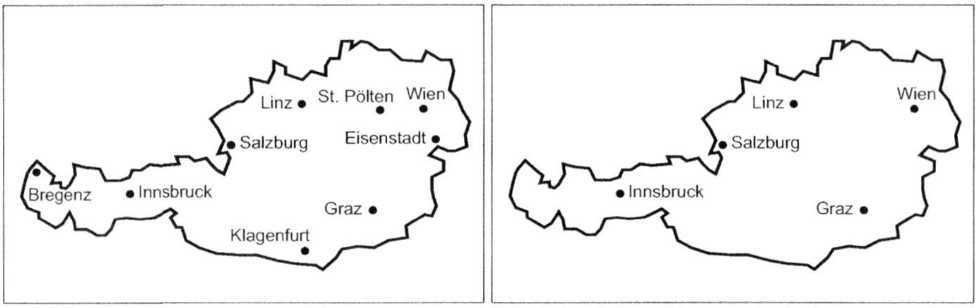

Fig. 11.1. Same design, different appearance

Fig. 11.2. Three of many different ways of visualising terrain: contour lines, hill shading, and altitudinal zones

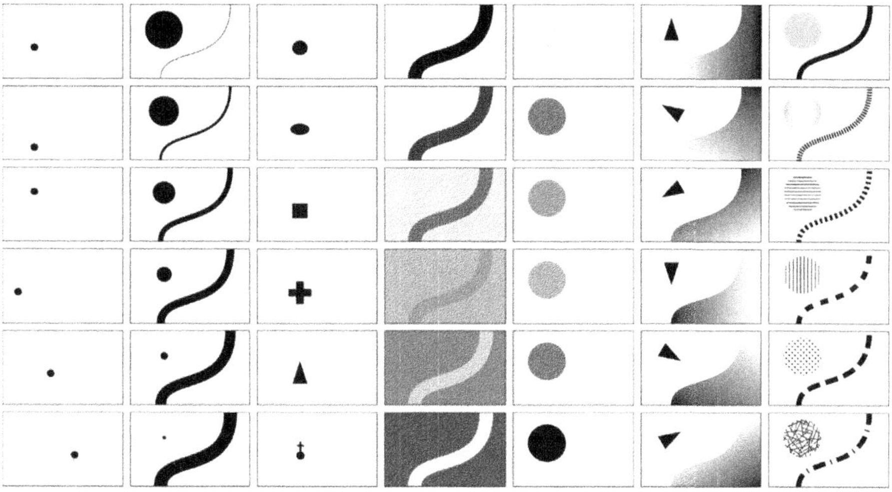

Fig. 11.3. Bertin's (1967) Visual Variables: position, size, shape, value, colour, orientation, and texture (after Carpendale 2003)

widely used and taught in cartographic education and rules covering all these variables can be found in cartographic literature. In respect of choosing colours Imhof (1965) writes that it was often tried to cover graphic arts with scientific methods. Imhof states that in this field it is not possible to find provable laws but only more or less generally accepted feelings, experiences, and habits. For the harmony of colours, for instance, there are some rules: The combination of two colours is harmonic if they are neighbouring colours in the colour circle or if they are complementary colours. This applies analogously to three colours. The more general rule is that a composition of colours is harmonic if the additive mixture of the used colours results in white or the subtractive mixture produces black or grey. The reasons why this is working are not known – there are only speculations – it could be that humans are used to the white colour of daylight or that we have a subconscious desire to order everything.

11.4 Experiment

In the experiment described below three topographic maps with slightly different designs were compared and people had to give their opinion and rating about the beauty of these maps. This method is rather imprecise – as there are so many variables and factors in these maps that vary it is not possible to find the most beautiful one, but this is also not the project's aim. Rather, the basis for future narrower research shall be found.

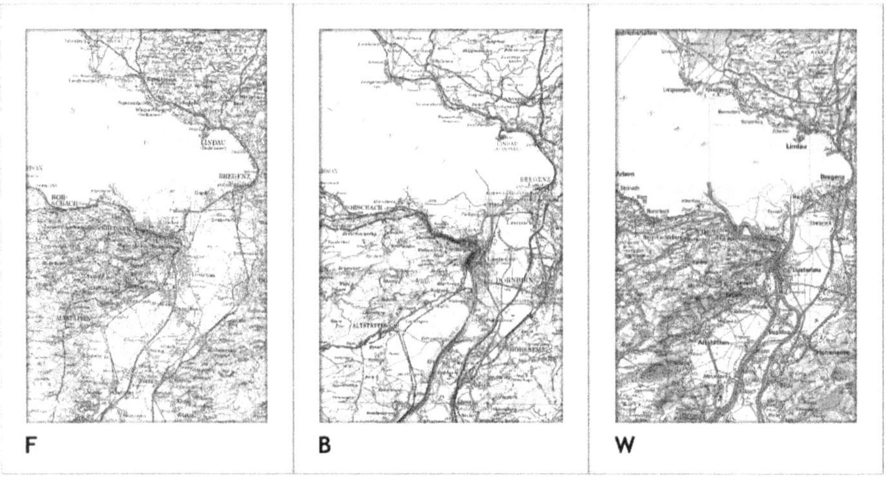

Fig. 11.4. Maps hanging on a wall, as presented to the experts

11.4.1 Setup

For the experiment test persons where shown three different topographic maps of the same spatial region. They were asked which one pleases them the most in respect of beauty and aesthetics. And they had to give reasons for their answer. The three exemplarily used maps depicted an area near Lake Constance in a scale of 1:200,000 and were produced by the national mapping agencies of Germany (Federal Agency for Cartography and Geodesy[1]), Switzerland (Federal Office of Topography swisstopo[2]), and Austria (Bundesamt für Eich- und Vermessungswesen[3]). The interviewees did not see the map border or the issuing company as the maps where masked. The main differences in the three maps are fonts, line styles for transportation network and border lines, density of content, relief representation, and colour of forested areas. The Swiss map uses mainly sans-serif fonts, the Austrian map mainly serif fonts and the German map only serif fonts. The content density is highest in the German map and lowest in the Swiss map. Not surprisingly the Swiss map uses the well known Swiss type of hill shading, in the German map the contrast between shaded and illuminated areas is relatively high and the Austrian shading is comparatively light.

In a first run the maps where shown at the *Cartography and Art – Art and Cartography Symposium* held in Vienna, Austria in February 2008. The maps where mounted on a wall *(Figure 11.4)* and participants had to fill in a short questionnaire, which 52 people did. Due to the topic of the conference the audience consisted of cartographers, artists and the like. Therefore this could be seen as the expert run.

1 www.bkg.bund.de (June 2, 2008)
2 www.swisstopo.ch (June 2, 2008)
3 www.bev.gv.at (June 2, 2008)

Fig. 11.5. Map presentation to pedestrians

In a second run the 101 interviewees were chosen randomly from pedestrians in Vienna, Austria. This time an interviewer kept a log *(Figure 11.5)*.

11.4.2 Results

Among the 51 experts the German map was the preferred one, followed by the Swiss map, whereas among the 101 pedestrians in Vienna the Swiss map was clearly preferred; see all results in *Table 11.1*. The explicit difference between experts and pedestrians could be a result of the different origin of the test persons. The experts came from various different nations whereas probably most of the pedestrians were Austrians. Or do cartographers and artists consider different aspects as important as pedestrians do? This cannot be detected with this experiment setup.

Table 11.1. Preferred maps

Map	Pedestrians (%)	Experts (%)	Total (%)
German	20	45	28
Swiss	50	33	44
Austrian	31	22	28

People were asked to give reasons for their decision; there were no predetermined reasons. The answers to this open-ended question were counted, similar statements were grouped. The results are shown in *Table 11.2*. *Colour* was undisputedly the most mentioned statement with 51 %. For the variable *content* there is a noticeable difference between experts and pedestrians; 24 % of the experts mentioned it, but only 3 % of the pedestrians did. It's similar with *realistic, natural:* only 1 % of pedestrians mentioned it, but every tenth expert.

Table 11.2. Mentioned reasons for calling a map beautiful (relative occurrence, only reasons with at least 5 % in any group are shown)

Reason	Pedestrians (%)	Experts (%)	Total (%)
Colour	47	61	51
Readability, Clearness	25	25	25
Relief, 3D impression	10	33	18
Font	14	12	13
Contrast	12	8	11
Overall impression	8	14	10
Content	3	24	10
Lines	5	2	4
Realistic, Natural	1	10	4
Brightness	5	0	3

11.5 Conclusions

The first result concerning the most mentioned map should not get too much attention, as it is only a conclusion to the specific maps that where tested and cannot be generalised. The difference between experts and pedestrians could be caused by cultural differences in taste, as the participating experts were from all over the world, whereas asked pedestrians were mostly Austrians.

People were asked to evaluate the beauty and aesthetics of the maps, interestingly 25 % mentioned readability and clearness as important factors for their decision anyhow. This shows that beauty/design and cartographic quality of a map can not be considered separately and brings back to mind Imhof's quote from above that the beauty of a map is the sum of its cartographic qualities. Some of the mentioned reasons are rather vague and imprecise and cannot be tested easily in future research (like *readability, clearness, overall impression, content, realistic,* and *natural).* Whereas the remaining variables *colour, relief, 3D impression, font, contrast, lines,* and *brightness* sound promising for further investigations. The same method described above could be used to test these variables, but this time by comparing maps having the same design except for one variable which is varied. This could allow finding design guidelines for beautiful maps, possibly with respect to cultural or personal differences in taste. Then the investigation would move away from syntactics to concentrate on the pragmatic aspect of map aesthetics. And only then the value of beautiful maps can be evaluated.

References

Bertin J (1967) Semiology of graphics: diagrams, networks, maps. University of Wisconsin Press, 1983 (first published in French in 1967, translated to English by Berg WJ in 1983)

Carpendale MST (2003) Considering Visual Variables as a Basis for Information Visualisation. Department of Computer Science, The University of Calgary. http://pharos.cpsc.ucalgary.ca/Dienst/UI/2.0/Describe/ncstrl.ucalgary_cs/2001-693-16. Accessed 9 June 2008

Di Dio C, Macaluso E, Rizzolatti G (2007) The Golden Beauty: Brain Response to Classical and Renaissance Sculptures. PLoS ONE 2(11):e1201. doi:10.1371/journal.pone.0001201

Franck G (1998) Ökonomie der Aufmerksamkeit. Carl Hanser Verlag, München

Freitag U (1971) Semiotik und Kartographie. Kartographische Nachrichten 21:171–182

Imhof E (1965) Kartographische Geländedarstellung. De Gruyter, Berlin

Karvonen K (2000) The Beauty of Simplicity. Proceedings on the 2000 conference on Universal Usability. ACM, New York

López FJA, Balboa JLG (2008) Approximating Cartography to the Customer's Expectations: Applying the „House of Quality" to Map Design. Cartographica 43:107–123. doi:10.3138/carto.43.2.107

Zipf A (2002) User-Adaptive Maps for Location-Based Services (LBS) for Tourism. Proceedings of ENTER Communications Technologies in Tourism, Innsbruck, Austria. Springer Computer Science, Heidelberg, Berlin

Section III: Art and the History of Cartography

12 World-Views: Art and Canonical Images in the Geographical and Life Sciences

Peter Vujaković

Department of Geographical and Life Sciences
Canterbury Christ Church University, United Kingdom

Abstract

Alan MacEachern, in his seminal 'How maps work', notes how certain representations, which he refers to as 'master images', have the power to popularize key scientific concepts and theories that are otherwise difficult to comprehend; he quotes the image of the double-helix as critical in the science of genetics. From a somewhat different perspective, the evolutionary paleontologist, Stephen J. Gould, discusses the construction of 'canonical' images of 'deep time' (past environments) and classic evolutionary iconographies, such as the 'tree of life'. Gould points to the hidden agendas in many of these images, for example, the anthropocentric concept of evolution as inherently 'progressive'. Maps provide examples of similar prejudices and power relations, for example, between global 'north and south', or Eurocentrism. This chapter examines the central importance of maps and other graphics as 'canonical images' in the geographical and life sciences, and the role of art in their making and propagation of a message. The key role of artistic input in creating apparently concrete or naturalised representations of abstract concepts, and in particular their role in pedagogic and popular representations, is explored. This study suggests there are fruitful connections to be made in studying the representations across a range of disciplines in the geographic, earth and life sciences.

12.1 Introduction: Canonical Images and the Role of Art

Images of the evolution of life and maps of the world show interesting parallels in the way that they ostensibly denote 'scientific facts', but are also capable of creating connotations which privilege certain world-views, for example, placing Europe or European 'man' in a dominant position. The aim of this paper is not to present detailed analyses of each of these themes, but to draw attention to some of similarities in the connotative power of images and the role of art in their construction. This power of pictures to present hegemonic concepts or 'worldviews' can be referred to by the generic term 'canonical images' – images representing an authorised viewpoint. The term originates from religious 'canon law', but may now be taken to represent any authoritative position, whether religious or, for example, scientific. Stephen J Gould adopted this term for key images that 'trigger' associations with specific scientific viewpoints; especially those associated with the evolutionary sciences (Gould 1996);

> *All scientific theories call upon canonical icons for their illustration (and often, even for their definition), but my field of evolution and the history of life has been more dependent than most* (p. 250)

Gould's particular concern was the manner in which such images can be used to explain scientific ideas, but also to impose socio-cultural meanings, such as the view that 'progression' is inherent in evolution.

In this chapter the term 'canonical' image is adopted, although it does have similarities with MacEachren's use of the term 'master image' (MacEachren 1995, p. 455). For MacEachren, 'master images' have the power to popularize key scientific concepts and theories that are otherwise difficult to comprehend. He uses the images of the double-helix and the Bohr atom as examples. He describes these images as providing 'concrete representations' that generate confidence in science as 'real knowledge'. He does not, however, make explicit mention of these as authoritative (and potentially hegemonic) images which reinforce a specific cultural viewpoint. Also, he does not give a direct cartographic analogue to the double-helix, yet certain world maps could be characterized as 'master images', at the macro- as opposed to the micro-scale. Scientifically generated world map projections can reinforce geopolitical worldviews (see discussion below), so the term 'canonical' is deemed more useful that 'master image'. MacEachren does, however, make the valuable point, when discussing master images, that "[p]art of their power probably stems from their three-dimensionality, a feature that brings the images closer to real objects, that "objectifies" them" (p. 455). This 'objectification' is very important to the following discussion of the role of 'art' in scientific representation. Art is used to drape 'skin' on the bare bones of scientific images, whether of the evolution of humans, past ecosystems, or maps of the earth.

Gould's 'deconstruction' of canonical images in the evolutionary sciences (see particularly Gould (1993) *'Reconstructing (and deconstructing) the past')* to has much in common with the cultural-turn in geography and cartography, for example, the influential work of Brian Harley (see Harley 2001; posthumous anthology). Both Gould and these cartographers understood that scientific images are often taken at face-value and the underlying ideological messages are thereby naturalized.

> *Iconography, in my view, provides the best domain for grasping th[e] interplay of social and intellectual factors in the growth of knowledge…Pictorial imagery catches us unawares because, as intellectuals, we are trained to analyze text and to treat drawings or photographs as trifling adjuncts.* (Gould 1993, p. 7)

Despite this critical stance it can be argued that publishers, science writers/journalists, graphic artists, *and* academics, are still, perhaps unintentionally, recycling ideologically loaded images, and certainly play a key role in naturalising ideology behind a veil of 'versimilitude'.

12.2 Canonical Images: Infection Via Educational and Popular Media

The term 'infection' is used in part for dramatic effect, but also to impress the importance of continual exposure to canonical images. Infection can mean *'to influence the mood or emotions of people'*, but its broader meaning *'to taint; to corrupt'* is also relevant to the subliminal, 'viral', influence of messages contained in many images. It is also a side-ways reference to Richard Dawkins', ideas concerning 'memes' (as opposed to 'genes') – self-replicating units of cultural imitation – that thrive in brains or the products of brains (books, computers, etc.) (Dawkins 1986, 2006). Dawkins regards memes as a form of cultural DNA and uses this concept to understand how a particular set of ideas about the world might come to flourish in specific populations (although he is careful to state that the 'meme pool' is not organized in same consistent manner as the 'gene pool') (Dawkins 2006).

Culturally embedded world-views such as 'Eurocentrism', reinforced by centuries of world-mapping conventions, or evolution as 'progession' can be seen as such self-replicating and invasive of other cultures. The imperialist legacy of 'Eurocentric' mapping, for instance, retains a potent hold on post-olonial cultures as evinced by a number of studies, the most important of which is Saarrinen's (1988) analysis of 3863 sketch maps drawn by students from 49 countries across the world. His findings confirm the persistence of a Europe-centred world-view, for example, 'developing' states with a marked colonial heritage displayed high levels of Eurocentrism (e.g. India and Brazil 95%).

The popular media and educational resources often reinforcing such concepts, for example Vujakovic (2002) shows how both the UK 'quality press' (serious newspapers) and university text books continue to reinforce a Eurocentric world-view, and to use world map projections that over-emphasise higher latitudes at the expense of the nations of the equatorial regions; this, despite years of effort by geographers and cartographers to counter this biased world-view and suggest diverse alternatives (Vujakovic 2004). The following sections explore several world-views and their representations in the geographical and life sciences.

12.3 Evolution as 'Progress'

Stephen Gould (1997) has done much to popularize the issue of bias in 'canonical images', particularly images linked to explanations of evolution and of past environments. While some of the images he discusses are often graphically simple (evolutionary trees), a large number involve elaborate input from artists or illustrators, for example, pictorial reconstructions of extinct faunal 'communities' and their environments. In many of these cases, the role of the artist is to 'imagine' the past and to breathe life into the scientific reconstructions of scientists.

Many of the images that Gould critiques provide a skewed presentation of life, focused on the rise of invertebrates, through mammals to modern humans. He notes "...the bias thus introduced is the worst and most harmful of our conventional mistakes about the history of our planet – the arrogant notion that evolution has a predictable direction leading towards human life" (1994, p. 10). These supposedly scientific images of 'progress' have clear links to the classical and religiously inspired medieval iconography of the 'Great Chain of Being'; the chain linking, but distinguishing humans from lower life forms, and from the angels and God above them.

The following case studies provide some insight into the nature of the anthropocentric bias inherent in much scientific imagery.

12.3.1 'Parade of Life' – Progress in Evolution and the Ascent of 'Man'

The key canonical representation of evolution as progress is the tendency to parade life as it inevitably leading to modern 'man' (sic) (*Homo sapiens*) (Gould 1993). Gould notes that this image is entirely misleading, with supposedly primitive/lower life being replaced in tableaux as the 'torch of novelty passed to higher vertebrates' (p. 10). In this classic imagery, the Age of Fish is replaced by the Age of Reptiles, which are in turn displaced by the Mammals, with 'man' at the apex. This ignores the fact that many major life forms have continued to evolve and diversify. Any honest representation based on diversity could legitimately represent the current age as the Age of Fish, with over 50% of living vertebrates represented by the teleosti (higher bony fish) (Gould 1996), while humans represent the *only* extant hominid. Unfortunately, these biased images abound in books of popular science as well as more formal educational formats.

Some representations of the 'parade of life' provide individual images over many pages, while others build these into one image, often with a version of the geological column alongside. A good example of the latter is found in *'The How and Why Wonder Book of prehistoric Mammals'* (Keen 1964) (author's collection) which shows the various ages of the earth, from the oldest at the bottom of the page to the youngest at the top. The top image includes 'man', who is shown (with clothes and weapons) observing a variety of now extinct mega-fauna from the a bluff – three important connotations of superiority, 'man' is placed higher than the other animals, 'man' is the only animal which is still extant, and 'man' is the only creature capable of technology. The classic example of a series of images is Charles R. Knight's (1942) 'Parade of life Through the Ages' in *The National Geographic Magazine* (author's collection) in which four of the twenty colour plates are devoted to modern hominids.

This format persists, whether in diagrams, the lay-out of museums, or the chapter order and content of books. It is intriguing to note that even Gould's own book (General Editor of *The Book of Life* (1993)) does not markedly move away from convention. As Gould himself states "the present book, though a great improvement in extended coverage of invertebrates, does not violate [the] tradition of tableaux parading towards Homo sapiens." (p. 10). One has to ask why this should be given his editorial role – perhaps knowledge that this dominant iconography is so embedded that no other format would be accepted in the market-place? A quick review of the book shows that the chapter entitled 'The primate's progress' (the title does not appear to be used ironically!) is disproportionately large, represents just under one-sixth of the total material directly related to life on earth. There are six large colour tableaux devoted to the hominids, yet only seven or eight similar images focused on the marine environment, but with not one devoted to the teleosti (see discussion above).

12.3.2 Anthropocentrism – 'The March of Progress'

Art also works to naturalise specific images of human evolution as a story of 'progess' or 'march of advance'. The classic image of human evolution shows ancestral apes slowly morphing into upright, modern 'man' – almost inevitably a male Caucasian (often blond, compared to the preceding black haired stage(s) that signify or connote a more primitive condition) and in some cases clothed in European fashion. The danger of this world-view has been noted by other popularisers of evolutionary science; for example, Richard Dawkins (2005) notes that the "dominant icon of evolution [shows] a shambling file of simian ancestors, rising progressively in the wake of the erect, striding, majestic figure of Homo sapiens: man as evolution's last word (and in this context it is always man rather than woman)" (pp. 1–2). This imagery has clear echoes in cartography's conventions of centring and orientation which privilege Europe on many world maps (Eurocentrism: see below).

A classic example from the author's own collection of popular science texts, '*The How and Why Wonder Book of Primitive Man*' (Barr 1964) shows an erroneous developmental schema, with European 'man' emerging (dressed in suit and hat and carrying a brief-case!) at the end of sequence from amoeba and jellyfish, through various modern vertebrates, including birds (certainly not linearly related to humans) and a gorilla. The sequence contains a switch-back to one extinct species of hominid (Neanderthal), before reaching 'man', who is set literally above all the other species on the page.

Over a decade after Gould's (1989) excellent deconstruction of the iconographies of progress, examples of the classic image continue to abound in popular media. The following examples illustrate its current use in a variety of media, including popular education and the press.

The first example is taken from a popular science text '*Animals: an illustrated A–Z*' (Cheshire 2006, pp. 200–201). This profusely illustrated book appears designed for both older children and a general adult audience. The diagram illustrating human evolution *(re)*presents the classic format through four stages, *Australopithecines* (represented by a single generic individual despite being varied group of related species) – *Homo habilis* – *H. erectus* – *H. sapiens*, as a direct line of ascent. There is no attempt made to suggest the line after the *Australopithecines* is a mere 'twig' on a larger branch, with hominids as lacking the diversity of other successful groups.

The second example is from *The Scotsman* (9-8-2007; p. 20) newspaper. The article describes recent evidence that *Homo habilis* and *H. erectus* may have been contemporaries rather than ancestral, but the paper still persists in using the classic canonical image, ending in a clearly blond European male human striding purposely forward.

The 'viral' reproduction of this image appears to be due to uncritical, and perhaps lazy, adoption of the canonical image, with only slight changes between images (this is mirrored in a similar uncritical adoption of the Peters world map by development agencies, see Vujakovic 1987). The 'progressive' message, however, is clearly understood and has been adopted by advertising and fashion industries *(Figure 12.1)*. The image has, however, also been used to lampoon human 'development'; with the stone tools of pre-historic humans being replaced by industrial and military technologies, with the final stage having humans return to a crouching position over a computer or games consol! Several other variants show grossly obese modern humans eating fast food. A web-search (e.g. using Google-Image) will generate as many satirical as serious examples of this 'canonical image'.

An interesting example of how the canonical image influences other representations, is a series of photographs in Jacob Bronowski's (1973) classic '*The Ascent of Man'* In a sequence of six images a 14-month old child is shown crawling, gradually extending its limbs, and eventually walking – the caption 'The human baby... is a mosaic of animal and angel ...The child has entered the human commitment to walk upright' (p. 31). Each image is larger than the pervious, until the photo fills the whole page (note textual reference to the 'Great Chain of Being').

Gould's (1989) other classic example of a false image of 'progress', the evolution of the horse, is also still circulating as a key image despite Gould's very clear message that this and human evolution are special cases, not representative of the bush-like nature of most successful animal groups (see for example, CGP 2006: a GCSE Biology revision guide currently in use in UK schools).

12.3.3 Eurocentrism – Art, Power and Cartography

Like the European bias in some images of human evolution, a specifically *Euro*centric bias in world mapping remains a significant feature of late twentieth and early twenty-first century mapping. This is regarded by some commentators as crucial in maintaining 'western' political and economic hegemony within the global international system (evinced by Saarinen 1988). Eurocentric imagery still dominates many popular and scholarly media (Vujakovic 2002), this despite a *de facto* shift of power across the Atlantic after WWII, and critical studies of the relationship between cultural representations and imperialism (e.g. Said 1993). In Brian Harley's words, Eurocentric maps reinforce, despite their apparent mathematical objectivity, "the manifest destiny of European overseas conquest and colonization" (Harley 1989, p. 10).

Throughout history there has been a tendency for map makers to place their own territory or some ideologically significant place (e.g. Jerusalem, or Mecca) at the centre of the map (Tuan 1974). Artistry works in two main ways to 'naturalise' and fix such positions. First, by the manner in which supporting images in the margins often reinforce centrism and other hierarchical relationships, and secondly, by draping the map with seemingly natural, but artistically generated, topographic or land-cover information. The later connotes – *'this is what the world looks like, and this is the natural position of view'*, and has become more important with the advent of satellite images that appear to provide 'real' images of the world, *seemingly* without artifice.

In the eighteenth and early twentieth century Eurocentric maps predominated, as major publishing houses in Europe supplied maps both to the metropolitan centres and the colonial peripheries. European educational models ensured that a Eurocentric world view was inculcated in children across the globe. The centrality of the metropolitan power was crucial to myth of a benign, paternal imperialism. This 'Euro-omphalism' is exemplified by a range of maps that reinforced the world-view of the metropolitan centre as benevolent parent (e.g. world map in P.H. and A.C. Kerr's primer *The Growth of the British Empire* (1911) "...in the middle is tiny England, which is the Mother-Country of them all" (cited in Lavezzo 2006, p. 2)).

A good example of the artistic use of marginal imagery to reinforce centrism *and* hierarchy is the map supplement to the '*Graphic*' (July 1886) showing the 'Imperial Federation – the British Empire in 1886'. This map's margins display a range of the people's of empire gazing on Britannia; she sits centrally, astride the globe, her trident a north arrow drawing the eye to the British Isles at *centre stage*. Art confirms the 'natural order'. Further detailed investigation unearths other forms of visual hierarchy, for example, a British soldier/explorer holds chained a clearly submissive tiger representing both European dominion over nature and Asia, while behind him, bent under bales of produce, are two indigenous people of the sub-continent. This latter aspect is all too redolent of the 'march of evolution' imagery discussed above, with the 'superior' European at the fore-front; in this case, of economic and social progress.

A more recent example of the importance of marginalia is the 'United Nations Map of the World' published in 'Great Britain' in 1945 (Bartholomew and Sons). This is particularly interesting given the geopolitical situation at the end of WWII. It retains a Eurocentric projection with Britain centre stage, and uses marginal images (crests/coats of arms of members states) to reinforce this. Britain retains the dominant position at top-centre, with the USA and Soviet Union (as supporters?) at the top corners. European state emblems dominate the top row, with Asian, South American, and African states largely relegated to the sides and lower margin. Even here, prominence is given to key Commonwealth states. This represents an interesting graphic attempt to maintain Britain's place in a radically changed world.

World maps, where the artistic input provides representations of the natural topography or land cover, abound, but these work to reinforce Eurocentrism in a different manner to the cultural hierarchies discussed above. These seemingly 'natural' views work, in part, because the Euro-centred view is already subliminally accepted as 'the norm'. They also work by connotation – 'this is what the world looks like, and this is the natural position of view'. This connotation is embedded in the seemingly 'scientific' *denotation* of the terrain (the reader rarely questions the cultural positioning inherent in the choice of projection, centre and orientation of such physical maps). An interesting example is the main world maps in the 1962 edition of the *Reader's Digest Great World Atlas*. Rather than a standard world map projection, the atlas opens with two '3-D style' views of the globe entitled 'South and North Pole', with terrain 'naturally' coloured according to land cover (e.g. orange for desert, and blue-green for coniferous forests) and topography suggested by hill-shading. On closer inspection these are clearly highly contrived viewpoints that 'suggest' a view from space. The larger of the two views (37 cm in diameter) is centred on England, not the North pole. Most of southern Asia, South America and southern Africa are excluded from either view (the second is centred on a mid-point between Antarctica and Australia (diameter c. 16 cm)). This is a world visually dominated by Britain and its 'white' commonwealth partners. Similar pseudo-globe views have been used elsewhere (see for example the *Collins World Atlas* (illus. ed. 2003); this atlas does however, provide a wide range of different points of perspective using this format, although the majority of its standard world maps are still based on a Eurocentric, non-equal area projection).

There have been some innovative and alternative world-views, although these have generally failed to displace Eurocentrism. For example, Americans, during WWII, quickly recognised the importance of air power, and of the polar region to hemispheric defense. Graphic designers, notably Richard Edes Harrison, began to create maps which made sense of this new strategic situation. Their artistic conventions created apparently concrete or naturalised, but deliberately subjective representations of the globe – views from various perspectives. As Henriksen (1980) has points out, the old cylindrical projections failed to show the continuity of the 'worldwide arena' and were replaced by more effective forms, notably North Pole centred projections. Harrison's *'One world, one war'* map for *Fortune Magazine* provides the classic example. A separate challenge to the Eurocentric bias emerged in the late twentieth century, in part as a response to international development issues, and in part as defiance against negative images associated with 'north-south as top-bottom' connotations. An early example was McArthur's polemic 'Universal Corrective Map of the World', published in 1979, which placed Australia top-centre (Black 1997). Other 'turn-about maps', often Sino-centric or Amero-centric versions, were also designed to challenge 'northern' hegemony. Despite such challenges traditional Eurocentric projections continue to be used across a wide range of media into the present century (see educational and news media examples, Vujakovic 2002).

The availability of satellite imagery has furthered opportunities for the creation of hegemonic viewpoints hidden behind a veil of natural versimilitude. Wood (1992) provides a trenchant critique of the Van Sant map, described at the time of its original publication as a 'portrait map' of the world. As he notes, the image, far from being a natural view of the earth from space, is highly contrived, from the development of the original satellite and sensor technology, through the scanning, filtering, and correction procedures used to create the image, to the choices regarding orientation, projection (non-equal area), centring (Eurocentric), colour, and other key attributes. He describes this cloudless, clean and totally day-lit image as the "... map, that would like us not to notice the fact that it is a map ... [but] a picture photographed from ... *life*" (emphasis in original, p. 49). Because we are infected with the virus of the rectangular, Eurocentric, northern orientated world-view, we do not question the artifice in this representation and many other similar satellite based images circulating in scholarly and popular media.

12.4 Conclusions: Good, Bad or Merely Ugly Science?

Canonical images of evolution and the earth provide potent examples of the parochial and partial nature of much scientific iconography. Whether a Euro-centred world-view or the chauvinism of anthropocentrism is unearthed, a close study of scientific images can alter us to the underlying agenda of supposedly objective images. It is unlikely that these are the only forms of imagery that lend themselves to this form of deconstruction. There are certainly potential synergies from cross-disciplinary work in this area, and it may be that study of other disciplines has much to offer (see MacEachren's (1995) discussion of images in sociobiology). The lessons are only valuable, however, if they lead to positive change and if alternative 'visionings' of the earth and its life forms develop from this debate. Alternative world-views (e.g. turn-about or Sino-centric map) may be not be any more objective than those discussed above, but a diversity of imagery, rather than being confusing, may be liberating in terms of a fuller understanding of the complexity of the real world and a cleansing of the viruses of hegemonic views. These, perhaps, are hard lessons to learn while some people continue to regard themselves as the topmost twig on the tree.

References

Please note: Later editions are sometimes provided for pagination purposes.

Barr D (1964) The How and Why Wonder Book of Primitive Man. London: Transworld

Black J (2003) Visions of the world: a history of maps. London: Mitchell Beazley

Bronowski J (1976) The Ascent of Man. London: British Broadcasting Corporation, first published 1973

Cheshire G (General Ed.) (2006) Animals: an illustrated A-Z. London: Starfire/Foundary Creative Media

CGP 2006, GCSE OCR Gateway Biology Higher Revision Guide, Coordinate Group Publications Ltd. (CGP)

Dawkins R (1986) The Blind Watchmaker. London: Longman

Dawkins R (2005) The Ancestor's Tale. London: Phoenix (first published by Weidenfeld & Nicolson, 2004)

Dawkins R (2006) The God Delusion. London: Bantam Press

Gould SJ (1989) Wonderful Life. London: Hutchinson Radius

Gould SJ (1993) Reconstructing (and deconstructing) the past. In Gould SJ (ed) (1993) The Book of Life (pp. 6–21). London: Ebury-Hutchinson

Gould SJ (1994) Eight Little Piggies: reflections in natural history, London: Penguin edition (first published by Jonathan Cape, 1993)

Gould SJ (1997) Dinosaur in a Haystack, London: Penguin edition (first published by Jonathan Cape, 1996)

Harley JB (2001) The New Nature of Maps: Essays in the history of cartography (posthumous collection edited by Laxton P). Baltimore: John Hopkins Univ. Press

Keen ML (1964) The How and Why Wonder Book of Prehistoric Mammals. London: Transworld

Knight CR (1942) Parade of life through the ages. National Geographic Magazine, LXXXI (2), 141–184

Lavezzo K (2006) Angels on the edge of the world: geography, literature and English Community, 1000–1534. Ithaca: Cornell Univ. Press

MacEachren AM (1995) How Maps Work: representation, visualization, and design. New York: The Guildford Press

Saarinen TF (1988) Centering of Mental Maps of the World. National Geographical Research 4:112–127

Said E (1993) Culture and Imperialism. London: Chatto & Windus

Tuan YF (1974) Topophilia: A study of Environmental Perception, Attitudes, and Values. Englewood Cliffs: Prentice-Hall

Vujakovic P (2002) What ever happened to the 'New Cartography'?. Journal of Geography in Higher Education 26:369–380

Vujakovic P (2004) World maps: a plea for diversity. Teaching Geography, 29 (2), 77–79

Vujakovic P (1987) The extent of the adoption of the Peters projection by 'Third World' organisations in the UK. Bulletin of the Society of University Cartographers, 21(1): 11–16

Wood D (1992) The Power of maps. London: Routledge.

13 Map Style

Alexandra Benová[1] and Ján Pravda[2]

[1] Faculty of Natural Sciences, Comenius University in Bratislava, Slovak Republic
[2] Institute of Geography, Slovak Academy of Sciences, Bratislava, Slovak Republic

Abstract

Cartography is defined not only as specific activity to execute maps, but also as art to create maps. Usage of maps as pictures in public buildings, at workplaces, and even in privacy is an apparent evidence that people are looking at maps as artistic works, which can influence them not only in the sphere of specific knowledge, but also in an emotional sphere. If we come from the character of map as a picture, then it is evident that in the set of various maps it is possible to differentiate various map styles. These styles are possible to identify by map stylistical features. In this article style-forming map factors are presented, which influence the selection of map stylistical features. We tried to classify the map styles on the base of these style-forming factors. We had distinguished historical and contemporary map styles, also individual and regional map styles.

13.1 Introduction

A map is a complex phenomenon and therefore has several interpretations. It is considered as the model, projection, but most frequently is defined as an image (reduced, generalized, sign, etc.). If we agree with the definition of a map as a picture, then we must accept that we can take note of such map properties, which evoke in humans some feelings, emotions – like paintings. It means that we can search in maps for properties like map style. Evidence of this is seen in the fact that many maps (mainly historical) are often published, displayed in the same way as paintings.

For over 10 years lectures in the 'map language' subject are offered within the geography specialisation area at Comenius University in Bratislava, Faculty of Natural Sciences. Under this theme a map is considered as a result of the map-sign system. This system has a number of levels: the symbol inventory level; the morphographical level; the syntax level; and the stylistics level (Pravda 1990). Map style is the main term of this level. Map style is a summary of characteristic traits, where maps gain stylistical features according to the theme, relevant map functions and the actual map application.

We distinguish two groups of map stylistical features:

- Intracompositional (inner map) graphical elements (e.g. individual map signs) or components, understood as groups of graphical elements (e.g. set of signs) in the map field, which participate in map style by their shapes, sizes, colours and other characteristics; and
- Extracompositional elements or components in the surrounding of map field, which add the finishing touches with their presence to the design and determining the actual style of every concrete map.

Stylistical features act not only as map signs and determine sign groupings, but they also determine the various map components (like the title above the border, graphic scale, legend, framework, etc.) and supplementary map components (diagrams, photos, text, etc.). However, there exist other factors which influence map creation and concrete conditions that force the map creator (or processor) to choose only some map stylistical features out of many that may be presented.

13.2 Map Style-Forming Factors

Map style-forming factors can be either objective or subjective. Objective style-forming factors are mainly map theme and composition, map purpose or technical facilities. These are briefly described below.

Map theme is the objective style-forming factor, which is "given from outside", like the theme of a painting. Map theme (content) influences the map composition, which generally determines the selection and arrangement, leading to the actual map appearance. For example, the composition of a world map is guided by the objective distribution of continents and oceans. A road map or a climatic map would show the same area, but express the theme using different expressing features, leading to a different map style as an outcome. Analogue maps (synoptic, geophysical, demographic, etc.) are predetermined, as their themes must accord to what is expected from this type of map, like topographic maps.

Map purpose requires the submission of the map theme (content) expression to a given specific intent, the goal of map production and whether this is a new edition of the map. For example, schematic or panoramic maps of the Alps in tourist guides markedly vary in style from topographic maps of the same area. A school wall map or an atlas map, even though it accords to the same content (theme) will appear differently.

Technical facilities (the set of tools for map creation) available to individual creators differ between those used for manual production and automated production. Different printing processes change the style or look of a map. So does the use of colour as opposed to a monochromatic map. These map style-forming factors can relate to the various painter's techniques, which have considerable influence on a particular painter's style.

Subjective style factors are mainly scientific ability, the approach to the map theme and subjective proclivities. These elements are now explained.

Scientific ability (competence, forwardness) is significant as in the case of creator, author (originator) of a map, as well as others who may also participate (e.g. map editor, compiler, draftsman, printer, etc.), but also appraiser or reviewer. It is a set of knowledge and experiences, which is necessary in the current historical-social period for map creation as an artefact, or as a graphical or technical work. Previously, it was enough if the map creator knew how to draw well, with experience of construction and the graphical formation of a map. Now the process of map creation is considerably different, and focuses on the use of computer technologies for map production. But just using a current computer and applicable software does not guarantee the creation of a stylishly distinctive map – here knowledge of cartographic interpretation and design and production experience are necessary.

Approach to map theme can be simple, involved, static, dynamic, analytical, complex, synthetic, etc. Some map creators like to express many map elements as individual entities (and thus evade expression of a complete 'set' of elements). However, successful depiction of many elements together depends upon generalisation and the expression of the whole gambit of elements, not omitting or suppressing any essential detail. This approach influences the selection of graphical features, and thus the resultant map style as well.

Subjective proclivities of map creators are presented generally by preferring of some types of expressing features (e.g. sheer or thick lines, geometric, abstract or to the contrary, illustrative, associative shapes of map signs), warm or cold colours, certain composition etc. To this group belong elements and components of map style, which come from the field of artistic (mainly painting) styles.

13.3 Map Style Classification

Looking at maps from a historical perspective, it is possible to distinguish historical and contemporary map styles (Pravda 1997, 2003). Historical map styles (i.e. from chronological viewpoint) can be classified as belonging to certain styles:

- Primitive
- Antique

- O-T (Orbis Terrarum)
- Arabian
- Chinese
- Portolan
- Topographic
- Georelief hachuring
- Other styles (georelief shading and others)

Within contemporary map styles (mainly from the viewpoint of map purpose) it is possible to classify them as belonging to the styles of:

- Utility
- Scientific
- Popular
- Artistic
- Other

There exists one more viewpoint from which it is possible to distinguish map styles. It is the viewpoint of dominance of individual or collective traits. From the conservation viewpoint of individual or collective traits as well as contemporary and historical maps, it is possible to distinguish:

- Author style
- Editorial style
- National style
- Regional style
- Others styles

The classification of map styles is open. It is possible to specify and earmark other styles or style variants to subvariants – in accordance with progress of our knowledge about this topic.

13.3.1 Historical Map Styles

Historical map styles sequence is a very demonstrative evidence of the evolution of creative cartographic thinking and interpretative art as reflection of that time. Research of individual map styles provides the means for revealing the various coherences related to the evolution of human thinking and knowledge – mainly to reveal the change from concrete thinking to abstract thinking, and the role that technical developments play in determining the way in which maps are expressed, etc.

Primitive style (style of prehistory and ancient maps) is characterised in the oldest maps (related to map expressions). For this style the predominant character of map drawing is the line. Within this style it is possible to distinguish style variants, which differ mainly by the design of lines and the actual map composition. The design of lines depends mainly on the material on which the map was carved or drawn, the tools used for carving or drawing and on the skills that the individual cartographer had when drawing the map. These maps were carved or drawn onto stone, bone or etched into clay tablets, etc. (e.g. a map carved on the curved surface of mammoth tusks, an archaeological discovery found near the village of Pavlov in South Moravia, or the map image carved in flat bone from an archaeological site near Mezhirichi in the Ukraine). Because there are not many maps from this period, more accurate specification of style variants can only be found as a consequence of deeper research. However, this style of map is supported from evidence found on prehistoric and ancient artefacts with map expressions in museum collections.

Antique style (style of ancient maps) is characterised by maps mainly written by Greek scholars, as the originals of ancient maps have not been preserved. For example we can only conclude indirectly how Ptolemy's own map of the world looked like on the basis of its verbal description in *Geographike Hyfegesis* or after its subsequent reconstruction. These reconstructions were redrawn many times (e.g. the reconstruction of Ptolemy's map of the world from the 2nd century A.D. redrawn at the end of the 15th century), so the resemblance to the original is unknown. We can only conclude that style-forming features were mainly line drawings (on papyrus) representing coastlines, lakes and rivers, signs of settlements, mountains and later the system of meridians. Another significant style-forming feature of ancient maps were names of countries, rivers, settlements and other geographical objects that were related to naming discoveries. This distinguished them from prehistoric maps, but the actual look of ancient maps remains unknown.

O-T (Orbis Terrarum) style is considered to be the heritage of the knowledge and culture from the Roman Empire. This style is characterised by maps from the period of the Christian Middle Ages, which are characterised mostly by their orientation to the East and positioning of the centre of the world at Palestine (Jerusalem). Letters O and T were not just the abbreviation of expression Orbis Terrarum (from lat. circle, disc of Earth = world, and simultaneously the Roman Empire). The letter O symbolised the shape of Earth in the shape of circle or ellipse and letter T symbolised the shape of the Mediterranean Sea with its side 'waters' – the Nile on one side and the Bosporus to the Don on the other. Main style-forming features of these maps are the simple linear drawing of coastlines of seas, lakes, rivers, pictorial drawings of settlements (forts), characteristic drawing of mountains and the abundance of textual descriptions of features (sometimes as full text in the map field). Examples of these maps are the schematic O-T map of Izidore of Seville (at the beginning of the 7th century) or the St. Beatus map from 776. Within this style we can earmark the existence of at least two style variants: schematised and approximate topographic.

Arabian style (a style of medieval Arabian map) can be characterised mainly by the south or west orientation of the maps as significant style-forming traits. The role of style-forming features on these maps are peculiarly signs of settlements, the drawing of rivers, seas, mountains and the mainly Arabic type used. Some examples of these maps are the geometric stylised map of the Mediterranean Sea (south-west orientation) from the *Atlas of Islam* (10th century) or the so-called *Rectangular World Map* of Al-Idrisih. He made this map in 1154 for the Sicilian ruler Roger II. The original Arabic type is substituted with the Roman alphabet. On the basis of generalised knowledge about Arabic maps it is possible to define at least two variants in this style: topographic (real information mapped) and geometrically schematized information.

Chinese style (the style of antique and medieval Chinese maps) is characterised by how the drawing of rivers, settlements, georelief is done and the mainly hieroglyphic type employed. Single hieroglyph names of geographical objects are in many cases also portrayed using localised cartographic signs. Some maps use a regular square grid.

Portolan style is characterised by the detailed drawing of coastlines, the richness of the names of coastal settlements (there is much other textual information), but the most typical style-forming feature of portolans is the net of rhumb lines showing the direction of main and incidental cardinal points. The portolan map is in principle a Middle Ages' utility (maritime navigation) map (*porto* from ital. – port). The style of portolans is very simple, but further study would be needed to determine style variants.

Topographic style is characterised by its evolution from iconic-topography variants to regional variants (still unnamed variants to conventional-topography variant). The most recent variant can be considered to be the ancestor of topographic maps (using a style variant which is part of contemporary maps).

The most important style-forming features are the representation of mountains and the gradual transition from its perspective (hillock) representation to ground-plan expression, the expression of rivers by gradually magnified lines, the increased representation of settlements by their patterns of towns and, on small scale maps, the noticeable transition from iconic and figural signs to round signs. This style is characterised as well by the application of regular net of meridians (connected to improved map accuracy) and the standardization of type styles used mainly for naming of represented objects.

Hachuring was used on maps mainly from the 17th till the 19th century. The technique was created from the necessity to express georelief on maps as a plan view – first just as a visual effect and later with more precision. The hachure is its most important and typical style-forming feature. It is possible to earmark some types of hachuring by the types of hatches used, which are: nonregular *orthographic-ichnographic hachuring,* for example the technique used by S. Mikovini on maps of Hungarian regions (Bel 1735–1742), *slope (orthogonal) hachuring* and *shadow hachuring,* which also create harmonic style variants. Hachuring was the forerunner of the expression of land relief by isohypses (contour lines) and seabed, lake and river relief using isobates.

13.3.2 Contemporary Map Styles

The development of contemporary map styles is linked mainly with progress in thematic cartography. The number of map types proliferated, which is inseparably linked with the expansion of map use. The development of contemporary map styles are reflected more by modern techniques and technologies of map creation, processing and reproduction than historic styles.

Utility style is the characteristic for a considerable part of contemporary maps. They are mostly maps of large and medium scales (1:1000 to 1:200 000) and they depict technical economic, topographical, aviation and marine navigational, water economy, forestry, road, geological, soil, botanic and various other themes. They are created to meet the needs of different organisations and institutions. Characteristic style-forming features are, strictly speaking, all types of figural, linear and areal signs, localised topographically or schematically.

Within this style it is possible to define style variants by using the localisation principle:

- Topographic,
- Schematic, and
- Combined (topographical-schematic).

In each category it is possible to further define sub-variants by the predominant type of signs used (figural, linear, discrete-areal, continuously-areal) and their mutual combinations.

Scientific style uses as style-forming features mainly non-motivated (geometrical) signs. In comparison with the utility style it is characterised either by detailed (analytical), complete (complex) or integrated (synthetic) approaches to a theme, a higher grade of abstract understanding of theme and a higher level of conventionality by map signing. Within this style it is possible to define style variants:

- *Magazine,* typically monochromatic (usually black and white) expressions and mainly mediated explanations as map signs;
- *Polychromatic (many coloured),* for which it is typical to use many colours (or coloured shades) to express complicated content;
- *Photomosaic,* for which it is typical to use expressional tools of aerial photographs, space, radar and other similar images (analogous or digital); and
- *Anamorphous,* characterised by a high level of deformation of map content (in comparison with the topographic style variant), which comes from the dependence of shape (or size) of the displayed area from a selected indicator.

Popular style is characterised by clearness, lucidity, often extreme simplicity, but mainly effectiveness, which is achieved by the use of associative elements (illustrative, symbolising, motivated) to icons (picture) signs as style-forming features, by engaging design and colour of graphical compositions, as well as practical binding or other elaborations. Within this style it is possible to define these style variants:

Popular-scientific (educational, aimed at cognition), characteristic for school, homeland study maps, tourist maps, road maps, city maps etc., which simultaneously represent style sub-variants:

- Pictorial
- Perspective
- Advertising/poster
- Map miniature (e.g. on postage stamps)
- Strip map
- Physiographic
- Image
- Others

Artistic style is characterised by proceedings of graphical to art painting elements in the role of stylistical features. Within this style it is possible to define these style variants:

Swiss, for which it is typical to use many coloured reproduction techniques with the application of the principles of areal perspective as georelief representation on topographic maps at large, medium and small scales, mainly in school atlases;

Painting, mosaic and others, for which it is typical to use traditional and modern (mainly combined) creative techniques on maps that can be considered in respect to their creative value as a work of art.

13.3.3 Individual to Regional Map Styles

The main differentiating criteria in this group of map styles are specialities, which depend on the knowledge and experience of their creators (individuals to collaborations). They are dependent on the status of social and technical-economic technological development in various parts of world.

Author style was prevalent in the past, when the author was also the creator and therefore the only one who influenced the final look of the map. Over time draftsmen, engravers, lithographers and several specialists in the process of map creation participated as well and each contributor left his characteristic marks which determined the final look of the map. Their work needed to be managed, so other specialists were involved – map editors (chief editor, responsible map editors, editors-compilers, technical, thematic specialist, art editors, etc.). Their influence was and is often so strong that they could significantly reduce or eliminate the individual style traits of the map author.

Now many maps are published without a distinct author and their final look (and style) depends on the software employed, editors' inputs or on others involved in using specialised map reproduction technologies.

Editorial style can be characterised as the result of the influence of editors and other specialists in map publishing or exterior cartographically-oriented groups that may influence the output (e.g. geographical, geological, meteorological and other institutes), which effect map creation.

Examples of these characteristic style divergences are map series from various publishers like Bertelsmann (atlases), Falk (city plans), Touring Club, Westermann, Freytag und Berndt, Kümmerly und Frey and others. Analogous maps from various publishers often differ at first sight due to special graphical design, colouring, fonts, signs, method of georelief shading etc.

National style is easily identifiable, when comparing similar cartographic products produced in different countries. Tourist maps, road maps, city maps, school atlases, but also newspaper, popular-scientific and sometimes scientific maps (cartographic products) published in one country can differ from similar maps having the same basic characteristics with regard to design, colouring either by single components of content (hydrographical components, vegetation, figural signs, fonts etc.) or general composition. These differences are conveyed irrespective of international standards.

Regional style can be noticed despite the existence of editorial and national styles on maps, which are published in different regions. Style-forming features at a regional level are elements like general map colour schemes, composition (graphical design) and map script (e.g. Arabic, Indian, Chinese, Japanese, etc.).

13.4 Conclusions

The fact that maps have unique image characteristics is indisputable. It is a specific image, which is not subordinate to an author's ideas about how a particular map should look like. We know that it is necessary to accomplish some conditions for map creation. Now, and in the past many topographic and thematic maps have been created by individuals or collectives in various projections using different techniques and also different methodologies. As in case of paintings we can also talk about maps and the art involved in their creation.

It is possible to distinguish various styles of maps. Considerations about the classification of these styles can be made according to basic map division as topographic and thematic did not lead to an acceptable result. Considerations about their division made on the basis of cartographic projections did not seem to be right. We decided to distinguish historical and contemporary map styles, within which it is possible to also distinguish individual to regional map styles. Within these styles there exist many variants and sub-variants. The investigations of one or two cartographers are not enough for definite conclusions about map styles, but knowledge from a bigger group of researchers is needed.

Maps used as pictures in public buildings, at workplaces, and even in private, provide the evidence that users view maps as artistic works, which can influence not only how they use maps to find specific geographical knowledge, but also emotionally.

References

Bel M (1735–1742) Notitia Hungariae, Novae historico-geographica, Teils 1–4, Vienna

Pravda J (1990) Základy koncepcie mapového jazyka (Foundation of map language conception), Geografický ústav Slovenskej akadémie vied, Bratislava

Pravda J (1997, 2003) Mapový jazyk (Map Language). 1st and 2nd edition, Univerzita Komenského, Bratislava

14 The World Image in Maps – From the Old Ages to Mercator

Mirjanka Lechthaler

Institute of Geoinformation and Cartography
Vienna University of Technology, Austria

Abstract

Studying the Australian aborigines' 'dreamtime' maps or engravings from Dutch cartographers of the 16th century, one can lose oneself in their beauty. Casually, cartography is a kind of art. Visualization techniques, precision and compliance with reality are of main interest. The centuries of great expeditions led to today's view and mapping of the world. This chapter gives an overview on the milestones in the history of cartography, from the old ages to Mercator's map collections. Each map presented is a work of art, which acts as a substitute for its era, allowing us to re-live the circumstances at that time.

14.1 Introduction

Long before people were able to write, maps have been used to visualise reality or fantasy. Their content influenced how people saw the world. From studying maps conclusions can be drawn about how visualized regions are experienced, imagined, or meant to be perceived. Often this is influenced by social and political objectives. Cartography is an essential instrument in mapping and therefore preserving cultural heritage.

Map contents are expressed by means of graphical language. Only techniques changed – from cuneiform writing to modern digital techniques. From the beginnings of cartography until now, this language remained similar: clearly perceptible graphics that represented real world objects.

The chapter features the brief and concise history of the appearance and development of topographic representations from Mercator's time (1512–1594), which was an important period for the development of cartography.

14.2 A Walk Through the History of Cartographic Representations

Activities such as surveying the Earth's surface, mapping gathered data and the production of cartographic representations (in analogous and digital form), as well as defining the coordinates of an arbitrary location anywhere on the surface of the Earth belong to current day cartography. But, historically, the determination of coordinates was the utopia of many astronomers, geographers and cartographers, and it was accompanied by much speculation, delusion and ingenious inventions. The next sections outline these historical events.

14.2.1 From the Stone Age to the Greek Antique Period

From the Stone Age there are no preserved traces, which could be related to attempts of representing an area. Later graphic representations of important elements of culture were engraved in caves, etched in wood, stone, bones, leather or horns (Leithäuser 1958).

The oldest preserved old-Babylonian map of the world from the 6th century B.C. was carved into a clay plate *(Figure 14.1,* British Museum, London). The world was represented schematically as a circle – the Babylon Empire, is surrounded by

Fig. 14.1. The whole world on an old-Babylonian clay plate

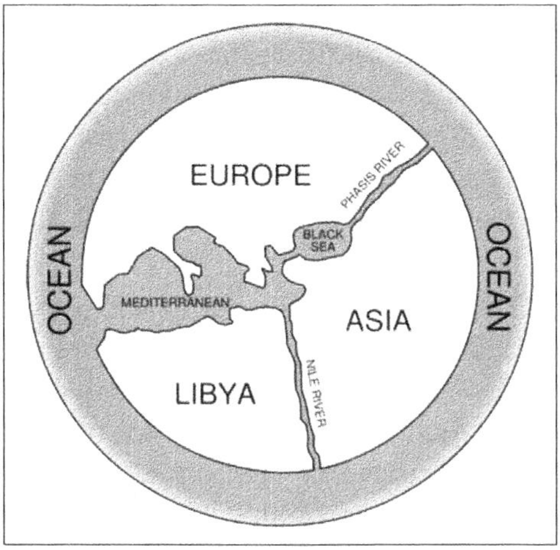

Fig. 14.2. Possible rendering of Anaximander's world map

an ocean. The river Euphrates is of central importance, flowing through the city of Babylon and into the Persian Gulf.

Geography and Cartography became scientific disciplines during the Greek Antique period. The philosopher, astronomer and astrophysicist Anaximander of Miletus (611–546 B.C.) gave a representation of the position of known land and sea of that time in the first known map of the world *(Figure 14.2)*. The original of his map of the world is not preserved, only the author's descriptions in his work *Philosophical Cosmology* (Leithäuser 1958, see also "Possible rendering of Anaximander's world map" in Robinson 1968).

14.2.2 Claudius Ptolemy

Several centuries later, the scientist Eratosthenes of Alexandria (276–195 B.C.) introduced geography as a scientific discipline to scholarly circles, associating it closely with mathematics. He determined the circumference of the Earth using astronomic measurements. Measuring the distance from Alexandria to Aswan (Assuan) in an ingenious way, he reached a result which is very similar to the currently determined one. He believed that measuring only the length of a single degree (out of total 360 degrees in a sphere) would be sufficient.

Claudius Ptolemy (87–150), one of the greatest scientists of the Greek Antique period, presented in his lifetime work *Megale syntaxis* the Earth in the center of the planetary system, relying on the theory by Plato and Aristotle. In his work *Geographia* (eight volumes), Ptolemy provided a list of all known settlements, their positions defined using geographic coordinates, *Map of the World* as well as maps of various areas. The first printed edition of Ptolemy's work without maps was published in Venice in 1475, the first printed editions with maps were done in Bologna (Italy) in 1477, and the prettiest one in Ulm (Germany) in 1482 (see "First edition of Ptolemy's *Map of the World* Ulm 1482", in Clark and Black 2005, p. 181) (note that the Cape of Good Hope is not presented). Unfortunately, only copies, no originals of Ptolemy's maps were preserved *(Figure 14.3,* National Digital Library of Poland*)*.

From Ptolemy and Marin of Tyre sprang the revolutionary idea (around the year 100) of introducing a geographical grid into cartographic representations, as well as the notion about how important mapping is adjusted to scale. His *Map of the World* was the first attempt to project the surface of a sphere onto a plane – the mantle of a cone. The projection grid consisted of meridians which cut the equator into equal segments and converged towards the North Pole. Parallels were depicted as circles of varying circumferences, which had a mutual center in the same pole. By changing the direction of the meridians in the southern hemisphere, the author attempted to represent the equator. He drew three continents: Asia, Europe and Libya to the best of his knowledge about geography at that time, within this grid, using auxiliary lines. Furthermore, the Indian Ocean, Mediterranean and Caspian Sea are also shown.

Ptolemy's scientific works affected cartographic activities authoritatively for centuries. Namely, in the lack of knowledge and the verification of mathematical

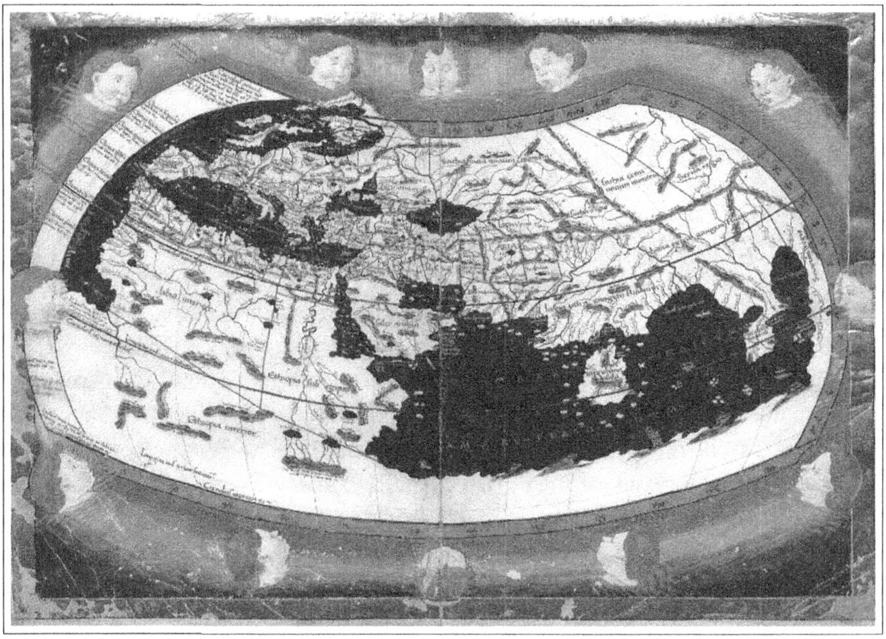

Fig. 14.3. Nicolaus Germanius: Copy of Ptolemy's *Map of the World (Geographia* $1^{st}/2^{nd}$ century A.D.) from 1467, size: 25.3 x 34.5 cm

sources and computing technique the author did not used the correct value of the Earth's circumference, which was already provided in the Greek Antique period from the geographer and historian Eratosthenes. However, Ptolemy employed the measurement and computation results of the voyager, geographer and historian Strabo of Amaseia (60–20 B.C.). Ptolemy's degree was shorter by almost a third. Consequently, his world image was therefore a third shorter in a north-south direction and the equator was drawn too far north. Due to the incorrect length of the Mediterranean Sea (62 instead of 42 degrees) and stretching of Asia to the east (50 degrees more than its true value) there was not enough space left between western European coast and the Asian eastern coast. This would later deceive Columbus (1451–1506) in his conclusions that he could reach Asia relatively quickly by sea, sailing in a westerly direction from Europe.

There is no doubt that Ptolomy's work advanced cartography, however his assumptions inhibited its further development. Up until the end of the Middle Ages, geographers and cartographers overlooked his fatal error.

14.2.3 Roman Antique Age

The further development of the mathematical foundations of maps stagnated during the Roman period. Emperor Augustus (63 B.C.–14 A.D.) was satisfied with approximate map geometry, but he was especially interested in a detailed survey of all objects, roads and paths of his empire. Marcus Vipsanius Agrippa (64 B.C.–12 A.D.) was in charge of the survey and started producing the general map of the world *Orbis terrarum,* which encompassed the Mediterranean area. It was unfortunately thought for a long time that Agrippa's large map of the world was the template for the well known map *Tabula Peutingeriana (Figure 14.4,* Austrian National Library, Vienna).

The Viennese humanist Conradus Celtis (1459–1508) discovered it in the library of a Benedictine abbey in Tegernsee (Germany) at end of 15th century (1494) and gave it as a gift to his colleague Konrad Peutinger (1465–1547) who collected old maps and manuscripts (Clark and Black 2005). After carefully considering this unique record of the past, it was concluded that the original map was produced around the year 250 (Leithäuser 1958, Clark and Black 2005) and that it was most likely one of the copies of a detailed road map of the Roman Empire from the first century.

Tabula Peutingeriana consists of 11 maps, which are assembled to form a strap (680 cm x 43 cm). Therefore, the whole painted area is stretched west to east. This map of the world is exceptionally important for geography and not so much for cartography, because its geometry of rivers, road networks, settlements and the most important mountain chains do not reflect reality. Its special value lies in the 3,500 toponyms on the map which describe the world of that time.

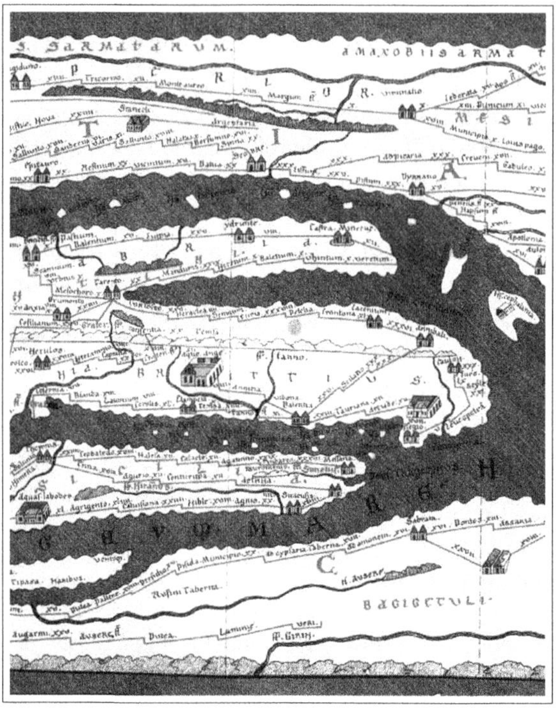

Fig. 14.4. Part of the section VII of the map *Tabula Peuntingerina*

14.2.4 The Middle Ages

During the Middle Ages, cartographic issues were addressed in particular by the Arabs who directed Christian scientists to Aristotle's works. In the 9th century they calculated the circumference of the Earth more precisely than Eratosthenes. They required good maps, and produced quite a large number of them in order to successfully expand the Arabian Empire and trade. The humanist, geographer and cartographer Abu Abdallah Mohammad al-Sharif al-Idrisi (1100–1172) contributed to the development of cartography with his works. His greatest work was a globe illustrating the newest geographical knowledge of that time, engraved into a 400 kg heavy silver ball. All seven continents are represented, as well as a river and lake network, important settlements and trade routes (Clark and Black 2005). Unfortunately, the globe wasn't preserved. However, its detailed description is featured in the work *Tabula Rogeriana* by Al-Idrisi, whose collection also contains a *Map of the World* from 1154 *(Figure 14.5)*.

The application of circular parallels was a great novelty in cartography of that time. Al-Idrisi measured the topographic information of object positions, altitude

Fig. 14.5. Muhammad al-Idrisi: *Map of the World* from 1154. Note that south is at the top of the map.

and included new geographical information from Spanish travelers or navigators. Ptolemy's (87–150) influence in his work is undoubted. (In the 9th century Ptolemy's works were translated to Arabic.)

In the Western culture it was primarily the work of monks in monasteries, which dealt with cartography during this time. This resulted in maps of high artistic value, like so-called T-O sketch-maps (1). *Map of the World* from Sallust's manuscript from the 14th century (see also *"Map of the World from Sallust's manuscript from the 14th century"*, in Leithäuser 1958, p. 63) dated from that period. The Mediterranean Sea and the Don and Nile rivers are represented in the form of the letter *T*, which divided the known world at that time in three continents. As the largest one, Asia lies in the north, Europe and Africa are west and east from the *T*. The *O* shape symbolizes the border of the known geography of the Earth, enclosed by the ocean.

Geographically and cartographically, works produced during this period do not possess great value.

14.2.5 Centuries of Great Discoveries and the Development of Cartography

The survival of the Christian image of the world became doubtful due to the information being added by explorers about newly discovered lands and the new natural-scientific understandings in the 15th and 16th centuries. Sailors believed the Earth had the shape of a ball and they often used globes for orientation in absence of good maritime charts (Leithäuser 1958).

14.2.5.1 Martin Behaim

After the Castilian King Alfons the 10th (1221–1284) wrote about how and which materials were needed to produce a globe, many scientists tried to model the known world in that form (Leithäuser 1958). In 1475, Pope Sixto the 4th entrusted Martin Behaim, a renowned cartographer and astronomer, to produce a globe-shaped model of the Earth. The so-called *Behaim's Apple* (German Museum, Munich, Germany), i.e. a globe of the Earth features the then-known inhabited world – Eurasia, which, he thought, stretches over 234 degrees of longitude (see *"Behaim's Apple, globe from 1492",* in Muris and Saarmann 1961, p. 48). The real value is only 131 degrees. This error, caused by Ptolemy's (87–150) influence, would later deceive Columbus (1451–1506), who died believing he discovered a nautical way to eastern Asia.

But back to Martin Behaim. Behaim, a sailor himself, gathered his information by sailing a third of the world known at that time. He acquired all other content he needed for representing the Earth by means of a globe by inspecting books of various geographers and scientists, such as Ptolemy, as well as the Venetian world traveler and writer Marco Polo (1254–1324). The circumference of the globe at the equator is 159.5 cm, diameter 50.7 cm, and scale 1:25.2 mil. The poles are connected with a metal bar which is a sign that people of that time thought the Earth rotated around its axis. To represent unknown areas, Behaim didn't use empty surfaces or write 'unexplored area' like other globe manufacturers. He filled those areas with emblems and characters from fairy tales, stories and legends.

Behaim's Globe is a unique and surpassingly valuable historical and cultural document, the oldest document representing the Earth as a globe and also the last globe produced before America was discovered.

14.2.5.2 Portolan Charts

The late 14th as well as the 15th and 16th century was the time of great discoveries. Sailors, surveyors and scientists (astronomers, geographers, cartographers and humanists) sailed and investigated the unknown areas of the Earth, exposing themselves and their crew to unbelievable strain, as they wished to explore the newly discovered parts of the world and record and bring back exotic objects. Income

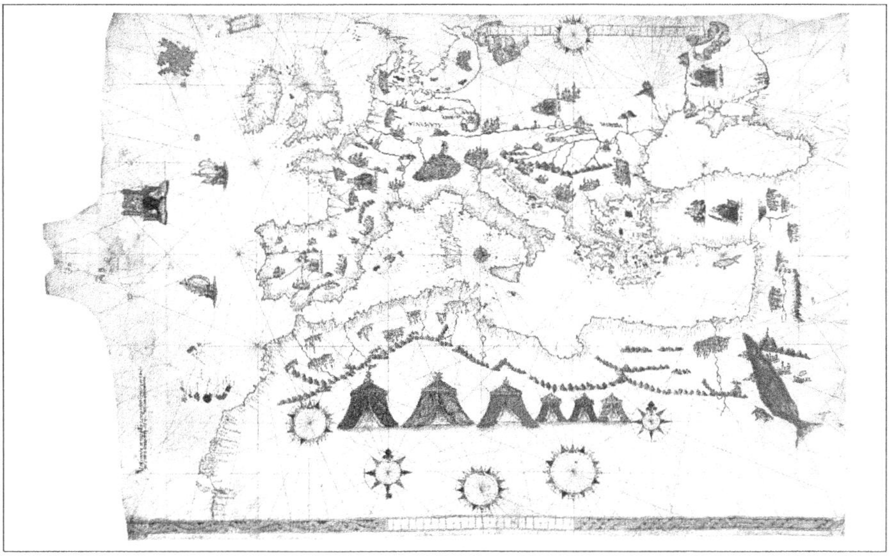

Fig. 14.6. Vesconte Maggiolo: *Portolan chart* from 1541

was not high for filling the 'white areas' on existing, so-called *Portolan charts (Figure 14.6,* Maps department of the Berlin State Library).

As well as noting compasses bearings (which were known already in 12th century) they also included written navigation descriptions. A grid of compass directions was first drawn onto a sheep or goat skin (vellum). Capes and bays, especially important objects for navigation, are disproportionably large in relation to the coastline. Names are positioned clockwise on the land and perpendicular to the coastline. Map geometry resulted from measuring lengths and directions obtained from astronomic observations. The map was decorated with emblems, compass roses and stylized paintings of known cities.

14.2.5.3 Christopher Columbus

The Spanish and the Portuguese shared and continued to conquer the world of that time. They were very interested in itineraries, records, surveys and, especially, maps. Christopher Columbus (1451–1506), an eminent sailor who discovered America, and his brother Bartholomew Columbus (1460–1514), also a sailor and cartographer, employed their own measurements for their maps, but also discoveries, itineraries, reports and fantasies of other sailors. For example Marco Polo (1254–1324) annotated information about unheard and unseen valuables of Asia and China (silver, gold, precious stones, temples, palaces, ...).

In 1487–1488 Bartolomeo Diaz (1450–1500), a Portuguese sailor, was the first to sail the south of the African continent – The Cape of Good Hope. Those discoveries enabled the inclusion of new geographical content in the maps, but they still refuted the theory and authority of Ptolemy (87–150) *(Figure 14.7,* British Library).

Back to Christopher Columbus: In August 1492, he set sail with his Spanish fleet west to the Atlantic, seeking a sea route to (magically described) India. In the same year he discovered a land in the far west – the Bahamas *(Figure 14.8).* He thought that it was just an island before the Indian coast was reached. Thus Columbus was the first Spaniard to step on American soil. He set sail for the third time in 1498, sailing around the Caribbean Islands and following the eastern coast of South America.

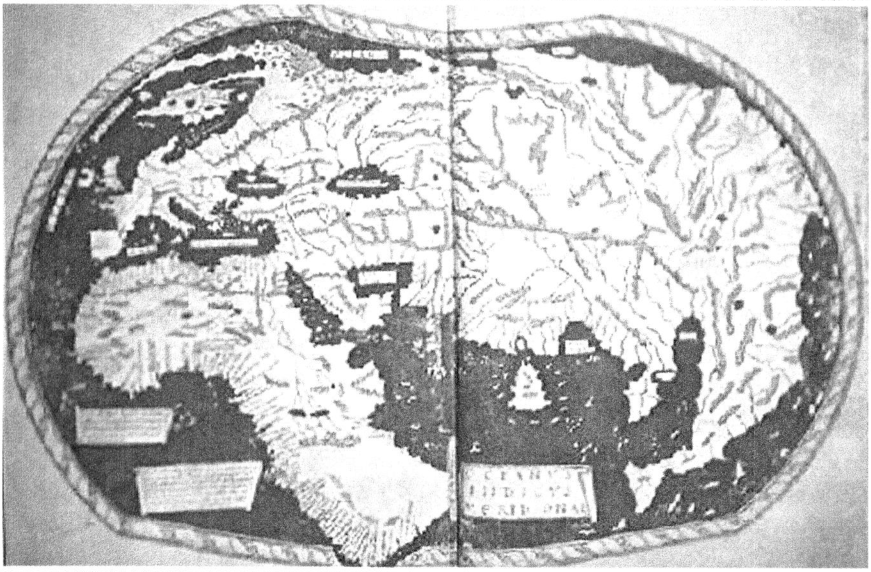

Fig. 14.7. Henricus Martellus Germanus: *Map of the world* from 1489

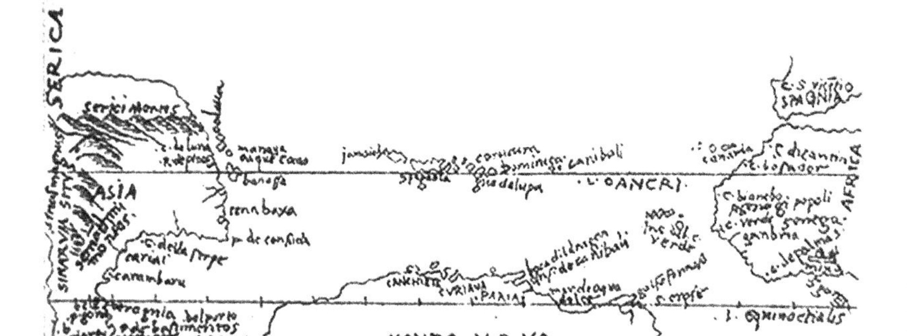

Fig. 14.8. Bartholomew Columbus (Christopher's brother): *Sketch of the Atlantic and coast of America* from 1503

14.2.5.4 Quarta Orbis Pars

Vasco da Gama (1469–1524), a Portuguese sailor and captain, discovered a sea route to India in 1497–1498 by sailing around the Cape of Good Hope. He discovered and explored the last part of the route to the spices of India. When the Italian sailor Amerigo Vespucci (1451–1512) returned from the west in 1504, he wrote that Columbus' discovery wasn't related to India and Asia, which were already known continents, but to a new continent.

In 1507, the cartographer Martin Waldseemüller (1470–1518) published a *Map of the World* from 1507 in his work *Cosmographiae Universalis Introductio* (Library of Congress, Washington, *Figure 14.9).* Presenting the fourth continent, he named it *America sive Americi terra,* celebrating Vespucci's discovery.

That image of the world lasts for a long time. The beginning of the 16th century brought great contradictions, which also affected maps. Vespucci's image of the world was again overshadowed by Ptolemy's influence. Thus America was not a new continent anymore but only *Terra Nova* – a newly discovered part of Asia (see *"Map of the World* by Hieronymus Marini from 1512", in Leithäuser 1958, p. 227. *India Nova* and *Brasil* represent the fact that America was discovered already).

The image of the world changed again when the Pacific was discovered. This was contributed to by discoveries by the Portuguese sailor Ferdinand Magellan (1480–1521). He sailed around South America, proving that 'beyond the horizon' there is another new ocean not included in maps of that time, and that the Earth was round. He also proved that *Terra Nova, Terra Incognita* or *Brasil* wasn't a part of Asia, but indeed a new continent, *America.* This gave true meaning to Columbus' discovery in 1492.

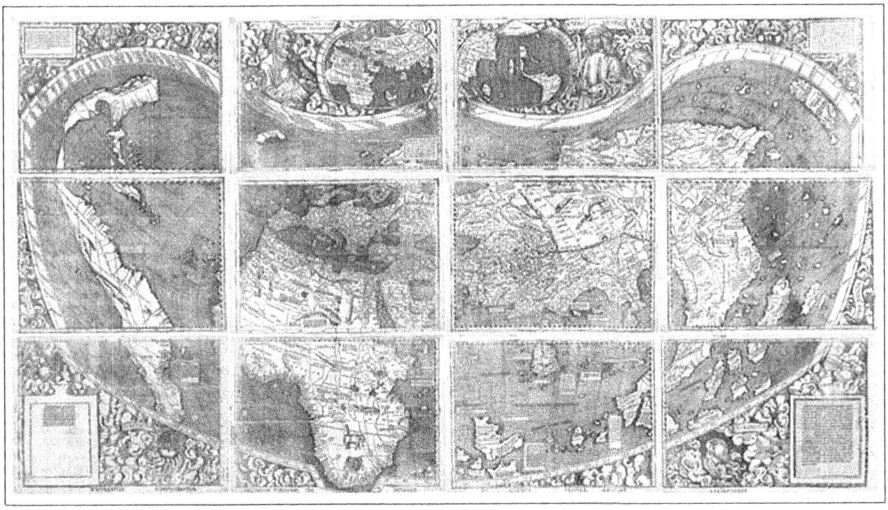

Fig. 14.9. Martin Waldseemüller: *Map of the World* from 1507

In the next year, 1523, Johannes Schöner (1477–1547), a professor of mathematics at the University of Nuremberg engraved a map into copper for the world globe (see "*Globe-map of the world* by Johannes Schöner from 1523", in Leithäuser 1958, p. 232) with travel routes shown and parts of Ferdinand Magellan's newly-discovered world. Schöner's *Map of the World* represents the turning point in the history of cartography. The map features the whole world known at that time, and the eastern Indian archipelago drawn on the left and right sides of the map. A full line gave America the shape of a continent, and the size of the Pacific became evident.

Thus, in only a few years, the mapped world doubled in size and lost its old borders. The need for correct and detailed survey grew daily. Cartography was disentangled from the model of Ptolemy, whose theory and cartographic representations of the world had deceived cartographers for centuries. Captains requested more complete and accurate maps for their journeys and surveyors wanted new and more efficient measuring instruments. During this period, a young scientist, humanist, astronomer and cartographer appeared, desiring knowledge and full of new ideas that revolutionized cartography as science. His name was Gerhard Mercator.

14.3 Gerhard Mercator – Ingenious Cartographer

Gerhard Mercator (actually Gerhard Kremer) was born in the small town of Rupelmonde in Belgium in 1512. At the age of 20 he completed a Master of Philosophy degree. He quickly changed his profession, wanting to work on map production, globe construction and building of astronomical instruments for surveying the Earth, applying his great knowledge of mathematics. Mercator was a cartographer, drawer and copper engraver, penman, publisher, printer of his maps, constructor of his own globes of the Earth and the celestial sphere and a precise machinist in production of measuring instruments. His lifetime opus and deserved fame of a great cartography reformer resulted from the combination of practical experience and ingenious mathematical talent.

14.3.1 Lifetime Opus of Gerhard Mercator

In 1538, under the influence of Frisius' school and his work *Cosmography*, Mercator published his first *Map of the World* in a very complex "heart-like projection" (see "*Map of the World* from 1538 by Gerhard Mercator", in Leithäuser 1958, p. 280), dividing the whole world at the equator into two heart-like forms. The representation and name of South and North America are historically important. The area of the South Pole contains textual annotations indicating that there are certainly many lands in that area, but it is not defined which lands and where their borders were.

Shortly after he published a book (in Antwerp) about map lettering, entitled *Lettering Booklet.* This revolutionized cartographic lettering, suggesting the use of very fine, small and legible italic lettering, which takes less space on maps, instead of using the so-called fractured lettering.

In 1541 Mercator constructed the largest globe at that time (diameter of 41 cm). Its 12 segments and 2 polar spherical caps were engraved in copper and printed by Mercator personally, with unprecedented precision (see "*Globe of the Earth* from 1541 by Gerhard Mercator" (The National Library in Vienna, Austria), in Muris and Saarmann 1961, p. 105). The circumference at the equator was 130 cm, which is a scale of 1:30,000,000. The geographic grid was especially precise. Mercator's globes differ from all previous ones because they represent rhumb lines – lines, which cut all meridians at the same angle, and are thus extremely important for navigation. Mercator had a vision that lines of geographic grid as well as rhumb lines could be mapped like straight lines on a map plane (Vermeulen 2006). At that time, Mercator also produced globes of the celestial sphere. Trading books with the publisher Plantin (1520–1589) showed that the author produced globes in pairs and gave him 24 to sell (Clark and Black 2005).

In 1554 he published his first *Map of Europe* (wall map consisting of fifteen sheets), the first ever to deserve that name (2). He had been working on it since 1538, precisely drawing positions of cities according to his calculations, carefully correcting incorrect positions that were attributed to Ptolemy, which were still present on maps.

In 1569, he finished working on his great *Map of the World* (wall map consisting of eighteen sheets) – *"ad usum navigantium" (Figures 14.10 and 14.11),* produced in the normal aspect of a cylindrical conformal projection – *Mercator's projection.* This started another new period in cartography, the period of mathematically correct maps, which formed the foundations of *new cartography.*

Mercator's cylindrical conformal projection is represented by a grid of meridians and parallels that intersect at an angle of 90°. The distance between the parallels gets larger, the closer they are to the poles. Therefore the represented area becomes more and more deformed. The poles can't be mapped. This projection was the first one that could be used to connect two points on the surface of the Earth by a straight line on the map – rhumb lines. A rhumb line cuts all meridians at the same angle. It is not difficult to imagine what kind of help that was in navigation. Even today, maritime and navigation charts are produced using this projection. But almost a hundred years had to pass until Mercator's projection gained the required credibility.

In the *Map of the World,* Mercator inserted the newest continent border geometry with accurate verification, detail and precision (the coasts of seas and oceans). For its production, he employed all his past works, for example his and not Ptolemy's geometry of the *Map of Europe,* and therefore the Mediterranean. He also included a huge, imaginary south continent *Continens australis.* The map is 1.31 m x 2.08 m

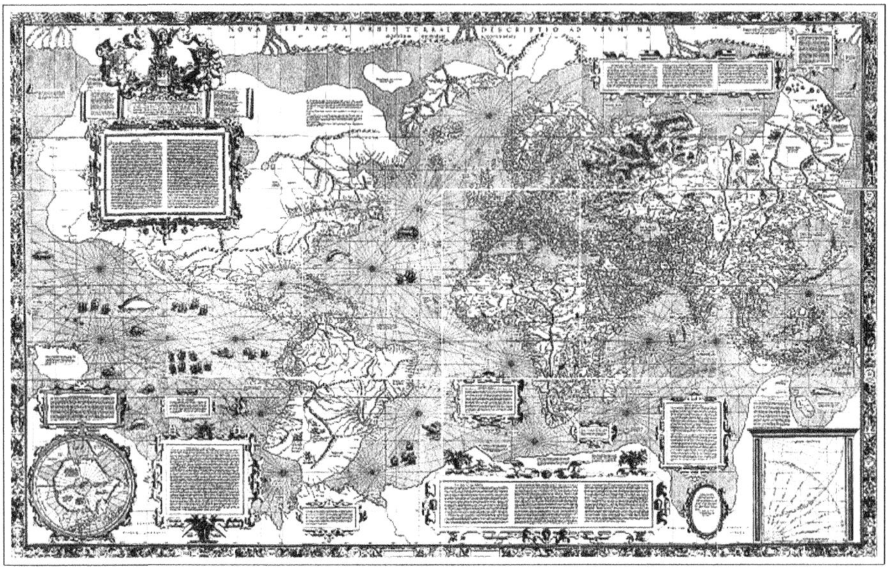

Fig. 14.10. Gerhard Mercator: *Map of the World* from 1569

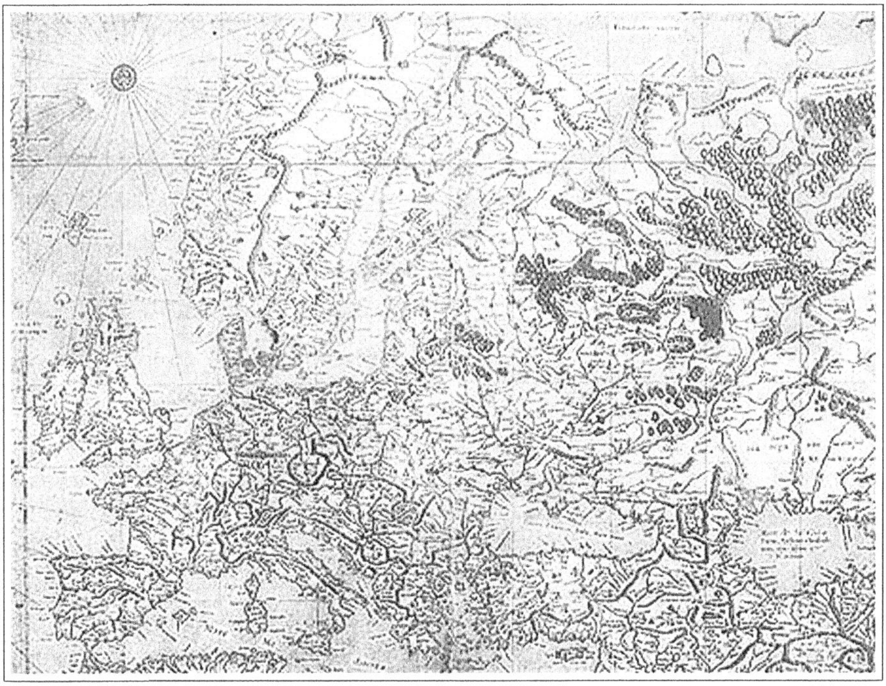

Fig. 14.11. *Europe*, as a part of the Mercator's wall map – *Map of the World* from 1569 (Scandinavia is disproportionately large)

in size and consists of 18 relatively large sheets, an unpractical format for sailors. An unknown sailor prepared three maps of Mercator reducing them to 29 sheets for his route (Leithäuser 1958). This opened the way for the first printed maritime atlas in a book.

Many cartographers adopted the idea. The first among them was Abraham Ortelius (1527–1598), Mercator's friend. His published atlas *Theatrum Orbis Terrarum* comprised 70 maps on 53 sheets *(Figure 14.12)*. This work continued to be updated even after the author's death.

For years, when Ortelius atlas continued to gain recognition, Mercator worked relentlessly and in much detail on his idea. He wanted to realize his lifetime idea, his *Opus magnum* in five parts (Leithäuser 1958, Monmonier 2004). In the first, he wanted to represent the creation of the world, in the second to describe the celestial sphere, in the third a representation of all countries and seas, in the fourth to represent the political history, and in the fifth to give a conclusive representation from the beginning of the world to his days.

The third part of his *Opus magnum,* he called *Tabula Geographicae.* He engraved a total of 102 new maps in copper, all in the same format. He controlled all astronomic calculations, determined geographic coordinates and distances between individual points, gathered newest geographic reports and transferred that data into his projection of 'increasing widths'. For the whole work, he chose the name *Atlas,* the name of the mythological Maori king Atlas, who was interested in astronomy

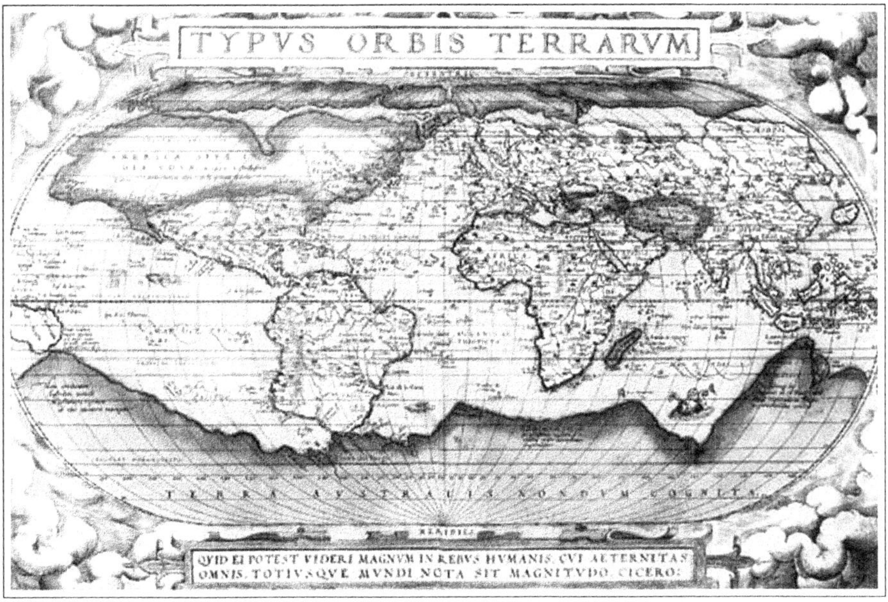

Fig. 14.12. Abraham Ortelius: *Map of the World* from his atlas *Theatrum Orbis Terrarum* from 1570

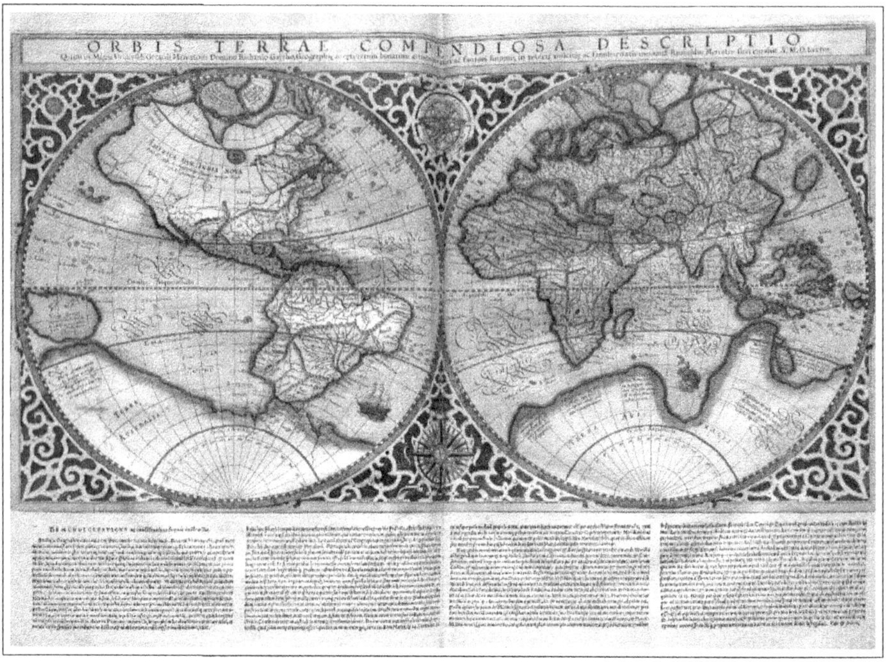

Fig. 14.13. Gerhard Mercator and Rumold Mercator: *Map of the World – Orbis terrae compendiosa descriptio quam ex magna universali Gerardi Mercatoris* from 1587

and allegedly produced the first celestial sphere globe (Leithäuser 1958). In reality, however, regardless of the model, the term 'atlas' as a collection of maps has its true source in Mercator's work (Monmonier 2004, Lechthaler and Stadler 2006).

In 1585, he published 51 maps belonging to *Tabulae Geographicae* of western and middle Europe (France, Belgium and Germany). Four years later, he published the last group of 23 maps of *Tabulae Geographicae* (Italy and Greece) (Monmonier 2004). In 1590, weak and partially paralyzed after suffering a heart attack, he couldn't do work on maps any more.

He worked without pause from 1568 until his death on December 2, 1594. Unfortunately, his *Opus magnum* with 123 books wasn't completed.

In 1587 Mercator's son Rumold finished and published the *Map of the world.* The map was engraved in copper and colored by hand *(Figure 14.13).*

In 1595 Rumold published the *Tabula Geographicae* as a complete work, with tied maps in the form of a book *(Figures 14.14 and 14.15).* He titled it *Atlas sive Cosmographicae Meditationes de Fabrica Mundi et Fabricati Figura (Figure 14.16).* Rumold stated that this was the title Mercator chose. At the beginning of the work he added a title sheet, genealogic table and a picture of god Atlas, who carries the entire celestial sphere on his back (Leithäuser 1958, Monmonier 2004).

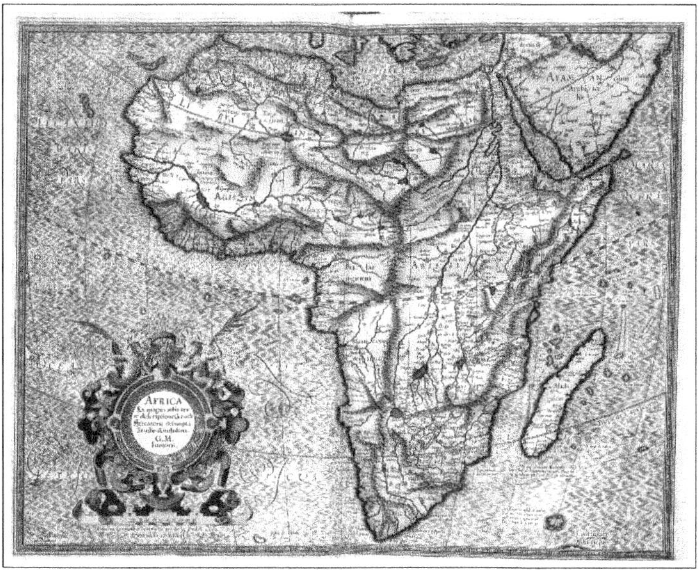

Fig. 14.14. Gerhard Mercator: Map of Africa from *Atlas sive Cosmographicae Meditationes de Fabrica Mundi et Fabricati Figura* from 1595

Fig. 14.15. Gerhard Mercator: Map of the North Pole *Septentrionalium Terrarum descriptio* from *Atlas sive Cosmographicae Meditationes de Fabrica Mundi et Fabricati Figura* from 1595

Fig. 14.16. Rumold Mercator: Title sheet of *Atlas sive Cosmographicae Meditationes de Fabrica Mundi et Fabricati Figura* from 1595

Fig. 14.17. Jodocus Hondius and Jan Janssonius: Gerardus Mercator from 1574

14.3.2 Mercator – The Greatest Cartographer

Because of the slow publishing progress as well as the development and use of Mercator's projection, which wasn't accepted at first, the *Atlas* didn't immediately gain the success it deserved. Neither Mercator himself nor his son Rumold got recognition during their lifetimes or learned the extraordinary work became famous and influencing the entire development of *modern cartography.* From 1595 to the present day, 31 editions of *Atlas Gerardi Mercatoris* have been published, translated to five languages, as well as the so called *Atlas minor* in a smaller format, which stayed at universities.

Undoubtedly, Gerhard Mercator *(Figure 14.17)* along with Marinus of Tyre and Claudius Ptolemy, can be considered as one of the greatest cartographers of all times (Mesenburg 2004). He was a cartographer who was more than a map drawer. Mercator was a teacher of the humankind and was the first who showed us by means of his globes and maps that the endless world could be so small.

14.4 Concluding Thoughts

Maps combine yesterday's, today's, and tomorrow's view of our world, other planets, and the whole universe – from the perspective of people, sciences, and the arts. The history of mapmaking (cartography) resembles the history of mankind. Long before people were able to write, maps were used to visualise human reality or fantasy. The centuries of great expeditions led to today's view and mapping of the world. One of the greatest cartographers of all times was undoubtedly Gerhard Mercator. His extraordinary life's work became famous and influenced the entire development of *new cartography*.

Casually, cartography is a kind of art. Each presented map is an artwork and a treasure chest for cultural heritage, acting as a substitute for its era making us relive the circumstances at that time.

References

Clark JOE, Black J (2005) Die faszinierende Welt der Kartographie. Wie die Karten die Welt verändert haben. Parragon Books Ltd, Bath, UK

Lechthaler M, Stadler A (2006) Output Media Adapted Cartographic Visualisation; in: Banassi E, Burkhard RA, Ursyn A et al. (eds): Information Visualization; Computer Society, IEEE, Los Alamitos, California, Washington, Tokyo, 304–309

Leithäuser JG (1958) Mappae mundi – Die geistige Eroberung der Welt. Safari-Verlag Berlin

Mesenburg P (2004) Abbildungen gestern und heute – Die Weltkarte des Gerhard Mercator aus dem Jahre 1569. In: Kartographische Schriften, Kirschbaum Verlag, Bonn. 9, 186–195

Monmonier M (2004) Rumb lines and map wars, a social history of the Mercator projection. University of Chicago Press, Chicago

Muris O, Saarmann G (1961) Der Globus im Wandel der Zeiten. Eine Geschichte der Globen. Columbus Verlag, Paul Oestergaard KG, Berlin

Robinson JM (1968) An Introduction to Early Greek Philosophy, Houghton and Mifflin. http://en.wikipedia.org/wiki/Anaximander#Cartography, accessed 13 October 2008

Vermeulen J (2006) Zwischen Gott und der See. Roman über das Leben und Werk des Gerhard Mercator. Diogenes Verlag, Zürich

URL

1. http://commons.wikimedia.org/wiki/Category:T_and_O_map accessed 13 October 2008
2. http://www.bl.uk/onlinegallery/ttp/mercator/accessible/introduction.html accessed 13 October 2008

Picture Credits

Note: In this article only images in public domain from http://commons.wikimedia.org/wiki/Image are included, where the copyright has expired or the author has referenced the adequate links in the text to give the reader the possibility to search for these cartographic mile stones of culture and art himself. The figures that are used in this chapter can be found at the following locations:

1. http://commons.wikimedia.org/wiki/Image:Baylonianmaps.JPG
2. http://commons.wikimedia.org/wiki/Image:Anaximander_world_map-en.svg
3. http://commons.wikimedia.org/wiki/Image:Ptolemy_Cosmographia_1467_-_world_map.jpg
4. http://commons.wikimedia.org/wiki/Image:Part_of_Tabula_Peutingeriana.jpg
5. http://en.wikipedia.org/wiki/Image:Al-Idrisi%27s_world_map.JPG
6. http://commons.wikimedia.org/wiki/Image:Maggiolo_-_Portolankarte_-_1541.png
7. http://en.wikipedia.org/wiki/Image:Henricus_Martellus_Germanus_(Wirkungsjahre_1480-1496).jpg
8. http://commons.wikimedia.org/wiki/Image:Bartolomeo_Columbus_map,_West_Indies.jpg
9. http://commons.wikimedia.org/wiki/Image:Martin_waldseemuller_map_1507_m_2.jpg
10. http://commons.wikimedia.org/wiki/Image:Mercator_1569.png
11. http://commons.wikimedia.org/wiki/Image:MercatormapFullEurope16thcentury.jpg
12. http://commons.wikimedia.org/wiki/Image:OrteliusWorldMap.jpeg
13. http://commons.wikimedia.org/wiki/Image:Mercator_World_Map.jpg
14. http://commons.wikimedia.org/wiki/Image:Mercator_Africa_037.jpg
15. http://commons.wikimedia.org/wiki/Image:Mer%D1%81ator_north_pole_1595.jpg
16. http://commons.wikimedia.org/wiki/Image:Mercator_-_Atlas_-_1595.png
17. http://commons.wikimedia.org/wiki/Image:GerhardMercator.jpg

Section IV: Integrated Media

15 Mapping Literature: Towards a Geography of Fiction

Barbara Piatti[1], Hans Rudolf Bär[1], Anne-Kathrin Reuschel[1], Lorenz Hurni[1] and William Cartwright[2]

[1] Institute of Cartography, ETH Zurich, Switzerland
[2] School of Mathematical and Geospatial Sciences, RMIT University, Australia

Abstract

Modern cartography has the ability to map almost any phenomenon for which spatial relationships are of primary relevance. While existing cartographic products cover already an enormous variety of topics, the visualisation of 'other' geographies gains more and more attention. These other geographies may not accord to the 'normal' spaces usually mapped, hence cartography is both challenged and forced to find uncommon solutions. Literature and its fictional spaces might serve as a fine example (but one could also think of soundscapes or emotions). Doubtlessly, the realm of fiction is defined by different 'rules' to the geography that cartography customarily addresses. This paper deals with two main questions: Firstly, how to map narratives and their complex spatial structure? Secondly, what do we achieve by mapping literature? By searching for some (provisional) answers, the horizon of a promising interdisciplinary research field – a future literary geography – becomes visible.

15.1 Introduction

Under the terms of literary theory a fictional or narrated world is made of three components: 1. characters , 2. plot/timeline, and 3. space. While characters and plot have been and are subject to countless philological studies, the topic of space and place in literature is rather neglected. Nevertheless, territorial and topographical aspects of literature have received renewed academic attention within the last decades (also labeled as the 'spatial turn' in humanities) – but so far no convincing definition of the research area, no concise glossary, no methods or tools have been developed in order to approach these matters in a systematic way.

The main focus of a future 'literary geography' are the manifold interactions between real and imaginary geographies in various literary genres. What happens when the 'literary world' and the real world meet or intersect? Which boundaries should be shown: one, both, or a hybrid real world / literature world demarcation?

This chapter aims to present some thoughts, occurring problems, and elements of a theory *(Section 15.2)* and points out that this emerging field demands for a genuinely interdisciplinary approach. It then moves on towards first cartographic renderings of fictional spaces *(Section 15.3)* and discusses the tools and techniques to prepare the texts for mapping *(Section 15.4)*. Finally *(Section 15.5)*, the paper outlines a larger research project that is currently entitled a 'Literary Atlas of Europe' (www.atlas-of-literature.eu).

15.2 Cartography and Literary Studies – A Meeting of Disciplines

It all starts with a surprisingly simple question: Where is fiction set? This question – as well as the variety of answers to it – leads quickly into 'a fugitive field' (Stableford 2003, p. xxxv), more precisely into a research area labeled as 'literary geography', but so far only established in its rudiments.

According to our own reading experience we know perfectly well that each literary work takes place somewhere. Moreover, it's impossible to even think literature without any spatial context. In many cases writers choose settings for their stories which have a 'realworld counterpart' – and they design these settings in a rather realistic way, sometimes even that realistic that one could as well use the novels in question as guidebooks to the described region or city. But literature has infinite options to deal with space (which makes it apparently so hard to map it in a sensible way). Of course, a setting doesn't have to be sketched realistically at all. It can be partly or completely invented (like Robert Louis Stevenson's *Treasure Island);* it can be a crossfading of two spaces (like in Julio Cortázar's short story *The other sky*, in which the setting consists of a confusing combination of two urban topographies, namely Buenos Aires and Paris); it can be an existing and known region combined with fictitious elements (like the Normandy with the towns of Balbec and Illiers in Marcel Proust's *A la Recherche du temps perdu),* it might be a likely place with an invented name, only vaguely localised (like Gottfried Keller's Seldwyla in the novella circle *The People from Seldwyla)* or an existing region remodelled (like Thomas Hardy's Wessex).

15.2.1 The Tradition of Literary Geography

Mapping literary spaces seems like a simple idea – but in fact it's a challenge for both critics and cartographers, a task which can be solved only in an interdisciplinary approach.

This becomes evident especially while looking at former stages of literary geography. Serving as an example of early literary geography is William Sharp's map from 1904.

The map displays Scotland and – in dark grey shade – the zone of action for several novels written by Walter Scott (among them the famous *Waverley)*. In other words: As a map user you're able to see immediately where real and fictional space overlap. There is a rather long tradition of literary geography, dating back to the beginning of the 20th century, with promising attempts in Great Britain, Germany, Switzerland and France. But the history of that (international) research field hasn't been written yet (for a draft and some map samples see Piatti 2008b). However, in the first stages of literary geography, scholars succeeded to visualise a couple of important aspects like the distribution of fictional settings ('gravity centres' vs. 'unwritten regions'). Nevertheless it has to be put on record, that most of these experiments remain on the

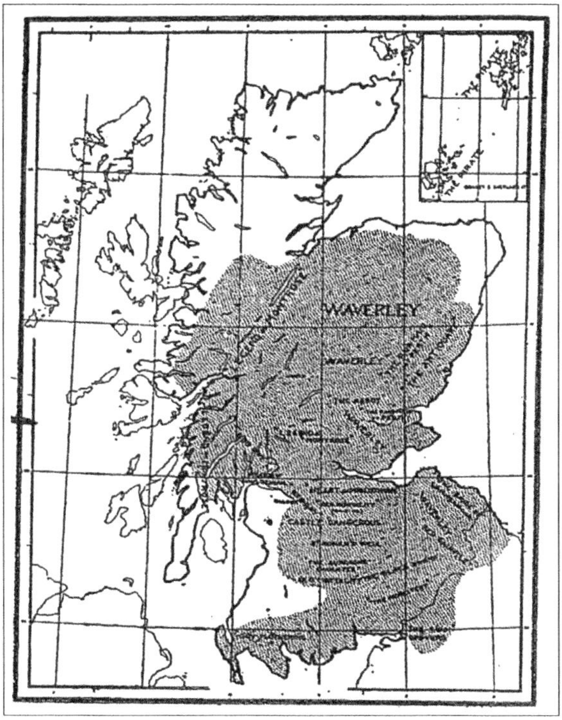

Fig. 15.1. Chief localities of Walter Scott's novels (Sharp 1904)

rudimentary level of an illustrative literary geography (that means: they basically support a written text without having a weight of their own).

Only with Franco Moretti's *Atlas of the European Novel* (1998) the beginning of a new era in literary geography is marked, where maps become truly tools of interpretation, allowing to see something which hasn't been evident before.

What is on display is a map of some of Jane Austen's most famous novels (like *Sense and Sensibility, Pride and Prejudice);* Moretti mapped the places, where the narration does start (triangles), and those where the plot comes to a happy or unhappy end (circles). You see again immediately that Austen could have had a whole lot more space on her disposal – but she didn't make use of this option. Writing her novels, she did deliberately not introduce colourful settings like London, the Lake District, or the Highlands in Scotland, and she neither turns west, to the "green Island" Ireland, nor east, to the continental geography. Her space, which supplies her with settings for many stories is quite narrow – it embraces only the pittoresque south of England, a peaceful area with fields and meadows, mansions and little villages, which are thoughtfully arranged in Austen's fictional worlds. Just think by comparison of plots with a significantly bigger geographical radius – like Herman Melville's *Moby Dick* or Mary Shelley's *Frankenstein,* the latter with settings situated in various European countries and a showdown located in the arctic regions. Moretti himself comments on this first map in his *Atlas,* that it already contains all the secrets of literary geography: You read books, focussing on spatial aspects, then you design a map – finally, you have a look at that map, and hopefully you discover things you didn't know before: New insights which are generated by that map.

Fig. 15.2. Jane Austen's England (Moretti 1998)

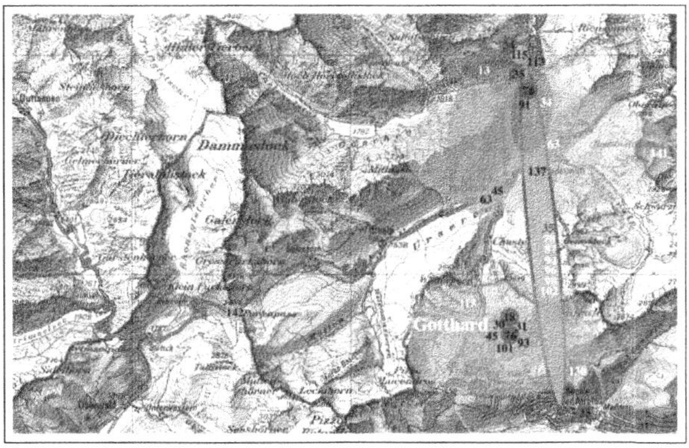

Fig. 15.3. Literary settings in the Gotthard region, Switzerland (Piatti 2008b)

Based on Moretti and developing his ideas further, Barbara Piatti wrote a monography about the exciting perspectives of literary geography, though still being restricted to static maps (Piatti 2008b).

The map (see *Figure 15.3)* presents a small extract of a famous and dense literary landscape: Mount Gotthard, situated in the middle of Switzerland. As a background in a variety of grey shades you see a topographical map of the Gotthard and its surroundings. Silhouetted in coloured ellipses and single dots the socalled literary metaspace becomes visible (ellipses = settings which can be indicated *vaguely*, as a zone; dots = settings which can be localised *precisely).* The prominent narrow, long shape symbolises the Gotthard tunnel. What we're able to spot immediately are the unwritten regions, covering distant valleys and mountain peaks, and the wealth of literature around Mount Gotthard, above and below ground. While the orange colour has been chosen for settings of fictional texts written by Swiss authors, the violet indicates settings described by foreign authors, among them Friedrich Schiller and August Strindberg. Although this last example offers a distinction between vague and precise localisation (ellipses vs. dots), and it also contains additional information about the settings (written by Swiss vs. foreign writers), it is obvious that literary geography needs refined, more flexible mapping solutions. It's also worth notifying that the majority of former approaches in the field of literary geography are made by non-mapmakers – surprisingly enough cooperation between critics and cartographers hasn't been established so far.

15.2.2 Towards a New Literary Geography

Moving to the next level means to work with interactive tools and mapping solutions. Moreover, a new literary geography should first of all follow some essential thoughts, which outline something like a philosophy of this emerging field. There are at least three aspects that make fictional spaces so unique – and consequently so difficult to map.

First of all, there must be an agreement about referentiality between geospace and textual spaces. That means: Fictional spaces don't have to, but can feature references towards the geospace (first space, actual space). Moreover, it's a fact, that both writers and readers are tempted by the option of anchoring texts somehow in the(ir) real world. On the other hand, authors also indulge in creating entirely new worlds with totally imaginary settings. These options – both faint and strong correspondences between geospace and textual space, as well as various connections that can be discovered to exist somewhere inbetween – have to be taken into account.

Second, narrated spaces don't have definite borders (with some very rare exceptions). Usually, cartography works with hard boundaries to show the edges of phenomenon mapped. A place, or the limits of one element mapped can be accurately described. But, when mapping literature, a 'soft' boundary might be more appropriate, whereby less accurate definitions about where exactly the 'edge' of a world of a particular piece of literature ends.

Finally (and connected with the two other aspects discussed above), fiction is sometimes hard to localise, whereby settings are located 'somewhere', with no precise correspondences to a given section of the geospace. This forms a problem for cartographers – where to locate settings when there are no geographic 'hooks' on which to 'hang' these elements of literature. Cartographers will need to hunt for clues about location, follow inferred routes and demarcate boundaries which refuse to accord to 'traditional' cartographic rules.

In short: The geography of fiction must be characterised as a rather *imprecise* geography. Additionally, it has to be pointed out, that there are virtually unlimited possibilities in literature – in literature you can create any space, hence there's the necessity to break these possibilities down into a coherent system. If you do so, be prepared to lose something – the complexity of a fictional text you encounter while performing a close reading (we lose style, vocabulary, psychology to name just a few elements) – but you also gain new insights (see *Sections 15.3* and *15.4)*. The explicit goal is to visibly render the specific geography of literature, but in a most sensible way, by paying respect to the logic of fictional space. In doing so, some parts of the geography of literature may prove to be unmappable. This is a different geography to map, requiring the exploration and development of different mapping techniques.

In order to design such an imprecise geography, visualisation models for single texts as well as for huge amounts of texts (statistical approach) have to be developed. In both cases literary geography should be much more than mere cartographic illustration or a support to scholarly literature in which everything is already said. Quite the contrary literary maps are meant to be tools of interpretation, powerful analytical instruments.

15.3 Imprecise Geography: Mapping Solutions for Single Texts

To map a fictional space, first of all the text (i.e. a novel, a novella, a short story) has to be read and prepared carefully, by breaking down the spatial structure into single elements and their respective functions. There are, according to our newly developed system, five main categories of which a fictional space consists. The following table offers an overview with some explications:

Table 15.1. Spatial elements of a fictional text

category	explication/definition
setting	where the action takes place (i.e. a house, a village)
zone of action	several settings combined (i.e. a whole city, a region)
projected space	characters are not present there, but are dreaming of, remembering, longing for a specific place
marker	a place which is mentioned, but not part of the categories above; markers indicate the geographical range and horizon of a fictional space
route	along which characters are moving: by foot, by train, on horseback etc.

Every topographical or geographical notion within the narrative belongs to one of these categories. This is already part of the preparation: The process of designing visualisations that depict geographies of literature is a process of translation. One set of symbols – text – is translated to another set of symbols – map symbols.

In the following a selected narrative will shortly be presented. It will serve as an example of how literary space might be visualised (with the exception of routes for which no symbolisation has been developed yet).

The exemplary text being analysed is Ernst Zahn's historical novel *Albin Indergand,* published in 1901 (Zahn 1981), set in the rocky regions and steep valleys at the north side of Mount Gotthard, Switzerland. The narrated events take place between 1789 und 1798, it's a rather simple story about the friendship between a devoted priest in a poor mountain farmer's village and an outcast, Albin, who struggles to find the right way in his life. During that process Albin has to deal with psychic (including a disastrous, tornamenting love affair) as well as physical catastrophes (landslide, war). But in the end he marries the love he had already met in his childhood, becomes the head of the village and lives happily ever after.

The map shown here (see *Figure 15.4)* makes use of visual information separation. The background consists of a topographical base map, reduced to a grey scale image. On top of the base map, thematic information from the analysed texts is superimposed and symbolised.

15.3.1 Topographic Markers

The large number of visually highlighted place names might strike the map reader first. These names are pointing to places that have been mentioned in the text but do not play otherwise a prominent role. Graphically, they imitate a highlighter whose fluorescent colours are occasionally used to visually mark important words or phrases in a text. They help a reader to reduce a lengthy text to a few key words and enable him or her to improve orientation or retrieve a text passage. Similarly, topographic markers assist a mapreader to visually grasp the geographic horizon of the analysed text. Typically, they are directly transferred from a text.

Topographic markers are commonly occurring and often surround the locations where actions take place and thus delimit the geographic space that is taken by a text. Considering specific collections of texts, the distribution of topographic markers will provide an idea of the geography of an author, a genre of literature, or a certain time period. This method might be appropriate for a relatively small number of texts but cannot be applied to large text corpora, since too many symbols would impair the readability of a map. For such cases, statistical methods would be more appropriate as will be alluded to in *Section 15.4.*

Colour is used to further categorise the geographic elements of a text and their relation to the geospace. The prevalent maroon colour tells that an existing place from the real world has been *imported* into a text. If a place (and/or its name) has been *transformed,* that is, it has been altered or renamed, the maroon colour changes to an orange shade, whereas completely invented settings will turn to a yellow colouring. As will be shown below, a colour coding scheme has been worked out to further categorise settings from literary texts.

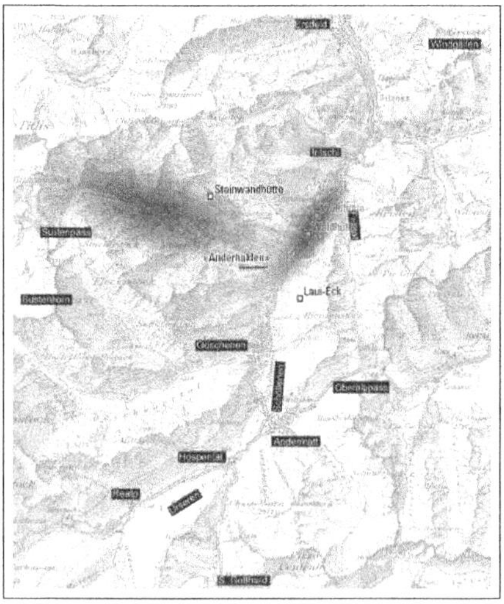

Fig. 15.4. The geography of Zahn's novel *Albin Indergand*

15.3.2 Fuzzy Shapes

The gesture that is related to topographic markers is that of pointing to a specific location. In many cases, however, there is much more uncertainty about the 'real' geography in literature. The blurred, diffuse and fuzzy coloured shapes on the map shown in *Figure 15.4* are intended to portray places, which are only vaguely known in their spatial dimensions. To put it in other words: A specific feature of a literary space is its numerous gaps – all you can detect on the spatial level is reduced, incomplete information. You have to imagine a fictional space – as presented in a text – as a foggy zone, with a few key elements, but often enough with no precise spatial relations between single settings and with no precise borders. These fuzzy shapes graphically express a typical gesture for imprecision, a circular or elliptical motion that forms a rather smudged shape. Size and proportion reflect the possible extent of the area of imprecision. For cartographical presentation, this gesture has been graphically translated to homogenous, smooth, and symmetrical shapes that avoid distinct borders. They do not force map editors to draw distinct lines and prevent mapreaders from measuring exact distances between, to, or from such shapes.

The fuzzy shapes in the maps above inform the mapreader about settings and zones of action that can only vaguely be located and whose extent is unclear. Again, the maroon colour tells that the geographic place is imported in the text. The orange coloured fuzzy shape points out that the place is renamed but that there is an agreement about the real existing place it stands for. The map from Zahn's text reveals the setting of the village of Wassen as being renamed to "Anderhalden". The crossed out and the newly placed name label indicates this place name transformation. While a set of warm colours ranging from red to yellow is used to show areas where action takes place, a different set of rather cold colours ranging from violet to cyan is used to denote projected spaces, a character thinks of, remembers, is longing for or imagines but is neither present nor acting at these places:

	Imported	Transformed	Invented
Settings	#660000	#FF9933	#FFFF33
Projected spaces	#330066	#3366CC	#66FFFF

Fig. 15.5. Colour coding scheme

Imprecision with regard to geographic places is commonly found in literature. From a cartographic point of view it would not be appropriate to more precisely locate and symbolise such places in a map than what is known. Ideally, imprecision should be shown as precisely as possible such that it provides the most appropriate impression of imprecision. Traditionally, cartography has not worried much about imprecision. Precision and accuracy has been regarded as a state, that could never be reached but at least aimed for.

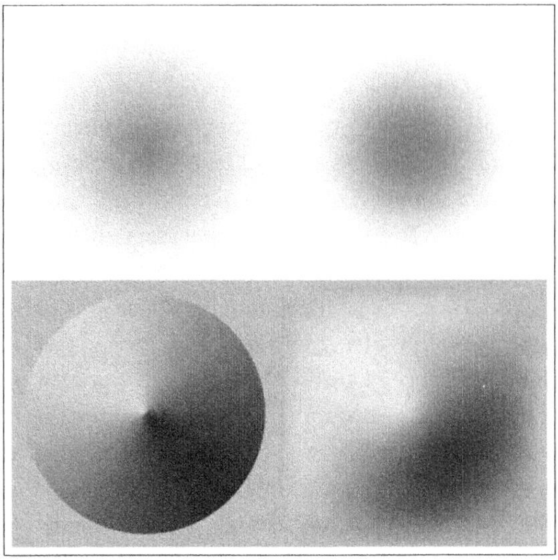

Fig. 15.6. The construction of fuzzy shapes. Upper row: fuzzy shapes, bottom row: shaded display (left column: linear function, right column: Bézier surface)

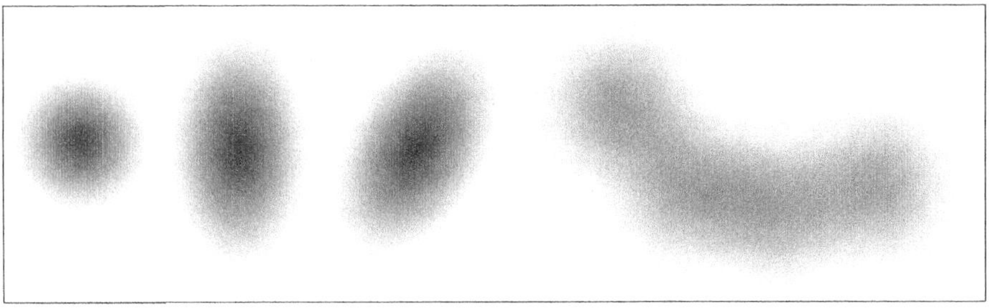

Fig. 15.7. Adaptation of fuzzy shapes. From left to right: Circular, elliptical, rotated, and multipart shape

The approach followed in the map example uses a shape that continuously reduces the opacity from the centre to the peripheral. A first attempt using a linear function reveals that the most intense part is a rather small area and a border can still be detected (see *Figure 15.6)*. Geometrically, this function describes the surface of a cone and is discontinuous at the centre and at the peripheral. This becomes evident when a shading is applied to the surface of the body. A more appropriate surface that avoids these shortcomings would be a rotated period of a sine wave or a Bézier surface with tangents laying on a horizontal plane. A Bézier surface also allows a variable flattening of the central part.

Among the properties of the proposed shapes are radial symmetry which makes the shapes independent of orientation, its smoothness and its simple parameterisation which takes automatic generation into account. If unequally scaled the presentation of imprecision can be oriented in space. Furthermore, due to the smoothness, single shapes can be combined to build more complex areas. Last, but not least, such shapes can be drawn sufficiently fast.

15.3.3 Animated Symbols

If settings denote a specific location rather than an area, fuzzy shapes are no longer an appropriate means of representation. The gesture that is related to present such settings is that of successively pointing at different locations. Or, to put it differently, it is a series of snapshots of the gesture of imprecision. Whereas the village of Anderhalden is shown as a fuzzy shape, the settings with the huts in the Reuss valley and the region of the Susten Pass are presented by animated symbols. The yellow colour of the point symbols suggest that these settings are invented locations and do not refer to existing real places.

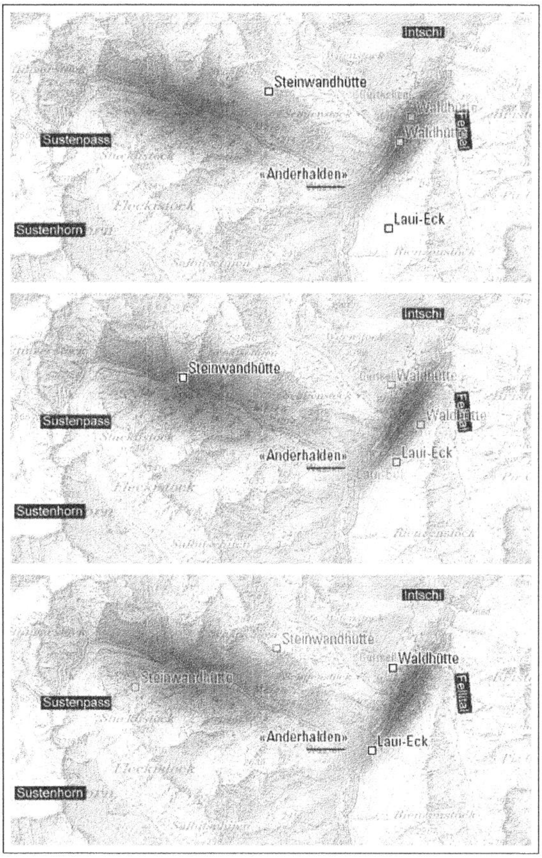

Fig. 15.8. Animated symbols from Zahn's novel shown as a series of snapshots

The animated symbols pop up equally distributed at random locations within a given area. This is mainly to express that there are different possible locations rather a movement or a dislocation of the setting (not a mobile setting). The place name is shown simultaneously with the symbol and uses a mask to enhance the contrast to the background map. In order not to overemphasise the meaning of such settings and to give them the appropriate visual weight a number of measures have been taken. Firstly, the animation period has been increased to up to eight seconds. Secondly, in order not to loose focus, the symbol softly fades in and out and thirdly, the animated symbols partly overlap such that the symbol at the new location appears before the previous symbol completely vanishes. Furthermore, to express the independency of the settings, the animations use slightly different period lengths. The smooth blending is achieved by a sine wave that modulates the transparency of the symbols.

15.3.4 Alternative Mapping Solutions

For some elements and aspects of the literary space conventional mapping clearly meets its limits. How to map journeys of the characters if the only information the text delivers are some indications about stopovers or intermediate stations – while the rest of the route stays literally in the dark? The structure of literary settings is (as discussed above) fragmentary. This does apply even more for dynamic elements within the narrated space like routes and paths. For one character's strolling through a city there are dozens of options, especially if the only hint the text offers is a vague direction (i.e. from the outskirts of a town towards the centre). Obviously it can't be the idea to reconstruct and map such a way in a pseudo-precise manner (this would clearly work against the intention of the text). Another major problem is caused by the category of completely imaginary settings without any references to the geospace. If there's no option at all to use a background map, only some schematic relations between the settings, arranged on a white surface, might be indicated.

In these cases a map need not necessarily be the best visualisation. Information graphics may 'paint' a better picture of geography and human movement through it than could maps. Take for instance the London Underground map. Its designer, Harry Beck, made the map more usable and an effective communicator about how to move about London (see Garland 1994). His original design moved away from the concept that the maps had to follow the actual geographical route of the lines. What were critical to Beck's design were connections: the interchanges between the various lines. It is these connections and the simple information graphics that make the map 'work', irrespective of whether the user is a seasoned London commuter or a tourist arriving at the capital for the first time. This example illustrates how reality is sometimes depicted better if non-realistic maps are used in preference to geographically correct maps.

Considering information graphics and their 'relaxed' considerations of geographical accuracy, can this concept be applied to visualising certain aspects of geographies of literature? Does the visualisation need to be faithful to geography? It might be worth investigating to ascertain if design strategies from this discipline area might be employed, particularly in regard to routes/journeys of characters and to completely imaginary settings.

15.4 The Quantitative Approach: A Database for Fictional Spaces

If we're aiming towards a literary atlas of Europe, methods of literary geography must be applicable to huge numbers of texts. So far, focus has been laid on an individual analysis and visualisation of the geographic space of single texts. While this is a valuable means for literary geography, a more systematic approach is envisioned that proceeds from a geography of single texts to a geography of literature as a whole. What is actually needed are consistent criteria for text selections and guidance of how to analyse such texts. A crucial point will be the identification of the type of information that is to store which also allows the application of statistical methods and the answering of questions on an upper level of abstraction.

The working process has to be imagined as a rather complex interplay between text hermeneutics and cartographic rendering. The core of the project and the crossing point of the two disciplines – literary studies and cartography – is a database that stores information about fictional spaces and is moreover the basis for automatically generated maps. As said before, it is necessary to refine topographical and geographical references of fictional texts to a consistent system – by assigning them to the main categories introduced above. But in addition to the function of the elements and their relation to the geospace there are other aspects to consider in each category. Via these so called metadata you will get a detailed, dense description of an individual fictional space. Some examples regarding that additional information:

- *Toponomy:* The place notation in the text might differentiate between actual and historical terms, between renamed and invented as well as paraphrased and nameless settings. This information is used to label the settings.
- *Spatial precision:* Quality indications show how precisely you can localise settings or projected spaces on a map, with the options precise, zonal and indefinite.
- *Function of the space:* This category indicates, if settings fulfil a specific task which goes beyond a simple background. This includes serving as a thematic background, but also the case of a possible protagonistic status of the setting, be it in a physical (i.e. avalanches, landslides, storm tide) or in a poetological sense (i.e. when an urban setting scares the characters, appears like a labyrinth or trap).

- *Special attributes:* If there are clues for a mythic or an allegorical/symbolical layer of meaning, it can be pointed up in this category.

Reading the texts, interpreting them and finally entering the data is defined as the work package of the literary experts. Meanwhile the cartographers are engaged in the next steps of the process: Maps at different scales have to be prepared. On one hand largescale maps are needed for fiction that is taking place at townlevel. It is possible to show important places and buildings, the author introduces for his story and to follow the ways characters are taking through the streets. On the other hand fiction can have a much wider geographic horizon, when acting in different areas, countries or continents. In that case mediumscale and smallscale maps provide an overview of the geographical background of the fiction.

To set up the automatic map generation process, further research has to be done regarding historical changes within the geospace. Firstly, important shifts and changes of political borders have to be taken into account. Secondly, the geospace is continuously subject of physical changes, for instance due to permanent variations in coastal and mountain regions or to the spread of a growing town.

Combining the entered data with the customised symbology and the cartographic material, the system is meant to generate literary maps automatically.

Once working, there will be options to manually adjust the symbology if necessary. These so called 'push-and-pull-tools' should enable the literary researcher to locate the objects more precisely or to change the shape of the symbols, according to the actual information the text is offering.

Besides creating any map automatically on request, working digitally offers other huge advantages, e.g.: The entered metadata and comments can be added as hyperlinks, extra tables or text next to the maps. Using an interactive device, the user is also able to adjust the information the way he or she needs it. Complexe questions might be asked and answered: In the interactive version of the atlas, the geography of a single author, a group of writers, or a particular historical epoch can be called up, as well as the literary geography of a minority or of an entire nation. One could ask: Are there epochspecific settings? Is there a typical Storm and Stress ('Sturm und Drang') landscape, a landscape of Romanticism, is there a site of Expressionism? Some questions could even be answered for the first time: are there works or types of works that are specifically set in imaginary places and others that are better suited for referential settings?

These requests can evaluate a relatively huge number of texts, making statistical methods necessary. Assuming a text corpus of a region might contain hundreds or even thousands of settings, traditional cartographic methods of portraying single settings will fail. Following a statistical approach, the focus of interest will be shifting from single settings to the visualisation of spatial distributions and patterns those settings describe. In order to answer questions as mentioned above it will be necessary to determine and visualise the 'densitiy' of the literary space. Such maps can be created automatically from database queries and might serve for comparison and animation.

15.5 Outlook: A Prototype of a Literary Atlas of Europe

Within the scope of our current project, entitled "A Literary Atlas of Europe" (www.atlas-of-literature.eu), the potential of such mapping, both for single texts and huge amounts of textual material, will be demonstrated via the example of three differently endowed model regions: a sublime Alpine landscape (Lake Lucerne/Mount Gotthard in Switzerland), a coastal border area, an amphibian landscape between water and land (North Frisia in Germany), and a multilayered urban space, a hot spot of world literature (Prague, Czech Republic). The literary permeation of these three spaces is examined within a time frame of about 250 years, from approx. 1750 to the present. A selection of 200 to 900 texts for each of the three model regions serves as a starting point for the critics. Through the interplay between these materials, the database, the mapping solution, the interpretations and comments a prototype of a literary atlas will be provided. In the end, as a last step, the resulting maps have to be analysed and commented by the literary scholars. In other words: They start to write single chapters of a future literary atlas of Europe. The idea is to develop methods which can be applied to any other geographical region and on any other selection of fictional texts. Once the system runs, the atlas could grow in an additive way: Other researchers and research groups may join the project by mapping selected literary landscapes and cities about which they could provide expert knowledge.

As a matter of fact, methods of literary geography could open a new dimension to literary studies. A literary geography as outlined above is a comparative and quantitative approach. Moreover, a broad range of questions is going to be dealt with, some of them indeed for the first time: Are there still any geographic areas entirely undocumented in literature? How densely settled by fictional works is a particular space? How internationally occupied is it? Or is the space inscribed almost exclusively by native authors? Under which conditions (political-historical not least) does the (imaginary) space of literature contract, and under which does it expand? As stated already above: The geography of a single author, a group of writers, or a particular literary-historical epoch can be called up, as well as the literary geography of a minority or of an entire nation. What comes in sight is Europe's (imaginary) space of literature, which has its own dimensions, which functions according to its own rules, but which is nevertheless anchored in the 'reality' of existing spaces and places. In the process, not only are the literary riches of single regions illuminated, but also, fictionalised landscapes and cities in all of Europe can be examined comparatively, in the sense of a literary-geographical system.

In the end the literary atlas project has more than one beneficial aspect: For the cartographers it's a new field with an extensive theory, it is both an invitation and a challenge to explore visualisations within the frame of an imprecise geography. For the critics it can be described as a chance to rewrite the histories of literature in a most attractive way as well as an option to combine a huge amount of single studies into an overarching project. For them and other users (e. g. pupils, students, teachers, the community of readers in a whole) it offers a spatially and no longer chronologically organised history of Europe's literary heritage and ongoing production – which does not stop at national or linguistic borders.

Acknowledgements

The following institutions are participants in the pilot version of the "Literary Atlas of Europe": Institute of Cartography, ETH Zurich (CH); Georg August University, Goettingen (D), Charles University, Prague (CZ). The project is fully funded by the Gebert Rüf Stiftung, Switzerland (duration: 2006–2009).

This article is based on an earlier version, see Piatti et al. 2008a.

References

Garland K (1994) Mr. Beck's Underground Map. Harrow Weald: Capital Transport

Moretti F (1998) Atlas of the European Novel 1800–1900. London: Verso

Piatti B et al. (2008a) Mapping Literature: Towards a Geography of Fiction, Proceedings of the Symposium 'Art & Cartography 2008', 31st January to February 2nd 2008, Vienna, Austria

Piatti B (2008b) Die Geographie der Literatur. Schauplätze, Handlungsräume, Raumphantasien mit Blick auf Vierwaldstättersee und Gotthard. Goettingen: Wallstein

Sharp W (1904) Literary Geography. London: Pall Mall Publications

Stableford B (2003) Introduction. In: Cyclopedia of Literary Places, vol. 1, Pasadena (CA): Salem Press, pp. xxxv–xlii

Zahn E (1981) Albin Indergand. Zurich: Ex Libris

16 Challenging the Digital Cartographic Continuity System: Lessons from Cinema

Sébastien Caquard[1] and Benjamin Wright[2]

[1] Département de Géographie, Université de Montréal, Canada
[2] Institute for Comparative Studies in Literature, Carleton University Ottawa, Canada

Abstract

This paper begins with the idea developed by critical cartographers that maps materialise and fix lines, stabilising meanings and identity, as well as economic and political power. Cartographic power derives from the impression that there is a natural continuity between objects and phenomena in the world and their cartographic materialization. As argued in a first section of this chapter, this cartographic continuity has been recently reinforced with technological maps and more specifically with virtual globes, such as *Google Earth,* and their smooth zooming and navigating capabilities. In a second section we turn to cinema in which the continuity system of narration has been widely challenged over the years by alternative filming strategies such as jump cuts, direct address, or self-reflexive narratives. The review of these reflexive strategies in cinema makes clear that filmmakers have attempted to undermine the conventional framework of cinematic form, seeking to break free from both the dominant cinematographic continuity narrative and its political implications. Finally, we argue that this cinematographic experience could serve as sources of inspiration for cinematic reflexivity in cartography through the use of montage, sound and human presence.

16.1 Introduction

The movie opens with a lengthy traveling shot. The camera captures what seems to be normal street activity and then follows at a distance a man exiting the subway. This otherwise normal, everyday situation comes to an abrupt end when a close-up shot showing the face of the director of the film, and of the film within the film, shouting: "cut!" As described by Jim Emerson (2006) in his blog, "The whole thing looks perfectly random and natural (...), as if the eye (camera) were just alighting upon one thing and then another as its interest is piqued. But we soon see how carefully and precisely it's all choreographed. Day for night. Illusion for reality." The movie *Day for Night* (François Truffaut 1973) is precisely about the making of the movie as disclosed by a subsequent wider shot revealing the soundstage filled with actors, technicians, and cameras (see *Figure 16.1*). Indeed, this opening sequence frames the entire narrative of the movie, which describes the technical, logistic, financial, social, and emotional dimensions that shape the construction of a film. In the opening scene, the distance between the constructed cinematographic image and that which it is supposed to represent is made apparent within the narrative and conveyed to the audience in a fictional mode.

Fig. 16.1. The opening shot of Day for Night (François Truffaut, 1973) presents a seemingly random and natural scene (a) before revealing its construction and choreography (b) (source: Emerson 2006, © Les Films du Carrosse)

This notion of image construction resonates in cartography. Critical cartographers have emphasised the constructed dimension of maps from a range of perspectives, thereby dismantling the myth of the map as a mimetic representation of the world and a mirror of reality (see Kitchin and Dodge 2007, for a recent review of the main ideas developed in critical cartography). As defined by Christian Jacob (2005, p. 99), the map is:

> *"The result of an intellectual operation as much as or more than a technical one. This intellectual operation allows a minimal schema or an image saturated with information to be projected on a medium. I prefer to consider the materialization and the construction of an image instead of the representation of the earth: in my view it is the only means of shielding the map from tautological definitions and of introducing a minimal set of questions in the positivist discourse that has sometimes dominated the history of cartography. A materialization and construction of an image of space: above all in the minds of the cartographer and of the society to which he or she belongs."*

Cartographers have long recognised the constructed dimension of maps (see for instance Wright 1942). Nevertheless, it is only with the development of critical cartography in the late 1980s that the persuasive and authoritative power of maps have been linked to their apparent objectivity and neutrality, and the illusion of reality they convey. In this chapter we explore how this apparent objectivity has been recently reinforced by the cartographic continuity system that has emerged with high-tech maps and more specifically virtual globes, as discussed in the next section. As cinema and cartography "share many of the same traits" (Conley 2006), and as they both "operate in spatial projection" (Möllmann 2004, p. 160), we then turn to cinema to review and illustrate some of the techniques developed that directly challenge the predominant continuity system of cinematographic narration. We conclude by discussing how cinema might serve to contribute to the development of a more reflexive approach to contemporary cartography.

16.2 Technological Maps and the Continuity System

A single map or even a set of maps cannot represent accurately a given space, as we are confronted with an unlimited multiplicity of social spaces (Lefebvre 1974). Therefore, the fixing of a line, a point or a name on a map is the result of multiple subjective choices. These choices will determine which, among the infinite number of possible maps, will be created for a given space. These choices can have some serious consequences on the way a space is portrayed and perceived. As John Pickles (2004, p. 3) points out: "Which lines we draw, how we draw them, the effects they have, and how they change are crucial questions." Following Gunnar Olsson's argument, Pickles (2004, pp. 4–5) asserts that "the world has literally been made, domesticated and ordered by drawing lines, distinctions, taxonomies and hierarchies." The way of fixing and ordering these lines on maps creates identities, differences, and hierarchies as illustrated by all the possible boundaries a map can materialise and fix. Maps have the enormous power of fixing and stabilising meanings and identities through lines.

Certain authors have pointed out the dangers inherent in the cartographic fixing of meanings, which is then translated into actions and decisions on the basis of the understandings created through these materialisations (Zook and Graham 2007). The cartographic fixing of meanings contributes to political stability, and often serves the existing power hierarchies (Wood 1993). To challenge this stability, Pickles argues that maps and mapping should be thought of in much broader terms, and that cartographers should move away from fixed lines, which stabilise particular perspectives and meanings. He calls for a "post-representational" cartography, one that is aware of its social consequences and open to alternative projects, while acknowledging multiple and changing identities. This approach moves us beyond objectivity, so that we can begin "to speak of cartographies as already and always involving imaginative open, contested and contradictory mapping" (Pickles 2004, p. 188).

Objectivity is deeply embedded in scientific maps. As James Corner (1999, p. 215) argues, "One can put one's finger on a map and trace out a particular route or itinerary, the map projecting a mental image into the spatial imagination. Because of this directness, maps are taken to be 'true' and 'objective' measures of the world, and are accorded a kind of benign neutrality". This apparent neutrality and the resulting impression of objectivity are at the center of a map's authority and power. Maps are authoritative because they look like they are objective, as if there is a naturalised continuity between the objects/phenomena in the world and their cartographic materialisation. The invisibility of the human presence in the mapping process reinforces the cartographic system of continuity. Maps are authoritative because they seem authorless, as argued by Brian Harley (cited in Wood 2002). They usually appear as autonomous documents totally independent from any context of production.

Dissociated from their context of production, maps cannot be perceived as the construction of an image of space (see Jacob in Introduction). Cleared of all references from the successive choices from which they result – choice of data, methods of data collection, of analysis, of representation, of diffusion, etc. – maps present in a mimetic form, a sum of choices that are subjective by definition. This dissociation reinforces the impression of continuity between the territory and its naturalised and miniaturised cartographic form. This continuity, which contributes to making the cartographic lines seem objective and authoritative, is only reinforced by technology.

While a map's objectivity and inherent authoritative power have been deconstructed and criticized by critical cartographers (e.g. Edney 1992), the impression of continuity has been perpetuated by the technological developments of the late 20th century, including automated cartography, GIS, GPS, satellite images, and the seemingly unlimited analytical potential they promise. The removal of direct human participation from many of the contemporary cartographic steps, including data gathering and map dissemination, contributes to a heightened impression of distance between the map and the context within which it was produced. Free from direct human manipulation, the high-tech map becomes the absolute reference, the direct projection of reality. This directness is reinforced by the smooth impression of continuity conveyed by the latest generation of virtual globes such as Google Earth.

Google Earth echoes earlier futuristic visions of the world, such as the one developed by David Gelernter (1992) in *Mirror Worlds.* Here, the author imagines a user looking into his computer screen and seeing reality, his town, company, hospital, etc. The representation of earth materialised through *Google Earth* combines an intuitive interface and a high level of interactivity with the unique commercial power of the Google company. This combination explains the worldwide instantaneous success of this application. A few months after its release, *Google Earth* had made its way into households, workplaces, education systems, hacking practices and mainstream media. People use it on a daily basis for personal, recreational, and professional purposes. In the context of this paper we focus more specifically on the cartographic continuity this application brings to the fore, both horizontally and vertically.

One of the major interests of Virtual Globes is that users can navigate interactively and indefinitely around the world. Horizontally the user can fly around the globe in any direction at a range of speeds and altitudes. Vertically the user can zoom from a view of the entire globe to the image of his/her house (and vice-versa) in a few seconds. These zooming capabilities reinforce the impression of continuity that exists between the world and its cartographic materialization (see *Figure 16.2)*. The link between the application and the world is made through the high resolution of the satellite images representing familiar environment such as streets, parks and houses. While the user is crossing layers of information recorded by different sources, at different periods of time, with different resolutions, these differences disappear through the smooth and fast visual crossing of these layers. The power of credibility of the local images/photos percolates smoothly through the different scales and layers of information and benefits the entire apparatus.

As emphasized by André Bazin (1975), photography benefits from a direct and automatic transfer of reality from the thing to its reproduction. This direct transfer and the inherent appearance of objectivity give photography a very strong power of credibility. Whatever the objections of our critical views, we must believe in the object represented in a photograph, effectively re-presented, meaning made present in time and space (Bazin 1975). Through the zoom capabilities of *Google Earth,* the entire world is re-presented; it is made present in time and space. The construction of a mosaic of images of the world, and the fragmented and partial vision of the world it offers are masked by the continuity system of navigation. In a sense, the barriers between the world and the materialization of the world have been removed. The power of credibility of photographs spans the entire *Google Earth* apparatus through its continuity system.

The fact that this cartographic application is controlled by a private sector company raises concerns about the limits of democratic practices (on this subject see Crampton 2003, Pickles 2004, Monmonier 2004). According to Zook and Graham (2007, p. 8), the Google mapping application "will not achieve plurality and unfixity because Google is the creator (and owner) of the code that creates these new representations." In other words, Google controls the information that is included and excluded in its maps, including how the information is rendered, ranked and made accessible to the users. This illustrates the importance of code in the production of space (Dodge and Kitchin 2005). According to these authors, the importance of political and economic imperatives is generally obscured by technological representations of place which are often accepted by non critical users. Instead of destabilising established lines and meanings, these technological maps have simply displaced the existing lines, from the visible line on the map to the invisible line of code.

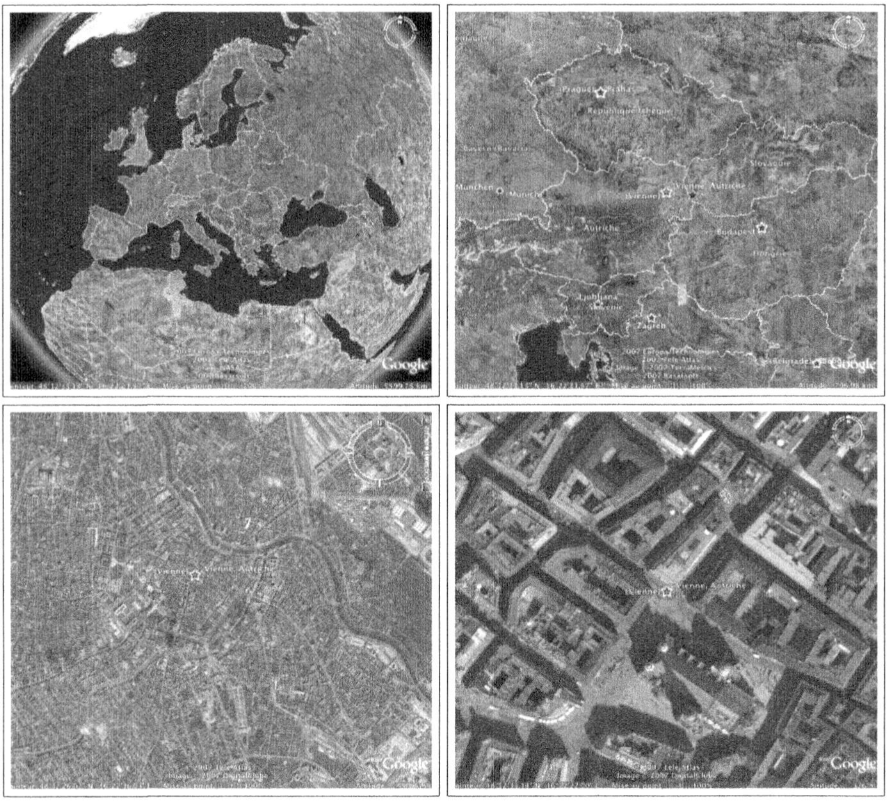

Fig. 16.2. With Google Earth the user can zoom from a view of the entire globe to the image of his/her house (and vice-versa) in a few seconds. These zooming capabilities reinforce the impression of continuity that exists between the world and its cartographic materialisation (© Google)

Finally it is important to emphasise the idea that smooth zooming capabilities of virtual globes and their inherent continuity system were first established in cinema, not in cartography. As argued elsewhere (Caquard, forthcoming), an early example of zooming capabilities can be found in the opening scene of *Casablanca* (Michael Curtiz 1942). This scene combines several contemporary navigation effects of virtual globes, except for the smooth crossing of different layers of satellite images and information. This smooth scale crossing, which is at the center of the continuity system, is be achieved by the 1970s in a very compelling and didactic film entitled *Powers of 10* (Charles and Ray Eames 1977, http://www.powersof10.com). This eight minute short film combines early satellite images with computer graphics to generate a fixed vertical smooth zoom in and out from a couple in a Park, to the entire universe and back to the couple in the park, down to the smallest cells in their bodies. Both of these films illustrate the potential of the cartographic continuity system that will be later materialised and made interactive in virtual globes.

The democratisation of mapping processes promised by recent cartographic applications remains far from delivering the plurality and unfixity required to stimulate the emergence of a post-representational cartography. On the contrary, technological advances made by Virtual Globes strengthen the apparent continuity between territories and their cartographic representations, which reinforces, in turn, the authority of the cartographic line. Challenging this apparent continuity by highlighting the ways in which it is a construction is necessary in order to stimulate user reflexivity and to destabilise cartographic authority. This is of utmost importance because millions of users rely on virtual globes to make a variety of everyday decisions. In the next section we argue that developments in cinema might offer suggestions for developing critically reflexive cartographic practices.

16.3 From Continuity to Reflexivity: The Cinematographic Approach

> *"Most narrative films tend toward a rhetoric of invisibility. They camouflage the ways they are made; their modes of production are most often seen, if at all, only indirectly in the field of the image. Some films make their process visible in the palpable difference between the technical aspects of the film – and its shooting style, its editing, its consciousness of the virtues of its apparatus – the more seductive effects of its plot, its editing, its psychology, or philosophies with which it may be affiliated. Others, self-conscious in style, mask the same relation by demonstrating allusively that they are conscious of ideology, but all the while they reproduce it simultaneously in a mosaic of fragments and self-reflexive gesture"* (Conley 2006, p. 4).

Not long after the advent of the motion picture, filmmakers shifted focus from simple visual spectacle to narrative storytelling. Indeed, the success of any cinematic narrative requires the use of specific stylistic and editing techniques to ensure that the story unfolded in a clear and coherent manner, and that the film's construction did not detract from the dramatic and emotional context of the story. This editing style, known as the 'continuity' system, emerged in the late teens and has since dominated Western film history. It remains the most fundamental system of cinematic construction around the world.

The continuity system does its best to mask graphic style and editing through a number of measures that work to mask transitions between shots. This approach is most evident in Hollywood films, where narrative clarity dominates filmic construction as the importance is placed on ensuring the audience follows the plot without distractions or confusions from shot constructions or editing techniques. To ensure smooth transitions from shot to shot, the continuity system advocates that actions unfold in the center of the screen, that lighting and sound remain constant, and that the camera remains objectively detached. To maintain spatial continuity, an establishing shot is used to provide an overall view of the scene before closer shots are utilised. The 'axis of action' – an invisible '180° line', or half-circle – is used to ensure the same angle of view across a series of shots. Eye-line matches ensure that characters in different shots appear to be looking at each other (this technique is evident in shot-reverse shot compositions that attempt to maintain fluidity during conversations between characters.) At the same time, the soundtrack (dialogue, music, and sound effects) remains constant across the different shots, even though camera angle and distance may vary.

Despite the hegemonic presence of the continuity system in film history, there are alternative editing strategies, which challenge the perceived 'invisibility' of the continuity model. The 'reflexive' nature of alternative editing strategies emerged in Soviet cinema of the 1920s, where disjunctive editing and disruptive visual style subverted the clean structure of the continuity system. Sergei Eisenstein's *Battleship Potemkin* (1925) and *October* (1928) incorporate jarring editing strategies that require audiences to understand shots as 'cells' of a much larger – and more political – organism. As Eisenstein theorised in "A Dialectic Approach to Film Form," montage is the 'nerve' of cinema (p. 45). In the 1960s, French filmmakers such as Jean Luc Godard (*À bout de souffle,* 1960) and Alain Resnais (*L'année dernière à Marienbad,* 1961) purposely rejected the strictures of the continuity approach and violated the smooth editing conventions and symmetrical *mise-en-scene* of Hollywood studio films. On a visual level, Godard's *À bout de souffle* incorporates jump cuts, unmatched shots that consciously break the 180° line. Godard inserts shots that are seemingly disassociated from the narrative action to purposely disorient the audience. With the jump cut, time and space is redefined, de-naturalised, and exposed as a construction of the medium. Characters do not finish their sentences, movements are left incomplete and action is presented in a choppy and de-realised manner. Experimental filmmakers such as Michael Snow have subverted the assumption that cinema is a transparent model of communication by playing with light exposure, filters, film stocks, degrees of processing in his 1968 film *Wavelength*.

While the continuity system is designed to hide its constructed nature, other editing styles have emphasised the artificial character of cinema for political means. The alienating spirit of Bertolt Brecht's theatrical productions in the 1920s pushed audiences to adopt a critical stance to the ideological nature of politics and entertainment. By denormalising the theatrical experience, Brecht's narratives exposed the artificiality of performance. Actors would speak to people in the audience, abandon the script, disrespect the proscenium and break the 'fourth wall'. Brecht's purpose was deceptively simple: persuade the audience to consider the constructed nature of theatrical performance. By showcasing the artifice of theatre, audiences could theoretically perceive everyday life (ideology) to be a performance in itself.

Robert Stam has argued that 'reflexivity' is not inherently political; it can be grounded in"art-for-art's sake aesthetics... It can be narcissistic or intersubjective, a sign of politically motivated urgency or nihilistic lassitude" (Stam 2000, p. 153). Although Stam's assessment remains relatively hyperbolic, it is important to consider the ways in which cinema deftly deals with reflexive style. The search for alternative aesthetic modes has led filmmakers like Godard in France and Michael Snow in Canada to challenge, more than anything else, the hierarchy of convention. Nevertheless, experimental and counter-cinema filmmakers continue to de-naturalise the cinematic environment for aesthetic *and* political means.

Films can play with the reflexive mode within the confines of the continuity system. Woody Allen's *Annie Hall* (1977) utilises direct address throughout the film, which works to provide the audience with a running commentary by Allen's character, Alvy Singer. This trait is not uncommon to the anarchic comedies of the Marx Brothers, who often turned towards the camera to address the audience. At other points in *Annie Hall*, Alvy defies the laws of cinematic convention by pulling Marshall McLuhan out of nowhere to confront an irritating pseudo-academic, who claims to know everything about McLuhan's work (see *Figure 16.3)*. McLuhan responds by saying, "You know nothing of my work." In this sense, Allen acknowledges the confines of the film frame by pulling someone into the screen space from a place 'outside' the cinematic world of *Annie Hall*. Again, Allen draws on Ingmar Bergman's self-conscious imagery in *Persona* (1966), where the film begins – quite literally – with an image of a projector revving up.

Fig. 16.3. Woody Allen's Annie Hall (1977) utilises direct address to defy the laws of cinematic convention by pulling Marshall McLuhan out of nowhere to confront an irritating pseudo-academic, who claims to know everything about McLuhan's work (© Rollins-Joffe Productions).

Atanarjuat (Zacharias Kunuk 2000) is a fictional tale of Canadian Inuit folklore and tradition that is at once an epic tale of narrative spectacle, and a personal story of self-representation. As a story about a community ravaged by outsider influences, *Atanarjuat* is set before the advent of European incursion, in a 'pre-contact' period. As such, Kunuk utilises the conventional logic of the continuity system to tell a story that seeks to preserve his own people's language and culture. In a highly reflexive moment at the end of the film, Kunuk's camera follows the cast and crew during the making of the film. To the surprise of some, Kunuk's crew wears contemporary Western garb, listens to popular music, and uses modern technology (CD players). This reflexive coda reinforces the film's representational project, which aims to disclose the history and tradition of the Inuit in the past *and* present.

In the documentary tradition, self-reflexive narratives consciously include the views and opinions of the filmmakers. Michael Moore's controversial films rarely if ever present two sides of the story, since Moore is the primary 'voice' of his films. Although he respects the continuity system, Moore often juxtaposes fairly innocuous images or sounds with violent ones, thereby producing a jarring, 'new' meaning. In *Bowling for Columbine* (2001), Moore ironically uses the song "What a Wonderful World" against images of failed American foreign policies. The effect is mildly Brechtian, since Moore does not break the fourth wall, but instead forces his audience to note the clash of meanings produced by the distinctive image-sound relationship. To be sure, reflexive strategies in cinema are not limited to particular historical periods or necessarily political narratives. Reflexivity simply provides a means by which filmmakers can acknowledge the conventional framework of cinematic form.

16.4 Perspectives: Toward a Cinematic Reflexivity in Cartography

This short review of reflexive strategies in cinema makes clear that certain filmmakers have attempted to challenge conventional frameworks of cinematic form and break free form the dominant continuity system and its political and aesthetic implications. This attempt could certainly be instructive in cartography as both disciplines "share many of the same traits" as argued by Tom Conley (2006). Central to his book *Cartographic Cinema* is the argument that "a film can be understood in a broad sense to be 'a map' that plots and colonizes the imagination of the public it is said to 'invent' and, as a result, to seek to control" (Conley 2006, p. 1). The control that both maps and films can have on those who take them without questioning their silent authority is a major theme of the book. The different techniques and strategies developed by filmmakers and discussed in the previous section could serve as a source of inspiration for exploring more reflexive forms of cinematic cartography that would, in turn, challenge the continuity system in cartography. These techniques and strategies can be regrouped in three major domains: montage, sound, and human presence.

Montage is a common practice in animated cartography, but has never really been studied. It is mostly used by default to show changes over time in a chronological order. Sequences that show snapshot maps of information at different periods of time (e.g. annually) are presented successively giving the impression that there is a smooth continuity between these situations. These sequences tend to reinforce the naturalisation of the temporal dimension of the cartographic message. As described in the previous section, montage has been used in cinema to challenge the continuity system. Instead of using cartographic continuity system in order to hide the constructed nature of maps, cartographers could explore the potential of editing styles to emphasise the artificial and constructed characters of maps and cartographic sequences, and to explore different ways of telling geospatial stories. As both cartography and films "operate with spatial projection" through scaling and framing (Möllmann 2004), these techniques could be used in cartographic montage to tell alternative geospatial stories as illustrated by some experimental sections of the *Cybercartographic Atlas of Canadian Cinema* (http://www.atlascine.org).

The use of sound in cartography is still rare (Brauen and Taylor 2007) and is generally either descriptive (e.g. a voice describing a geographic situation) or mimetic (e.g. a sound attracting the attention of the user). Musicians and artists have started to explore less conventional uses of sound with maps (see Caquard et al. 2008). In the context of this paper sound could serve to directly address the audience, to comment on the content of the map, on the data gathering, or on the analytical methods used to produce the map. It could be used descriptively or allegorically as illustrated by the audio-visual map on territorial claims of the *Cybercartographic Atlas of Antarctica* (https://gcrc.carleton.ca/confluence/x/XAc). In this interactive map different voices are triggered when the mouse is over a part of Antarctica claimed by different countries. The cacophonic overlaying of different voices serves to emphasise "the cartographic tension existing between the complexity of the world and the cartographic simplifications required to analyze, represent and communicate these phenomena" (Caquard et al. 2008, p. 1231). In this example, there is no direct continuity between image and sound, but an opposition between the messages conveyed by each of the mediums. Sound provides multiple opportunities to challenge the cartographic continuity system. It could serve to develop stimulating forms of experimental cartography and to make maps not only talk about the territory but also about the limits of cartographic lines, their inherent power, and the illusion of the continuity system. It could serve as a counterpoint of the image in order "to disrupt the cartographic conventions" (Sparke 1998, p. 473) instead of reinforcing them. This contrapuntal use of sounds and images echoes work done by filmmakers such as Michael Moore in *Bowling for Columbine* (2001) or Jean-Luc Godard in *Contempt* (1963). Adding sound through voices can also be a way of adding human presence to the map.

Cinema is shaped by human presence in the action as well as behind the camera as illustrated by the Truffaut sequence used to introduce this paper. In cartography human presence has been made invisible in both the construction process and the representational result. As discussed above, cartographers have historically concentrated most of their efforts on designing maps that resemble reality – maps in which human implication is as invisible as possible – without integrating elements of self-reflexivity in their creations. The invisibility of human presence is also obvious at the representation level. There are no human beings in maps. As emphasised by Brian Harley (1988) the dehumanization of the map makes it easier to avoid conscience issues related to the impact that any military or administrative decisions might have on the population of the mapped area. Freed of human presence, the map as a tool of decision dissociates the territories from its inhabitants, and the decisions from its consequences. Integrating human presence through audio, video, photos, and montage could serve to break the cartographic convention, and bring different perspectives on the mapped information as illustrated by the work of recent artists including Heather Rogers (2007) and her mapping of the entire network of the New-York garbage industry. Many contemporary artists are humanising the map in different ways for different purposes, but the importance of mapmaker in the cartographic construction process remains to be acknowledged. Examples from cinema including from the opening scene of Truffaut *Day for Night* and others described previously might be inspirational.

Montage, sound and human presence could be further explored in cartography, but the transfer of cinematographic techniques in cartography faces several challenges. Technically, there are no limits to the introduction of cinematographic effects in electronic atlases or virtual globes, as long as there are resources available for this purpose and a real interest from the mapmakers. This interest is far from being obvious as cartographers have historically concentrated most of their efforts in designing maps that resemble as much as possible reality – maps in which human implication is as invisible as possible – without integrating elements of self-reflexivity in their creations. The introduction of these elements that break cartographic continuity, and therefore cartographic conventions, will require a profound change in cartographic practices and approaches. In cinema, self-reflexive narratives and strategies have been explored mainly in independent cinema and from an artistic point of view. In cartography this approach will then require to rethink the place of the artist in the overall mapmaking process in order to design maps that are not only scientific and aesthetic, but also self-reflexive.

Acknowledgements

This work is part of a research project entitled "Un atlas cybercartographique du cinéma canadien" (http://www.atlascine.org) supported by a Social Sciences and Humanities Research Council (SSHRC) of Canada strategic grant under the Image, Text, Sound and Technology (ITST) program.

References

Bazin A (1975) Qu'est-ce que le cinéma?, Paris, Editions du Cerf

Brauen G, Taylor DRF (2007) A Cybercartographic Framework for Audible Mapping, Geomatica, 61, 2, 19–27

Caquard S (forthcoming 2009) Digital Cartography Originated in Cinema – Mapping the Origins of Modern Cartography in Films, The Cartographic Journal, Special issue on Cinematic Cartography

Caquard S, Taylor DRF (2005) Art, Maps and Cybercartography – Stimulating Reflexivity among Map Users. In Taylor DRF (ed) Cybercartography: Theory and Practice. Amsterdam, Elsevier

Caquard S, Brauen G, Wright B, Jasen P (2008) Designing Sound in Cybercartography; From cinematic structured narratives to unpredictable sound/image interaction, International Journal of Geographic Information Sciences, 22 (11): 1219–1245

Conley T (2006) Cartographic Cinema Minneapolis, London, University of Minnesota Press

Corner J (1999) The agency of mapping: speculation, critique and invention. In Cosgrove D (ed) Mappings. London, Reaktion Books

Crampton JW (2003) The political Mapping of Cyberspace, Chicago, University of Chicago Press

Dodge M, Kitchin R (2005) Code and the transduction of space. Annals of the Association of American Geographers, 95, 162–180

Edney MH (1992) Harley JB (1932–1991) Questioning Maps, Questioning Cartography, Questioning Cartographers. Cartography and Geographic Information Systems, 19, 3, 175–178

Emerson J (2006) Opening Shots: 'Day for Night', post on the blog Scanners. http://blogs.suntimes.com/scanners/2006/08/opening_shots_day_for_night.html, Accessed, Dec. 2007

Gelernter D (1992) Mirror Worlds, Oxford University Press

Harley JB (1988) Maps, Knowledge, and Power. In Cosgrove D, Daniels S (eds) The Iconography of Landscape. Essays on the Symbolic representation, design and use of past environments. Cambridge: Cambridge University Press, 277–312

Jacob C (2005) The Sovereign Map: Theoretical Approaches in Cartography throughout History, Chicago, University of Chicago Press

Kitchin R, Dodge M (2007) Rethinking maps. Progress in Human Geography, 31, 331–344

Lefebvre H (1974) La Production de l'espace, Paris, Anthropos

Möllmann D (2004) Mobile Camera. Cartography in Film and Video, In Möntmann N, Dziewior Y (eds) Mapping a City. Hamburg: Kartierung, Hatje Cantz

Monmonier M (2004) Spying With Maps – Surveillance Technologies and the Future of Privacy, Chicago, University of Chicago Press

Pickles J (2004) A History of Spaces: cartographic reason, mapping and the geo-coded world, London and New-York, Routledge

Rogers H (2007) The Power of Garbage, In Mogel L, Bhagat A (eds) An Atlas of Radical Cartography, Los Angeles, Journal of Aesthetics & Protest Press

Sparke M (1998) A Map that Roared and an Original Atlas: Canada, Cartography, and the Narration of Nation. Annals of the Association of American Geographers, 83, 8, 463–495

Stam R (2000) Film Theory: An Introduction, Oxford, Blackwell

Wood D (1993) Maps and Mapmaking. Cartographica, 30, 1–9

Wood D (2002) The map as a kind of talk: Brian Harley and the confabulation of the inner and outer voice. Visual Communication, 1, 139–161

Wright JK (1942) Map makers are human: Comments on the Subjective in Maps. The Geographical Review, 32, 527–544

Zook MA, Graham M (2007) The Creative Reconstruction of the Internet: Google and the Privatization of Cyberspace and DigiPlace. AAG 2007. San Francisco

17 On Written Language in Works of Art and Cartography

André Skupin[1] and Marinta Skupin[2]

[1] Department of Geography, San Diego State University, San Diego, USA
[2] San Diego Historical Society, San Diego, USA

Abstract

In what some have called *the visual turn,* visual imagery has become an increasingly pervasive element of contemporary society. As traditional boundaries between media are being dissolved and the respective roles of creators/users are changing, written language is not simply being replaced, but instead is appearing in new places and fulfilling new roles. The aim in this chapter is to highlight key parallels and differences in recent developments of cartography and art, with particular concern for the use of textual elements. On the cartographic side, influential trends include the paradigm shift from communication to exploration and the increasing recognition of the socially constructed nature of all cartographic artifacts. Meanwhile, over the course of the last century artists have increasingly questioned the role of written language, increasingly integrating it into their works. Some artists have also sought to integrate geographic depictions into their works, including the linguistic codes included on them. Finally, linguistic codes play a role as both cartographers and artists reflect on *space* and *abstraction* as shared concerns.

17.1 Introduction

What role does *text* play in cartography and the visual arts? The aim in this chapter is to illuminate some key parallels and differences in the recent history of these two fields. Specifically, the focus is on the use of text within the products of artistic and cartographic work. To the cartographer, at least one not fundamentally opposed to critical inspection of one's discipline, it would appear that such conflicting aims as truth, clarity, possession, and control can be identified in artifacts from both fields, though means of identifying and considering those aims will differ. Meanwhile, the artist may be interested in the physical conventions of text as it is employed in cartography and art, and in how these conventions interact with the semantic function of text. The visual nature of written language has subjected it to the scrutiny of artists' examination throughout the history of art. Ever since the invention of the alphabet, when writing became a largely non-pictorial mark-making system, artists have experimented with integrating writing and images in their work. Since the modern period, especially, artists have been using text in visual art, thereby challenging notions of representation, abstraction, the relationship between language and art, and examining how meaning is visually encoded and socially constructed.

We would argue that a successful mapping between these two perspectives could provide insights that are useful to either side. More importantly, the confluence of such powerful forces as global change (incl. climate, economy, etc.), the *visual turn* of society, and an environment increasingly filled with text (e.g., billboards) may demand exactly such exercises at venues like this symposium.

Both of the authors are at the same time creators and consumers of the artifacts of their respective fields, as well as heavily invested in educational activities. In addition to a review of the recent evolution and current state of the treatment of text, the authors will present some of their recent works. On the cartographer's end, there has been a focus on visualising text documents and present them in a form resembling cartographic artifacts on the basis of various geographic metaphors and GIS technology. Meanwhile, the artist has sought to integrate text into works on canvas.

We do not explicitly address the question of whether and to what degree cartography actually is art and vice versa, though that might be a critical issue, even when only the use of text is considered.

17.2 Text in Cartography

Cartographic works have included labels and annotation for millennia. Even the earliest surviving maps already consisted not only of geometric primitives, but also contained annotation regarding the depicted places (together with religious and broader cosmological elements). In fact, those early maps – produced in Mesopotamia – would likely not have been recognised as depicting geographic space, if the names of the early city states had not been included. Meanwhile, the stick charts of Polynesian Islanders include no written labels and the uninitiated would have a difficult time recognising them as highly useful geographic depictions that were derived through careful observation and hands-on experience of geographic phenomena.

The German language has reserved a special term for a map that is deliberately void of labels: *stumme Karte* (mute map), indicating that such a map does not speak to us the way a *normal* map would. Interestingly, such maps are often seen as particularly useful in elicitation of geographic factoids, where students are expected to come up with the missing elements in order to complete the map. It is as though they are inserting the missing words in a sentence. Such didactic uses of unlabeled maps can, according to Stams (1983), be traced back to the late 18th century.

In the cartographic literature there are plenty of statements pointing towards a special role of text appearing on maps, along the lines of David Fairbairn's proclamation:

> *"Text on a map is indispensable. ... a map without text is impossible, such a product being merely an image or a graphic."* (Fairbairn 1993)

In many ways, text is the easiest map element to illustrate what – at least in our view – constitutes the main characteristic of a map: *deliberate abstraction*. The purpose of a map is not to depict what is visible, but what *is* or *may be* relevant to the map user or the map creator. Of course, many of the items found on geographic maps are in fact visible as part of our experience of geographic space. An indication of this is that such features can now be extracted from satellite and aerial imagery through fairly automatic means. Good examples would be rivers, forests, and streets. Many other map features at first seem hidden from view, but are in fact delineated by visible features. For example, boundaries between municipalities often reveal themselves as rapid changes in land use patterns, due to different zoning ordinances.

Text on cartographic maps is profoundly different. The graphic marks generated by adding labels to a map do not correspond to any geometric feature visible from above. Among the rare exceptions are some encountered during ground travel, when a transportation route crosses a linear feature (like a river) or enters/exits an area feature (like a city). At such locations one finds text markers in situ, e.g. the name of a river indicated by a sign at the apex of a bridge. Interestingly, to a traveler these limited textual imprints are a welcome sight for orientation, together with other textual indicators of geometrical and topological relationships (e.g., road signs). Without them, orientation is much more difficult, as it is when faced with an unlabeled map of an unknown area. Again, in the orthogonal view of a map there are no textual imprints existing in geographic reality that could be delineated.

Cartographic labeling involves the following main steps:

1. Choice of elements to be labeled;
2. Determination of label content;
3. Label design; and
4. Label placement.

Successful labeling depends on a consideration of all of these steps. Unfortunately, when it comes to the scientific cartographic literature, labeling is mostly addressed as a label placement problem, and sometimes a typographic design problem, but questions of what to label and what the content of those labels should be – both of which precede placement – is rarely, if ever, addressed by those involved in the creation of maps. While text appearing on maps may be the easiest indicator of power relationships and social constructs (e.g., reflecting colonial and post-colonial histories), that has had limited impact on actual mapping practices. As a result it, cartographers have likewise failed to inform the broader realm of scientific and information visualisation, in which geographic and non-geographic maps are produced mostly by non-cartographers. Issues of feature choice and determination of label content are also where cartographic labeling could potentially meet up with certain tasks found in data mining, such as when labels of n-dimensional clusters are to be determined in an interactive setting. However, such pursuits are dependent on developing deeper conceptual understanding of the complete process by which labels eventually appear on a map. Such conceptualisation may also be a prerequisite for meaningful dialogue between cartographers and those artists interested in incorporation of text.

17.2.1 From Lettering to Typography and Beyond

Explicit concern over textual elements appearing on maps has changed over the years and so has the language used to describe language in maps. *Lettering* used to be the dominant term applied to the process of putting text elements on maps. This reflected the most common technological approaches. Whereas the spatial arrangement of text in a book is characterised by a regular pattern, line-by-line, cartographic text is distributed more unevenly and frequently curved to follow the shape of geographic features being labeled. Historically, this meant that cartographers had to carefully consider the placement of every single letter. Only in the last twenty years has fully electronic label placement become a reality, such that labels can be placed and iteratively adjusted on a computer monitor. Labels can be associated and placed alongside elongated features or stretched to fill the extent of area features; true-type fonts can be scaled to any size without pixilation effects; variations in color and such effects as drop shadows can be easily introduced. Consequently, one now encounters the term *typography* much more frequently when cartographic text is discussed.

To investigate the treatment of text by cartographers more thoroughly, we surveyed some of the leading cartographic textbooks of the last sixty years, beginning with Erwin Raisz' "General Cartography" (1948). The results are summarised in *Figure 17.1,* with a special focus on the relative amount of coverage devoted to text issues. For every book, the percentage of pages dealing with text was determined. Around 1950 we found significant variation ranging from 3.1% in Raisz (1948) to more than 30% in Arthur Robinson's map design monograph "The Look of Maps" (Robinson 1952, not shown in *Figure 17.1).* During the 1960s and 1970s the various volumes of Robinson's "Elements of Cartography" and Raisz' updated volume, now called "Principles of Cartography" (Raisz 1962) showed remarkable consistence, with coverage of text issues taking up between five and six percent of the total page count. After 1990, text books began to incorporate more and more of the results of cartographic research produced in the 1970s and 1980s, including such emerging topics as animation and interactive maps. Borden Dent's book is a bit of an exception, with 4.8% still devoted to typographic issues (Dent 1999). Interestingly, among traditional, print-focused cartographers this remains a favorite textbook. Meanwhile, the last edition of Robinson's book, now involving a total of five co-authors (Robinson et al. 1995), covered cartographic text with only 3.26%. Those cartographers [and other geovisualisers] looking for coverage of emerging, computer-based techniques tend to turn to the book project headed by Terry Slocum, in which coverage of typographic and labeling issues is limited to around 1% (Slocum 1999, Slocum et al. 2005, 2009)! Meanwhile, Cynthia Brewer (2005) dedicates almost 1/3 of her book to text issues, in recognition of the potential power of GIS and a growing user base that is however not cartographically trained (see also *Section 17.3.1)*.

As already mentioned, cartographers tend to focus on the typographic design and placement of labels. In terms of design, standard semiotic rules are extended from the traditional concerns over point, line, and area geometry towards driving the appearance of text labels on maps. For example, differences in terms of hierarchy and quan-

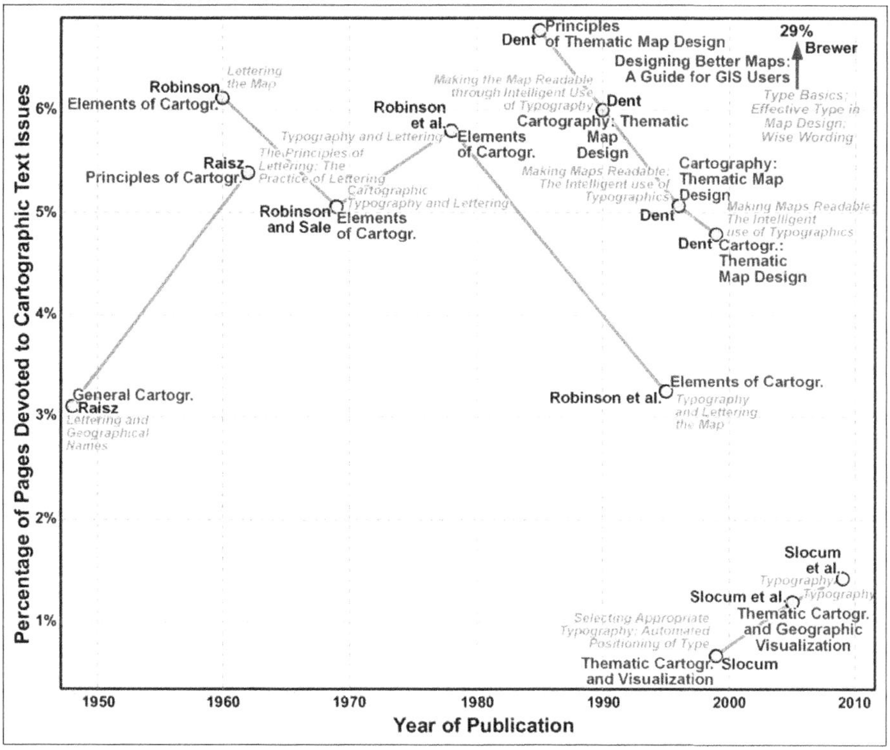

Fig. 17.1. Coverage of cartographic text in leading textbooks. Listed are authors, book titles, and headings of chapters or sections devoted to text issues.

tity have traditionally been expressed through such graphical differentiation as size or color value. These are usable with cartographic text as well, such as variation of font size or font color. However, text affords a number of additional variables to work with, including letter spacing and case *(Figure 17.2,* Kraak and Ormeling 1996).

As far as label placement is concerned, the main difficulties encountered hark back to our earlier argument of text not being found inscribed in geographic reality. On one hand, cartographic labels should be unequivocally associated with the features they refer to. On the other hand, they must conflict neither with other labels nor with the variety of other content found in a map, i.e. labels should not obscure actual feature geometry. Meanwhile, the text strings that will become labels vary tremendously in length, which adds complexity. For example, country names can be as short as "China" (and in that case be associated with a very large geographic feature) or as long as "Swaziland" (a very small country).

One sometimes underappreciated aspect of cartographic labels is that they convey much more than simply the name of a feature. First of all, to give something a name is a far less innocent operation than is sometimes thought (see the following section for more on that). In addition, quality label solutions are able to convey far more than the mere name of a feature. The shape of a label can mirror the shape of its respective feature, e.g. the name of a river running alongside it. Sometimes, the label can be used to indicate the geometry on its own, e.g. when "Mid-Atlantic Ridge" snakes its way across the space between South America and Africa, without any other indication of the location of the referenced feature. Another example for the intricate role of labels is the common rule of placing the name of a coastal city in the ocean, while a city near but not at the ocean is supposed be labeled on land. In other words, intelligent labeling can help to convey important geometrical and topological aspects of geographic space.

Fig. 17.2. Semiotic principles applied to text labels for cartographic maps (from Kraak and Ormeling 1996; courtesy of Menno-Jan Kraak)

17.2.2 What's in a Name?

Maps were for the longest time thought of as mere depictions of geographic space. Granted that abstraction was always seen as necessary – due to the reduced scale of a map – but cartographers thought of themselves mainly as faithful servants of geographic reality. There was much talk of selecting map projections that served the purpose of the map, to choose source data providing the best available geometric accuracy, and of other concerns conveying a sense of responsibility and innocence. However, powerful arguments put forward during the last two decades have undermined many of our most basic assumptions about maps and mapping. Postmodern critics have charged that maps are reflective of and, more importantly, instrumental in shaping relationships characterised by differences in power and control. While in-depth reflection on postmodern, postpositivist, feminist, and related positions on the role of maps is beyond the scope of this chapter, labels are in fact an element that lends itself most easily to such critique, because, compared to other cartographic content, they are such explicit expression of its creators' viewpoint.

Geographic names are also often more volatile than other geographic and therefore map features, and reveal much about a region's development as well as about the cartographer's disposition. For an example, consider how dramatically labels can sometimes change in editions of the same atlas published at different moments in time. Such is the case with the map "Africa – Southern" appearing in *Goode's World Atlas* in 1960 versus 2005 (Espenshade 1960, p. 166; Veregin 2005, p. 232). Geometric elements, expressed as point, line, and area symbols are virtually identical in both editions. It is only in the labels that the enormous political, economic, and social changes in this region are hinted at. "Bechuanaland (British Protectorate)" becomes "Botswana," "Southwest Africa" becomes "Namibia," and so forth. Yes, names and labels are important signifiers and they seem to operate quite differently from other geographic features. Changing the name of a country (e.g., from Burma to Myanmar) is seen as relatively unproblematic, but changing the geometry of a country's outline – remember that label placement is based on this – is typically referred to as *war!*

17.3 Text in Contemporary Visualisation

Increased concern about the use of text seems warranted not just in the light of postmodern critiques of mapping approaches, but also due to recent advances in the underlying concepts and technologies. The technology of mapping has changed dramatically in recent decades, beginning with the development of GIS and digital geographic databases and leading up to such current trends as Google mash-ups. One result has been that more and more people are actively engaged in mapping activities. On a conceptual level, the most fundamental change has been a shift in the dominant cartographic paradigm from *communication* to *exploration*. The goal of cartography has thus shifted from the creation of *the perfect map* to *the perfect interface* allowing users to discover patterns and generate new insights. This is one of the contextual elements behind the apparent lack of attention paid to labeling. If the map seizes to be the primary geographic database – which had been its main purpose for many centuries – and instead becomes something much more fleeting, ephemeral, then the role of the label is bound to change as well.

17.3.1 Geographic Information Systems

Based on much of the survey presented in *Section 17.2.1* one might think that cartographers have all but lost interest in how cartographic text is generated. To a classically trained cartographer, labeling is frequently reduced to issues of typographic design and label placement. As we've argued in *Section 17.2.2,* that is a somewhat naïve view, with respect to the very nature of attaching labels to geographic features. However, the fact that cartographers may feel quite comfortable with producing quality labeling solutions (e.g., using desktop publishing software, like Adobe Illustrator) obscures the fact that novices can feel overwhelmed by the complexities of label design and placement. That never used to be an issue, since only cartographers had simultaneously access to geographic source data and to the various technology solutions for data transformation, cartographic design and production. The "only" element cartographers had to add was their knowledge regarding what constitutes good map design.

All that has changed with the advent and maturation of geographic information systems (GIS), together with huge volumes of geographically referenced data now being easily obtainable. As a result, for the last decade or so, non-cartographers have been able to create visual output from geographic data. One hesitates to attach the label "cartographic" to much of what has been produced in this manner. GIS users were for many years able to violate even the most basic conventions of cartographic design, many of which had been derived from or supported by extensive research by cartographers, cognitive scientists, and psychologists. Not surprisingly, cartographers were for the longest time extremely apprehensive about GIS technology, basically associating them with "bad maps." In recent years this has begun to change. For example, the commercial software ArcGIS (www.esri.com) now enforces many of the fundamentals rules regarding the link between data and symbols that had been followed by cartographers for several hundred years and were finally formalised by Jacques Bertin (1967/1983).

Text labeling is a logical part of these efforts towards formalisation and algorithmisation of cartographic expertise by academia and industry alike. However, given the inherent complexity laid out earlier, automated labeling turns out to be quite complicated. Most labeling software approaches address only some portion of the labeling issue, e.g. point feature labeling, but there are very few examples of true integration of labeling within a comprehensive mapping system. One of the few exceptions is the Maplex extension to ArcGIS, which gives GIS users access to many of the methods used by traditional cartographers. *Figure 17.3* illustrates how the concerns and, frankly, tricks of cartographers are turned into parameters, which are then driving the automated placement of potentially very large numbers of labels. For example, the visualisation shown on *Figure 17.4* was generated in that manner.

Modern GIS software may provide the tools necessary to create impressive labeling solutions, but their effective use still depends on the operator having a fair amount of understanding of map design concepts. That largely explains the extensive coverage that Brewer (2005) dedicates to cartographic text.

17.3.2 Virtual Globes

It may be time to ask whether the role of cartographic abstraction is changing in the light of the kinds of geographic visualisations people are now surrounded by. The various implementations of digital globes (e.g., *Google Earth,* Microsoft *Virtual Earth*, NASA *World Wind*) are focused on what's visible through an orthogonal view from above. On one hand, the selective processes involved in generation of abstraction *seem* deemphasised, with so much of the display occupied by highly textured imagery. Some of the general attraction of digital globes are likely due to

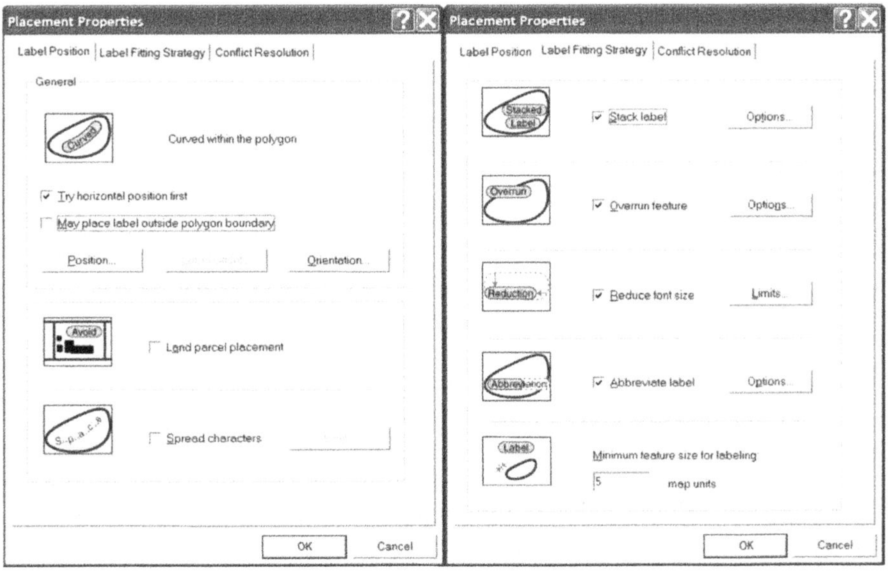

Fig. 17.3. Labeling controls in the Maplex® extension of the commercial GIS software ArcGIS®

the perception of a live depiction of reality, notwithstanding actual heterogeneity in spatial, temporal, and thematic characteristics. On the other hand, recent versions of such software have increasingly included more overtly abstract elements, and text is among them. In other words, the emergence of virtual globes is forcing us to pay more attention to text in visual displays as a subset of the broader issue of abstraction. What, when, and how to label in a virtual globe environment ought to be a major area of concern for cartographers.

17.3.3 Information Visualisation

In *information visualisation* – which deals with abstract data – there is a pronounced focus on generating novel geometries. All matters of rectangular, circular and network layouts of abstract data have been proposed (Card et al. 1999), but text elements are typically added as an after-thought, out of sheer necessity. Labeling is seen as something to be made available only on request, like when a user clicks on a geographic feature. Meanwhile, the broader cartographic view of labels as able to convey much more than just the name of a feature is largely ignored.

Naturally, text labels receive more attention where text documents themselves are visualised, such as in knowledge domain visualisation (Börner et al. 2002, Skupin 2002). There is even a place for GIS software, once *n*-dimensional document vectors are projected into two dimensions. For example, the first author has used ArcGIS and the aforementioned Maplex extension to generate large-format knowledge domain maps involving thousands of text documents and, ultimately, hundreds of labels *(Figure 17.4)*.

17.4 Text in Art

Writing and pictures are essentially two ways of visual representation on a flat surface. Artists who have combined elements of text and imagery in their work have explored ways in which these two systems of representation differ or correspond to inform perception and cognition.

17.4.1 Coexistence of Text and Image

Among the types of image-text relationships identified by Simon Morley are "multi-medial" (Morley 2003) relationships, in which either the text accompanies the image, in support or contradiction, or the image accompanies the text. In both cases, image and text remain distinct from each other. Examples of this kind of relationship include René Magritte's 1929 painting "The Betrayal of Images,"

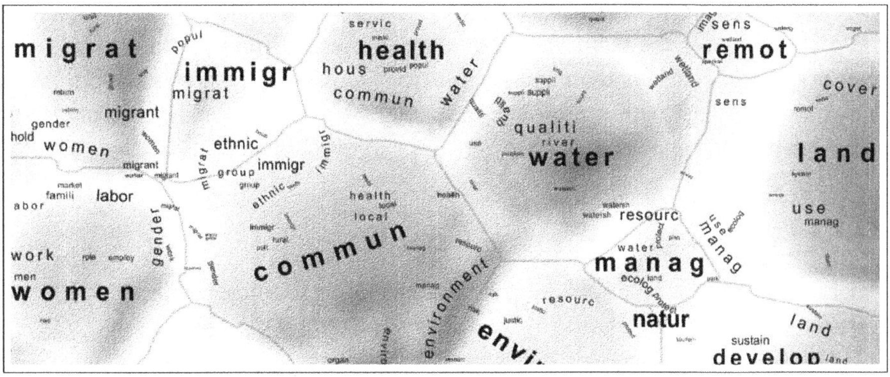

Fig. 17.4. Detailed portion of a visualization of the geographic knowledge domain. André Skupin, In Terms of Geography, 2005

where the words "Ceci n'est pas une pipe" are painted below an image of a pipe and Barbara Kruger's "I shop therefore I am," 1989, in which Kruger combines a caption reminiscent of advertising with a black-and-white photograph of a hand. From a formal perspective, an important issue arises in instances like these. It is impossible for text, being a flat medium that exists right on the surface of the picture plane, to co-exist with an image depicting 3-D realism, without disturbing the illusion of depth and the adherence to perspective in the composition. The combination of text with imagery resembling reality exposes the fact that the imagery in the painting or photograph is no closer to reality than are the words that we use. Pictures and words are therefore presented as two kinds of signs used to represent reality. By appropriating the look of commercial communication and advertising techniques, Kruger specifically comments on the use – and commodification – of both language and image as signs in contemporary society.

17.4.2 Text as Image

Instances in which the distinction between text and image is blurred, and where parts, or all, of one becomes the other, fall within what Morley terms "mixed-media" and "inter-media" relationships (Morley 2003). Examples include Robert Indiana's piece where the four letters spelling "LOVE" constitute the entire composition, and Ed Ruscha's gunpowder-drawing "Optics," in which the rendering of the word "Optics" makes it appear suspended and voluminous in space. Both of these examples take a single word as their main compositional element and subject it to formal treatment: Indiana by, amongst others, the unconventional arrangement of the letters and focus on the positive and negative spaces formed by the letters in this arrangement. Ruscha's drawing creates an optical illusion through his rendering of the image, which has the effect of making the word look like it is made of ribbon or bent paper. There is thus a deliberate play by the artist with the connection between the presentation of the word and its content. It could be argued that the relationship between the presentation and meaning of the word is more confrontational in Indiana's "LOVE," with a profound and enigmatic concept being presented in a very orderly and simplified manner.

17.4.3 Language as Art

Another use of text in art can be found in examples like Joseph Kosuth's "Titled (Art As Idea As Idea)" from 1967, and Jenny Holzer's series of "Truisms," started in 1977. These two examples differ from previous ones in that the content of the words, or the ideas encapsulated by the words, is the art. The presentation often consists of nothing else than the legible text in the form of standard-looking printed (or photocopied) words on paper. The artist Sol LeWitt's statement that "ideas alone can be works of art" (Harrison 2001) explains the aim of conceptual artists like Kosuth to change the very nature of art by making it more about thinking than about seeing. One way conceptual artists achieved this was by replacing the art object with language. Kosuth presents a photocopy of the dictionary definition of the word "art" on the gallery wall, while Jenny Holzer presents her "Truisms" statements in public spaces, like phone booths, as well as in galleries and museums.

17.4.4 Maps in Art

Artists have at various times incorporated elements of cartographic maps into their works. In this chapter, our interest lies with those to whom the role of text in such depictions was particularly important. The following two examples fit that requirement nicely.

In the case of the futurists of the early 20th century, it was imperative to escape from the structures imposed by linear text. That was exemplified by their motto "Parole-in-Libertá" (Words in Freedom). The futurists deliberately emulated structural elements of geographic space – especially urban space – and its cartographic depiction, including the use of visual hierarchies expressed through font size and spacing *(Figure 17.5)*. The creators as well as the audiences seem to have clearly understood these as maps, such as when Tristan Tzara refers to them as "cartes-poemes géographiques futuristes" (futurist geographic map poems, Tzara 1920). Incidentally, the view of two-dimensional space and mapping as providing a movement towards freedom, as an escape from the rigid structure of the traditional written word, should be worth further investigation in the light of current views on mapping as a mechanism of appropriation and control.

As a final example from the world of art informed by maps, consider "Nordrhein" by Guillermo Kuitca, 1992, in which the artist copied a portion of a map of Germany

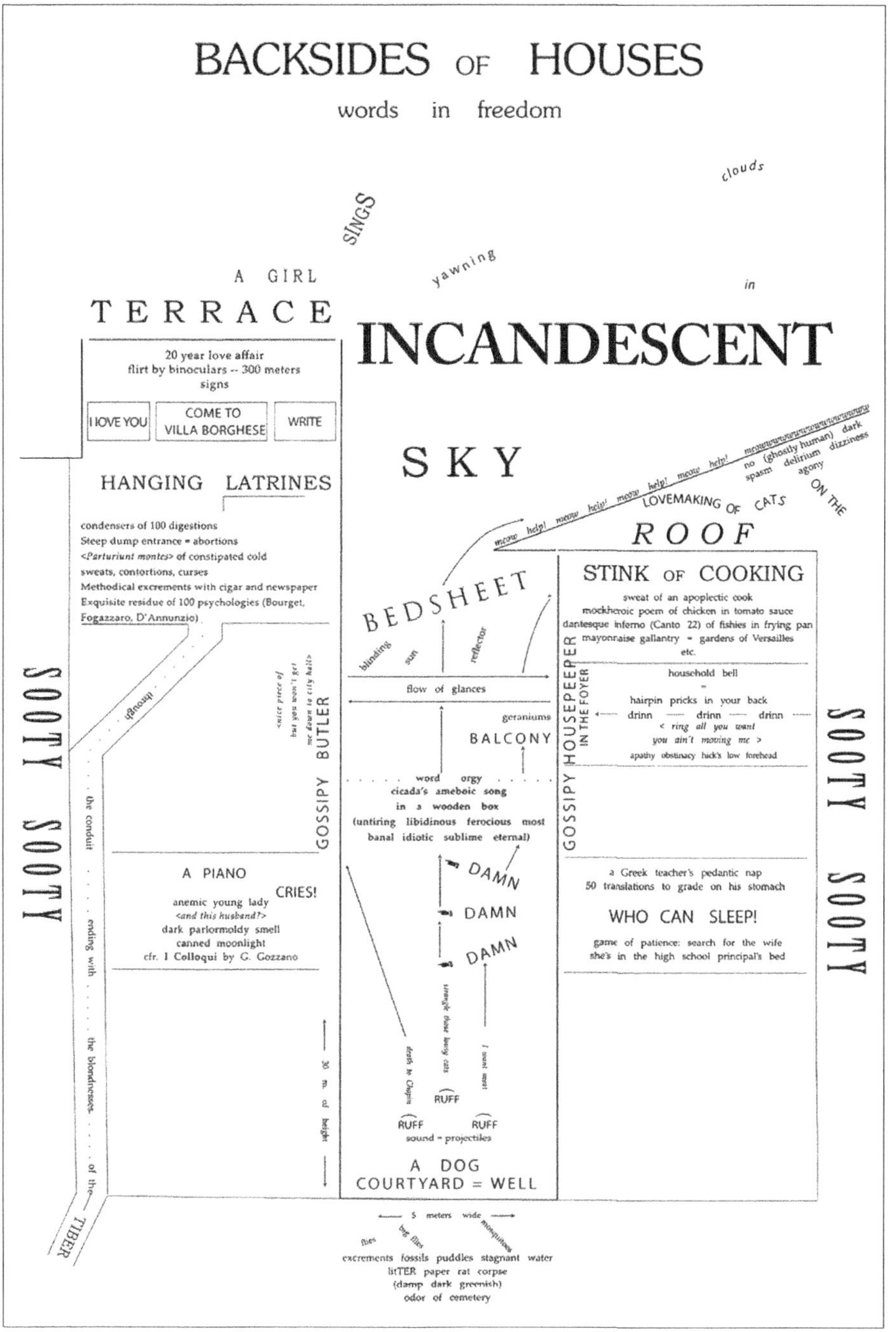

Fig. 17.5. Futurist geographic map poem. Vincenzo Volt, Deretani di Case, 1916 (Translation and translation design by The Getty Research Institute; Copyright Research Library, The Getty Research Institute, Los Angeles, California)

by painting it on a mattress. Kuitca, who is Argentinean, deliberately picked a map of an area with which he was completely unfamiliar, thereby reducing the labels and features to a maze of incomprehensible signs, and the map to a visual form devoid of geographic meaning. Keep in mind that, similarly, Magritte's complex message in "The Betrayal of Images" or the original version of Vincenco Volt's map poem *(Figure 17.6)* would be lost on anyone not able to read French or Italian, respectively.

17.4.5 Seeing Geographic Space in Art

A major intent of this chapter was to capture elements of the dialogue between the two authors, the cartographer and the artist, on the role of text in their respective fields. Any successful conversation depends on a certain amount of shared understanding and shared language and to that end, much of this chapter consisted of the authors exposing their own views of the role of text. We hope that readers from either field will find the result interesting. However, at this point not much of a synthesis of the two perspectives is contained. In order to give some hint at the direction of our conversation and collaboration, consider finally the differences between *Figure 17.6* and Peter Davies' 1998 work "Fun With the Animals: Joseph Beuys Text Painting" (http://www.saatchi-gallery.co.uk/artists/artpages/peter_davies_fun_with_animals.htm).

Davies constructed what one would typically call a semantic network. Starting with the artist Joseph Beuys, Davies linked together one concept/person/characteristic from the field of art history after another, in a playful manner. In the end, the work includes connections such as "De Kooning" to "Pollock" via "drinking."

Figure 17.6 contains recent work by the second author, aimed at deconstructing the appearance of language by exploiting the individual letter as its most basic primitive. Focusing solely on the appearance of letters, new arrangements are created, guided by visual effect, but nonetheless evocative of spoken and written language.

The conversation between the authors revealed that Davies' "Fun With the Animals ..." and *Figure 17.6* generated quite different ideas regarding the degree to which elements of geographic space were recognised. It turned out that earlier conversations regarding the so-called *First Law of Geography* were crucial in this context. Long held as a fundamental underpinning of how geographic space operates, it claims that everything is related to everything else, but closer things are more closely related (Tobler 1970).

The artist-author was reminded of geographic space in how Peter Davies had managed to visualise the interrelatedness of diverse elements. Meanwhile, the cartographer-author felt that the space Davies had generated and visually depicted looked far different from what one might expect in geographic space. One issue was that it was not clear whether planar enforcement was present or not, i.e. whether lines were ever crossing other lines or nodes without an actual link. The closest geographic

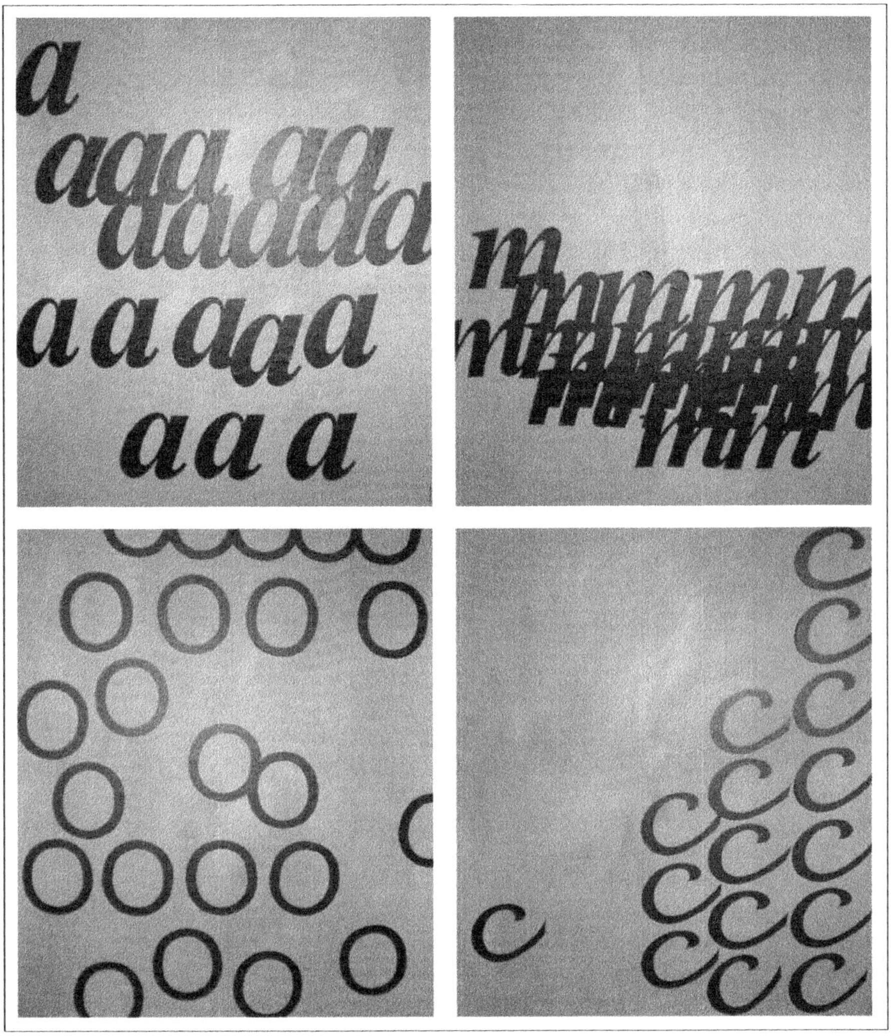

Fig. 17.6. Seeing geographic space in art: Marinta Skupin, Letter Paintings, 2007

network that this may resemble is that of an airline network. However, the distinct hub-spoke structure of today's airline network is absent. The cartographer-author also was [mistakenly] looking for meaning in the size and color of individual nodes.

Instead, *Figure 17.6* triggered in the cartographer a much stronger recognition of some elements of geographic space, quite unintended by the artist. Numerous structures reminiscent of geographic space appear to exist in the arrangement of letters, with distinct spatial distributions across the four canvasses. None of the paintings contains a regular distribution of letters. The "m-painting" shows not only pronounced clustering, but overplotting of multiple symbols (like is often found when plotting geocoded crime locations over long time periods). In the other letter paintings, there are other cluster structures, such as the apparent crowding of *c* letters towards the right edge of that painting. Overall, it was the *diversity* of spatial patterns that reminded the cartographer of the diversity encountered in geographic space.

Much more could be said about the subjective nature of the dialogue attempted in this chapter. There clearly is no such thing as "the artist" or "the cartographer," and neither can claim to speak for a whole discipline. Instead, both authors found the value of this exercise to lie in generating fresh perspectives on each other's fields and, ultimately, in informing their own scientific and/or creative ventures.

References

Börner K, Chen C, Boyack KW (2002) Visualizing knowledge domains. In Cronin B (ed) Annual Review of Information Science and Technology, 179–255

Brewer C (2005) Designing better maps: a guide for GIS users. Redlands, CA: ESRI Press

Card SK, Mackinlay JD, Shneiderman B (eds) (1999) Readings in Information Visualization: Using Vision to Think. San Francisco: Morgan Kaufmann Publishers, Inc

Dent B (1999) Cartography: thematic map design. 5th ed. Boston: WCB/McGraw-Hill

Espenshade EB (ed) (1960) Goode's world atlas. 11th ed. Chicago: Rand McNally & Company

Fairbairn DJ (1993) On the nature of cartographic text. The Cartographic Journal 30 (December):104–111

Harrison C (2001) Essays on Art and Language. Cambridge, MA and London: MIT Press

Kraak MJ, Ormeling FJ (1996) Cartography: visualization of spatial data. Harlow, England: Longman

Morley S (2003) Writing on the wall: word and image in modern art. Berkeley; Los Angeles: University of California Press

Raisz E (1948) General Cartography. 2nd ed. New York: McGraw-Hill Book Company

Raisz E (1962) Principles of Cartography. New York: McGraw-Hill Book Company

Robinson A (1952) The Look of Maps. Madison: University of Wisconsin Press

Robinson AH, Morrison JL, Muehrcke PC, Kimerling AJ, Guptill SC (1995) Elements of cartography. 6th ed. New York: Wiley

Skupin A (2002) A cartographic approach to visualizing conference abstracts. IEEE Computer Graphics and Applications 22 (1):50–58

Slocum T (1999) Thematic Cartography and Visualization. 1st ed: Prentice Hall

Slocum T, McMaster RB, Kessler FC, Howard HH (2005) Thematic Cartography and Geographic Visualization. 2nd ed: Prentice Hall

Slocum T, McMaster RB, Kessler FC, Howard HH (2009) Thematic Cartographic and Geovisualization. 3rd ed: Prentice Hall

Stams W (1983) Umrißkarte. In Ogrissek R (ed) Brockhaus abc Kartenkunde, Leipzig: VEB FA Brockhaus Verlag

Tobler W (1970) A Computer Model Simulating Urban Growth in the Detroit Region. Economic Geography 46 (2): 234–240

Tzara T (1920) Chronique Zurichois. In Huelsenbeck R (ed) Dada almanach, 10–29. Berlin: Reiss E

Veregin H (ed) (2005) Goode's world atlas. 21st ed. Chicago: Rand McNally & Company

18 Imagining Europe

Christian Spanring and Emanuel Andel

ÖIR-Informationsdienste GmbH, Vienna, Austria

Abstract

As part of the 50th year anniversary of the Austrian Institute for Regional Studies and Spatial Planning, the idea to present visitors different images and faces of Europe was born. Traditionally one of the main tasks within the institute is editing, scientific analysis and cartographic representation and visualization of regional data and spatial indicators.

The planned presentation should contain appealing entertainment and sophisticated regional information at the same time. A way to cover both aspects, embedded in an informal and pleasant setting, had to be found.

We concluded from the success of our developed project – a horizontal light sensitive and interactive map projection – while its first public presentation, that our defined approach was fulfilled very satisfying.

18.1 Projection

A map of Europe is projected onto a horizontal table. Having a room height of about 3m, the projected map and the interactive area would be around 150x150cm.

Currently the projection contains:

- a map of Europe
- a menu on both sides of the map
- and a text inlay for additional information.

The EU-27 are, mainly because of reasons of data availability, highlighted in the map. Outside the EU-27 were, at the time the spatial indicators have been defined, no usable data available.

Basically, interaction with either one of the two map menus leads to entirely different map types in each case:

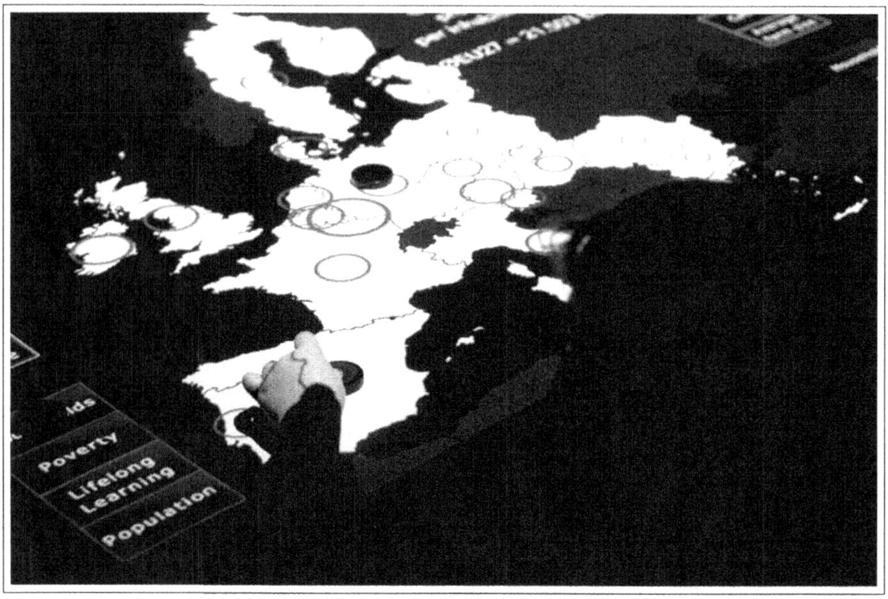

Fig. 18.1. Menu

- on one side, harmonized and pre-defined maps of EU-27 countries of several regional themes are shown
- on the other side countries can be compared directly among each other, regarding several regional themes.

While the maps are being displayed, the text inlay provides additional information such as map title, EU-27 average value or brief indicator descriptions right next to the cartographic image.

18.2 Interaction

Three game pieces, known from games like Backgammon, are available to map users for map interaction. By placing the pieces on different positions within the projection, map actions are triggered.

Putting a piece on the pre-defined map column calls and shows the corresponding map on the projection.

To use the "country comparison" all three pieces are necessary: the first piece chooses the regional theme. The next piece must be placed on a country. A popping up circle will indicate the selected country. The first selected country is used as index for any following country selection. A second circle will pop up if the third token is set to another country. The different sizes of the two circles represent the

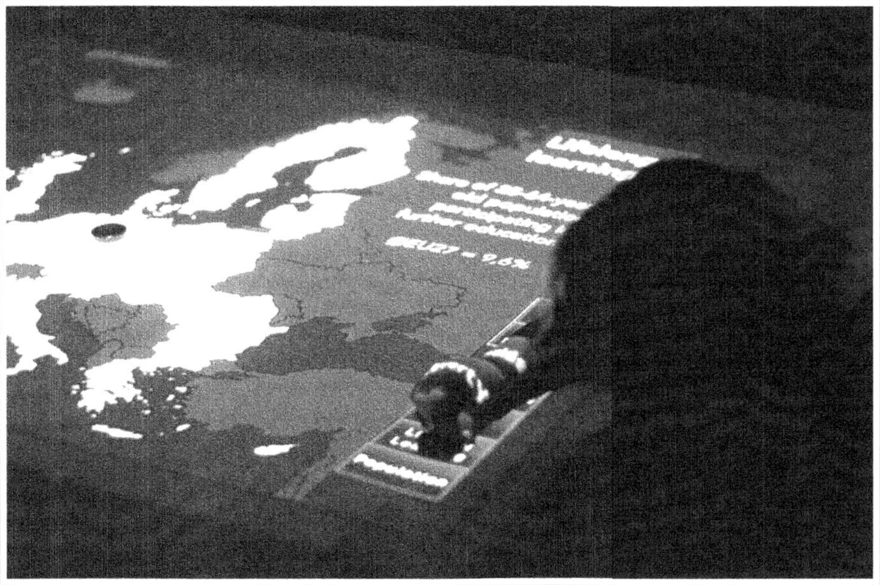

Fig. 18.2. Country comparison

relative relation between the two selected country values. The map user can now explore the map, Europe in our case, by placing one piece after another. The intention was to create different views of Europe, different views on different themes among European countries.

18.3 Cartographic Aspects

Unintentional but observed during the first public presentation was the ensued lively discussion around the projected map. From our point of view there were mainly two reasons responsible for it:

People gathered around the table where the map was projected on. Everyone was standing at exactly the same distance to the viewed cartographic image, which formed the basis for the discussion. Unlike watching a vertical projection as we know it, where every participant is at a different distance to the projection, the horizontal projection helps to attach all viewers the same way. No viewer is discriminated by standing behind another viewer, no view is blocked by anyone. On a horizontal projection every participant gains identical access to the information. Furthermore, around the table participants are already facing each other and are therefore in an excellent position to actively join the communication process and discuss about the map. A horizontal map projection is comparable to large paper plots, frequently used as tools on tables in planner working groups.

A second noteworthy point is the haptic aspect while interacting with the map. The very physical interaction – the touching, grabbing, moving, putting and pointing, using only the own hands and fingers without any instrument in between – provides a very familiar way, known from everyday life, of using handling objects. It supports and shortens the connection from the map to the user.

18.4 Technical Aspects

An usual projector is used to project the map image on top of a table. The movement of the three game pieces is tracked by an infrared-camera. An interpreter (an OpenSoundControl hack) processes the camera signals and passes it to the map application (developed with Flash® and ActionScript™). Based on the received parameters, the map application defines and sends a query containing country and theme information to a database (MySQL) where all map values are stored and retrieved dynamically. Finally a map image is rendered, sent back to the projector and displayed on top of the table again.

Section V: Non-Graphics

19 Dots, Lines, Areas and Words: Mapping Literature and Narration (With some Remarks on Kate Chopin's "The Awakening")

Armin von Ungern-Sternberg

Department of Philosophy and Philology, University of Mainz, Germany

"Oh! to be able to paint in color rather than in words!" exclaimed Miss Mayblunt.
Kate Chopin, The Awakening

Abstract

The so-called spatial turn of the humanities has led to an increasing interest in map-making. Looking at various trends in map-making the paper evaluates possible links and cooperation between cartography and literary studies. Differentiating between two central questions - which features of literature may be considered as spatial and in what way literature as a social practice could become spatially relevant – the paper argues that maps of literature should be based on specific features that constitute a literary text as opposed to other forms of art and discourse. Claiming that maps of literature should provide more than a visualisation of structure or an illustration of certain forms or content, the paper pleads for a general understanding about aims, terminology and standards of literary geography. As this requires a theory of space in literary texts, the paper distinguishes between various levels of spatiality in texts and different forms of constructing narrative space, suggesting outlines for future map-making of literary phenomena. Future endeavours of literary cartography and geography should reflect that in literature there is no space as such. Literary space is both sequential and discontinuous.

19.1 Maps and Literature

Generations of scholars have laboured and fallen out over the question of how to define and analyse literature. The implications of map-making have caused no less difficulty. To the literary critic this might come as a relief, except when having to deal with both. In this paper I will examine how spatial features of literature can be analysed and how mapping them might be possible or useful. As a literary critic I will not dwell on the issue of whether we can speak at all of something like an 'objective external reality' as I cannot – in the words of Woody Allen's alter ego Kleinman (in 'Shadows and Fog' 1992) – rule out for sure "that everything exists only in the dream of a dog".

In the broadest sense maps can be regarded as spatial data handling tools fulfilling the double function of storing knowledge and providing orientation, in order to depict what is known and to prompt further discoveries or understanding. The question of mapping is a new one to literary criticism and literary critics have become sceptical about ever new imports of methods from other disciplines and the 'exact' sciences in particular. Interestingly, geography and cartography seem to be meeting us half way in seeing maps increasingly in their social function and in a pragmatic and situative setting, the general view being that they organise rather than simply show space and spatial relations following a choice of what information to include and how to convey it. This is a choice any author also has to make when writing a work of literature. The "perceived level of 'accuracy' of a map is, to a great extent, dependent on the task for which it is used and on the subjective assessment of the user" (Dorling and Fairbairn 1997, p. 58) and similar remarks can be made about what is perceived as 'realism' in fiction. The mappae mundi of the Middle Ages are 'correct' in placing Jerusalem as the centre of a world forming the body of Christ. For them, location was an issue not in the sense of geographical positioning but in the grand order of the world. Literature, too, adds another dimension to its text. It forms not merely a relation to the world but creates realms of its own and on its very own terms.

The history of cartography in terms of both the content and appearance of maps is not simply the story of growing accuracy but also one of different world views. On both literature and map-making, religion and philosophy have always had a far-reaching influence – in European, Islamic, Asian and other traditions alike. Research has become increasingly interested in maps as instruments of power and control, in "persuasive cartography" (Tyner 1982) and "how to lie with maps" (Monmonier 1991). If maps are seen today as a "conceptual object, yet the interesting thing is the grotesquely token foot it keeps in the world of the physical" (Harbison 1977), this definition appears to borrow from our understanding of literature: We are used to acknowledging the dual nature of art as something segregating itself from empirical reality that still falls into this reality and social relationships (Adorno 1970, p. 374). If that is so, mapping and literature cannot simply be divided along the lines of reality and fiction. Interestingly, modern cartography has questioned the referentiality of maps while literary theory has moved away from the idea of literature as mimesis and towards concepts of structure, social practice and discourse, taking as the basis of art – if the concept of 'art' still holds – the complex of artistic norms, the artistic structure of a more than personal, social character (Mukařovský 1974, p. 10).

There are other and perhaps more obvious reasons why we should deal with maps and literature: Literature itself refers to maps or uses them as illustrations of a book if not as part of an artistic statement. We could think of literary traditions dealing with cartographic subjects. Modern literature rejecting omniscient points of view was for some time seen as "moving in the direction of spatial form" in the sense that, in order for a text to be "properly understood", its elements "must be juxtaposed with one another and perceived simultaneously" (Frank 1963, pp. 8, 12).

Maps have become a popular subject of scholarship, and ‚mapping' has become a metaphor increasingly used in various subjects in the humanities. Obviously maps appeal to recent thinking: They have the attraction of juxtaposing elements, simultaneously presenting, as it were, the whole world on a plate.

Yet when it comes to literary criticism, we should watch the terminology, even in an interdisciplinary approach. Many speak, for example, of mapping the human genome. If this is not just another word for classifying it, it is the expression of a growing trend, extending from, as it were, cellular activity to social and ecological data, which could more aptly be called scientific visualisation (McCormick 1987). Obviously cartographic metaphors lie deep in western thought (Conley 1996). But we might go too far if we try to map any spatial expression. We may speak of mental maps, an interesting and growing field of research; but they are best defined as structured concepts of a part of spatial environment (Werlen 2004, p. 287) and not as the structure of a belief or an argument, even if 'mind-mapping' has become a popular tool in workshops and discussion.

The 'spatial turn' of the humanities seems to be suited to our age of globalisation and its experience of different, maybe clashing cultures. On the other hand, the growing interest in mapping seems strangely out of time: It is probably "no exaggeration to say that the map as we understand it was effectively an invention of the sixteenth century" (Harvey 1993, p. 464). In this paper I will argue that in mapping literature we need to think of new forms of map-making that detach from that very everyday experience which seems to ask for research on spatial dimensions of literature.

In recent years cartographers have tried to overcome the traditional two-dimensional and fixed (printed) pictorial system of mapping in order to arrive at more dynamic and individual forms of representation, traditional cartography being seen as rigid, if not imperial or even masculinist in nature (Rose 1993), while alternative forms of mapping are upheld as modes for decentralised and democratic local empowerment (Aberley 1993). New imaging technologies and computerised maps spatialising social or environmental data and incorporating them in complex, interacting models have added enormously to our understanding of the world around us. It is understandable that "to the young practitioners of this new art, the old geographers believe in a flat earth – two-dimensional, static and Euclidean, akin to a page in a book" (Hitt 1995). What we see, may be indeed "a distinctly new visual system, a new way of rendering the world-as-picture" (Pickles 2006, p. 164), prompting further questions such as whether "vision and mapping have their own intrinsic geographies" (Pickles 2006, p. 181). But we can also ask whether, in a stricter sense, these new forms of image-processing should not be seen as a new *visual* system rather than as maps. As soon as we start to deal with moving pictures and literature we could, rather than think of maps, return to the problems of adaptation of literary texts for the screen.

It comes perhaps as no surprise that the ubiquity of 'mapping' appears to coincide with a remarkable crisis of cartography which has led some scholars to fear nothing less than an end of geography (Flusser 1992, p. 31). Others have spoken of a "cartographic anxiety" (Gregory 1994), pointing towards increasing doubts about once honoured cartographic traditions such as 'systematicity', 'boundedness' and 'totalization'. Such concerns are less prompted by discoveries and latest theories of science shaking common concepts of space and time but rather seem to be influenced by the overwhelming developments in telecommunication widely perceived as a disappearance of distance. What we are facing might be indeed a new paradigm of orientation by individual self-positioning through the enormous success of GPS, the users of which are less concerned with their position on a map than with what to do next. In effect we "may be returning to a world before formalised cartography, where many different people around the world developed their own, slightly different forms of mapping" (Dorling and Fairbairn 1997, p. 125).

So in dealing with the cartography of literature are we best to content ourselves with designing different maps for individual tasks? I think not. If we are to avoid a mere accumulation of individual pictures that do not say much when put side by side, if our endeavours are to prompt further discoveries and understanding, we have to agree on what it is we actually want to map and how we should do it. We have to develop a system of scale and meridians, as it were. Modern cartography may only be able to reach out beyond its once strict systems because it can look back on them as a common basis for academic discourse. When it comes to mapping literature there is much less we can rely on. As the issue of maps and literature is – all in all – a new one, we should particularly aim for clarity in order to structure both hypotheses and discussion.

So I doubt whether we should indeed follow some cartographers' decided move away from the map as the object of their studies and towards the "user's mental image of the map" (Kimerling 1989, p. 688) (if not the intended mental image of others), calling for a "distinctly post-modern reconceptualization of spatiality that connects the social production of space to the cultural politics of difference" (Paulston 1996, p. xvii). The aim of some recent studies is to "blur this distinction between map and territory [... in order] to acknowledge the socially constructed character of the mappings within which our lives are orientated" (King 1996, pp. 16–17). Following Baudrillard it became the objective of some studies to overcome "the sovereign difference [...] that was the abstraction's charm. For it is the difference which forms the poetry of the map and the charm of the territory, the magic of the concept and the charm of the real." (Baudrillard 1983, pp. 2–3). We could connect this to widespread ideas of the 'undecidability' or indeed 'unreadability' of texts or of a sheer infinite extension of intertextual readings. But as a basis for dealing with maps and literature, this is less than satisfactory.

It is equally bold and simple to suggest that "the map is a text, like any other" (Pickles 2006, p. 43), but it does not help where literature is concerned. Like literature, a map can be taken as a highly complex super-sign (Wood 1992, p. 132). They may both be, express or lead to cognitive concepts. But the observation that two things can be compared does not make them similar. They might share a similar structure but still consist of different elements. Though we generally speak of 'reading' maps and though their 'language' too follows a given and flexible set of conventions, though they contain and render 'information' of some sort, it is clear that they do so in different ways. Maps and texts do not let us see anything as such nor simply "how others have seen it and how they want us to see it" (Pickles 2006, p. 61) but through a medium – that is: different media. Literature works by linguistic means, and in the case of fiction, through the various ways and functions of narration. There is, for example, no first person narrative or indirect speech in map-making. And "unlike the author of a written text, the cartographer cannot express the limits of technique in the map itself" (Broek 1965, p. 64), whereas "literature is not just the only art that can criticize itself, it is the only art [...] that can criticize anything" (Carey 2005, p. 177), at least in the form of an argument or by voicing an opinion. Literature acts through sequences of words, more often than not in sentences that are grammatically correct.

A map, on the other hand, works through visual effects. Much has been written on text-picture relations and it might help to draw some distinctions: A map is not an illustration nor is it the same as a chart or a diagram, nor is it just a form of visualisation of various facts. It is rarely self-contained – as one might consider a work of art – but usually understood to be a tool for something else, its aim being to render space and location: Above all, maps rely on methods for determining an accurate position on the surface of the earth, mostly in terms of longitude, latitude and altitude. Even at the most general level "map-making is the process of winning data points from nature" (Hall 1992, p. 8), followed – in a stricter, geographical sense – by a 'projection', i.e. mathematical transformation. In our case we have to complete an even further transformation, as spatial (simultaneous) configurations fundamentally oppose the structure of narrative sequence (Pellegrino 1994, p. 32).

What then could we project from literature onto a map? And what forms of projection could we apply? Literature's subjects and plots are manifold and of course also include questions of positioning and cartography which have also inspired a number of recent acclaimed novels, e.g. by Pynchon and Kehlmann. It is safe to say that literature's relationship towards cartography may be described as oscillating between two counteracting tendencies, to share in and to elude a disposition of positioning (Stockhammer 2005, p. 338). But the possible use of cartography for literary criticism is not just a question of subject matter nor merely an issue of local flair and discourse but should equally reflect what constitutes a literary text as opposed to other forms of art or occupation with reality.

19.2 Spatial Aspects of Literature

Insofar as map-making is concerned with positioning and spatial relationships, two questions have to be raised: 1) Which features of literature can be considered as spatial and 2) in what way could literature as a social practice become spatially relevant? I begin with the latter.

Though "the human subject is difficult to map for numerous reasons" (Pile and Thrift 1995, p. 2), it is of course possible to record where authors came from, where they found their education, where they chose or had to live, where their works were published. While it is easier to map an individual life, it should be more interesting to try the same for a number of authors or works. We can expect to learn if there are patterns to be traced, if there are centres and connections that change over time, whether literary trends, genres and epochs could have social or regional roots in some places and areas. But of course it is not that simple.

Maps like those on modern data traffic or in the "State of the World Atlas" (Kidron and Smith 1995) can spatialise all data – from circulating copies to residential addresses – that are statistically available from the publishing industry or literary sociology. This tells us much about markets, centres of publication and the spread of works and translations, of censorship and perhaps public taste. It is less clear how much it tells us about literary history. In order to compile such maps we need not study literature.

We are used to associating certain regions or places with literature, as for example England with 'English' literature. The traditional concept of literary history, mainly one of national philologies, places texts in the context of a given area, its history, its economic structure, its social space if not its mentality. But relationships that matter more than the occasional anecdote are, though often assumed, very difficult to show. Genres, styles or plots cannot be tied to the place of the printing press or where the author had breakfast. We clearly cannot map literature in this way or on the basis of a political map, taking different colours to stand for individual literatures. This is not only because territories have changed throughout the centuries. It is unclear what such a map should actually signify: the location of authors, their influence on others (perhaps in similar colours?), the 'setting' of their works or the public they address. To draw a more detailed map or to blur and distort boundaries and continents might be an act of good will but obviously is no solution. It is characteristic of literature that it exerts influence across borders, but ideas and styles do not simply travel through 'places and paths of thought' (Holenstein 2004) as on an autobahn or by coach, passing through villages and cities one after the other. We might indeed be adhering to an antiquated model of the 16th century in trying to 'map' literature this way from a "drive to locate and implant oneself [and literature for that matter] in a named space" (Conley 1996, p. 303) in order to form a total relation to a total world (cf. Foucault 1974, p. 69): our world as we see it.

At a time when bilingual reading (if not writing) "becomes the norm rather than the exception" (Spencer 1994, p. 345), we may come to appreciate the modern cartographers' wish to deconstruct the map. We might even come to understand the unease of some geographers who argue against geoinformatic systems that may help "to create a world, in which people are not treated and judged by who they are and what they do, but more by where they live" (Dorling and Fairbairn 1997, p. 134), as so many important features of writing and literature obviously can be neither quantified nor localised. Literature "may be a cultural product, but it is never simply contained by a culture" (Attridge 2004, p. 6). If so, it defies any traditional mapping other than in its sociological or economic facts, and even those might be, though verifiable, ultimately misleading.

Literature obviously is part of our life. But that does not mean that it is possible or sensible to map it. The question of which spatial features may be identified in literature and how literature deals with space is less obvious, yet this is where literary cartography should begin. I have elaborated on spatiality of texts elsewhere (von Ungern-Sternberg 2003), but would like to repeat that for me there are several levels of spatiality in any text, the first being that to which I suggest we restrict the term 'space' of a text. In literature we do not encounter space as such but only mediated and shaped by various narrative functions. Yet it is this *'space'* that for me constitutes literary space and that I would like to distinguish from the *'atmosphere'* of a text or what used to be called the 'world' of a text or an author, e.g. the 'Kafkaesque' in Kafka. 'Space' and 'atmosphere', in my opinion, demand different ways of mapping. When it comes to comparative research – the backbone of literary criticism (as even the analysis of an individual work has to understand how it differs from others) – the analysis of literary 'space' provides a clearer, more objective basis than the analysis of 'atmosphere': Literary 'space' is formed by particular elements in a particular way that can be identified and clearly described. But when it comes to 'atmosphere', values and connotations, we are working on ground that is much more difficult to chart. On a third level the idea of the *'structure'* of a text is linked to the notion of spatiality; but structure is more an abstract concept of spatial elements; what we look upon are rather inherent groupings or codes of semantic fields. It is of course possible to clarify mental associations graphically, as we know from any mediocre powerpoint presentation or from sketches in the notebooks of great thinkers, yet this is obviously something to be distinguished from literary space in a stricter sense. Then again, on a fourth level, any text has its own *'spatiality'* – as an object in our life as well as consisting of one sentence and paragraph after the other, chapter after chapter – and on a fifth level its own *'area'* of communication, in which the text calculates with its readers' reception: We should not judge a book by its cover, but many of us do or are influenced by a title or by what we know about the author. Even before we start reading the first sentence of a text, we already see it through a web of spatial coordinates that are neither arbitrary nor entirely individual but that can be presupposed by the author. Although all this comes together to create the complex impression of 'space', it is better not to confuse these levels as many studies do.

In fact, it helps to be even more precise ourselves: These five levels may be extended to six: Through the first level, the actual text and its narrative shaping of literary 'space', there is some *'stage set'* that we do not perceive as such while reading but that might be (at least partially) reconstructed. It is the first three of these six levels of spatiality, this 'set' rendered through narrative techniques of literary 'space' – and shaded by the 'atmosphere' of a text – that we might attempt to map; if we think of visualising the other levels we should rather speak of charts or diagrams. In terms of geographical cartography, this distinction might still not be correct, but I think it helps to clarify the options of literary maps which, as I will try to show, will deviate significantly from what we perceive as the surface of our globe.

If literary 'space' is a construction, we ought to look at how it is constructed rather than any illusion of space that we think evolves from it. It has always struck me as peculiar, given the amount of research and clever remarks on tempus and structural organisation of narratives, that narrative theory has neglected the mechanisms of 'space' in literature. Some have felt that "the relationship between time and space is of importance for the narrative rhythm" (Bal 1997, p. 139) but even they have moved on quickly to interpreting it, mainly looking at structural groupings of locations or analysis of values, the depth and meaning of a text as juxtaposed with the sequential surfaced narrative (Bronfen 1987, p. 356). This idea pervades Lotman's influential remarks on literary space, whose most important topological feature, he claims, are an unsurpassable border dividing space into two disjunctive parts (Lotman 1993, p. 327). However, bipolar concepts fall short of the true complexity of space in literature, whose most striking feature I find to be less dividing lines than a net of elements, angles and vectors adding up to something ultimately perceived as space.

So I would deny that in literature "in principle, places can be mapped out, in the same way that the topological position of a city or a river can be indicated on a map. [...] Of course, in fiction, these places do not actually exist, as they do in reality. But our imaginative faculty dictates that they be included in the fabula" (Bal 1997, p. 133). Literature deals with our imaginative faculty, but the two are not the same.

19.3 Features of Literary Space

The problem is less the age-old dichotomy between reality and fiction. It has been said that "even a map of an imaginary country is objective, in the sense that the mountains, roads, towns, and so on that it pictures were suggested by corresponding objective things in the real world" (Wright 1966, p. 34). But literature is first of all a text, not an imaginary country. A text cannot form an area or surface but only name it:

> *"Mr. Pontellier, unable to read his newspaper with any degree of comfort, arose with an expression and an exclamation of disgust. He walked down the gallery and across the narrow "bridges" which connected the Lebrun cottages one with the other. He had been seated before the door of the main house. The parrot and the mockingbird were the property of Madame Lebrun, and they had the right to make all the noise they wished. Mr. Pontellier had the privilege of quitting their society when they ceased to be entertaining. / He stopped before the door of his own cottage, which was the fourth one from the main building and next to the last. Seating himself in a wicker rocker which was there, he once more applied himself to the task of reading the newspaper. The day was Sunday; the paper was a day old. The Sunday papers had not yet reached Grand Isle."* (Chopin 2000, p. 3)

We are given conceptual dots and various details that form a continuous narrative but not a continuous space. There are gaps that prevent us from mapping Mr. Pontellier's position at the beginning of the excerpt or his way to his cottage as well as from envisaging the architecture and outline of the place. "Fiction, in the final

Fig. 19.1. Hotel on Grand Island

analysis, presents us with strictly verbal information, and in spite of the popular saying, no amount of words will ever add up to a picture" (Marcotte 1974, p. 272), such as a contemporary photograph of one of the hotels on Grand Island showing one of the "plantation residences [that] have been converted into rustic hotels, and the negro-quarters remodelled into villages of cosy cottages for the reception of guests" (Hearn 1889, p. 11; *Figure 19.1).*

Contrary to maps, there is not only no continuous but also empty space in literature. If map-makers are afraid of unexplored and uncharted areas (which they used to cover with large ships or giant elephants in earlier centuries), the reader of any text must be prepared to overcome an even bigger horror vacui. Space in literature is always bordering on a final frontier of what is mentioned and what is not. It has been nicely put that reality is used like a reservoir to fill fictitious worlds (Zipfel 2001, p. 90) and it is correct (to a certain extent) that even those facts that are not described but only implied by a text are part of such fictional worlds (Zipfel 2001, p. 85). We can infer that Mr Pontellier sat somewhere, when he "arose", which is confirmed one sentence further, but whether on a deck chair, a bench or at a table is something any director of a screen adaptation would have to add. Nor do we know how many cottages there are, whether they stand in rows, circles or any other pattern, or what they look like. Texts do not consist of images nor do they paint a picture. Instead they consist of a sequence of statements, propositions and given opinions. They cannot render a recognisable skyline, landmarks or sights of a city but only name or describe them in less or greater detail.

Any text draws on its readers' "encyclopaedia" (Eco 1987, pp. 236–278) as it is impossible to create or describe a space completely. The act of reading has long been accepted as playing a both active and integral part in literary narratives. Any form of art and even normal everyday communications rely on a complex interplay of "presentative" und "suggestive factors" (Munro 1970, p. 39). It may be that literature displays a particularly extraordinary "ability to capitalise on language's disabilities" (Carey 2005, p. 239) and such interplays. But in dealing with this ability we must not simply fill in such fragmentary descriptions or else we would be mainly occupied with ourselves, painting our very own elephants on the map of literature. It is true that in reading we also react to what we have created ourselves (Iser 1984, p. 210), but academic analysis should try to do better than this and recognise the rudimentary nature and incomplete net of a text's very own "presentative factors". There has been much debate on whether items in fictional worlds could be characterised as complete or incomplete objects (Zipfel 2001, pp. 93–97). Perhaps it would be safe to argue that they are complete as concerns the text, its story and its interests, but incomplete with regard to 'our world' and in such a manner that is as important as characteristic for a given text.

It is understandable that in mapping literature we are tempted to 'fill' these gaps, be it in showing Mr. Pontellier's walk across – what? – or in placing his accommodation near a hotel which might have provided a model for it on Grand Isle in the Gulf of Mexico. But I doubt its heuristic benefits. Rather than turning words into areas of which we know no contours, it could be more interesting and perhaps more appropriate for a given text to create a map showing which places are linked to each other in a narrative and in what way.

Literature can name or hint at area, length and size. But no text enables us to pin down its space as if on a floor plan. If someone turns eastwards in a text, we still do not know whether she turns left or right. Trying to seat Edna's dinner guests at the correct sides of the table is a puzzle we cannot solve:

> *"Edna herself made the tenth, and at half-past eight they seated themselves at table, Arobin and Monsieur Ratignolle on either side of their hostess. Mrs. Highcamp sat between Arobin and Victor Lebrun. Then came Mrs. Merriman, Mr. Gouvernail, Miss Mayblunt, Mr. Merriman, and Mademoiselle Reisz next to Monsieur Ratignolle."* (Chopin 2000, p. 96)

The subject of literary illusion is complex, but it is safe to say that there are no dimensions in any space described by language (Assert 1973, p. 39). We cannot point to anything in a fictitious space, and transition theory fails in the realm of fiction (Hamburger 1968, p. 107). Classic topoi in literature, such as the locus amoenus, are specified by their elements not by their relative position; Arcadia is first of all a place of certain qualities. Location in literature is to be understood in its intellectual and linguistic context, its reference is at the very least as symbolic as realistic. When Mr. Pontellier reaches the door of his cottage, we do not associate at first a certain position in relation to others but simply recognise that he stands in front of a door and not in front of a swimming pool.

19.4 Functions of Literary Space

Texts do not simply replicate or represent objects of the world nor the world as such; they realise a world in its own entirety (Blumenberg 1964, p. 19). This literary world draws on our own experience – as language does – and refers to certain objects or places, yet it is separated from our own reality not only by the covers of a book but by forcing its own perspective onto things. Much has been written on point of view in literature, and it might be useful to think of an analogy here: Like a 'view', a literary world is complete in itself and on its own terms. We cannot map a beautiful view; we can only indicate that there is one if we stop the car and have a look. A landscape is not found in the world; rather the world is transformed into a landscape in opening up to an aesthetic experience (Lobsien 1981, p. 1). Literary texts do not only create space; they create something through space as an aesthetic function. In Kate Chopin's "Awakening" neither New Orleans nor Grand Isle are 'there' as such but under the story's own terms and headlines. In mapping literature it may be more fruitful not to take narrated space – in the sense of signifiant/signifié – simply as references that can be tied up with their real counterparts. It is less important that a text names New Orleans than what it makes out of it. In Kate Chopin's "The Awakening" the house of the Pontelliers is the one place in relation to which others are valued by Edna or indeed by the reader. It is significant – and not a sign of paranoia (as it might be in the 'real' world) – to imagine some 'deeper meaning' in the fact that Edna's other house, which she moves into in order to set up a life on her own, is situated "around the corner" (Chopin 2000, p. 88). Its location "around the block" (Chopin 2000, p. 91) is stressed several times in the text and obviously denotes more than just a setting. We are not given any detailed address. Yet a literary map would anyway have to try harder than just to colour in some bits on a street map of New Orleans, on whose long axes we might imagine the protagonists to live and move. It would also have to include a 'vertical' dimension in the text: Obviously it is a spatial relation and no coincidence that the Ratignolles, widely known for their soirées musicales, live "at no great distance from Edna's home, on the corner of a side street [...] in commodious apartments over the store" they own and conduct (Chopin 2000, p. 61), that the pianist "Mademoiselle Reisz always chose apartments up under the roof" (Chopin 2000, p. 68) and that Edna, trying herself at art, sets up "her atelier – a bright room in the top of the house" (Chopin 2000, p. 64).

Moreover, the "Awakening" is a text in which the semantic fields of borders, of indoors and outdoors, upstairs and downstairs, of coming and going play an important role. It has also been called "a novel of beginnings" (Dyer 1993). It is no coincidence that the story of an awakening starts and ends in the same place: a coastline, about which not much information is given. We need not know more; it is the symbol that works and is important. It is, however, a *spatial* symbol. So how do we map that? Maps are, in a way, all about positioning. This is why thought on literary maps turns first of all to the most obvious question of how to treat real as opposed to fictitious places/objects or those that, although somehow transformed, we can connect to real 'counterparts'. Yet there are other variants: Some texts deliberately blur or explicitly refuse to name their places to the reader. Occasionally we might almost feel excluded:

> *"Madame Lebrun knew where Mademoiselle Reisz lived. She gave Edna the address, regretting that she would not consent to stay and spend the remainder of the afternoon, and pay a visit to Mademoiselle Reisz some other day."* (Chopin 2000, p. 68)
>
> *"There was a garden out in the suburbs; a small, leafy corner, with a few green tables under the orange trees. [...] The place was too modest to attract the attention of people of fashion, and so quiet as to have escaped the notice of those in search of pleasure and dissipation."* (Chopin 2000, p. 116)

Even if we were given exact information, it seems strangely 'out of place' to try to identify space in literature by means of longitude, latitude and altitude. It would be similarly inapt to take the names of Edna's dinner guests – Mrs. Highcamp, Mr. and Mrs. Merriman, Mr. Gouvernail, Miss Mayblunt – and look them up in a contemporary New Orleans address book. Literary space is more complicated than a land register. The following two 'addresses' set up a space in which much is said and implied but few spatial relations are given.

"The Pontelliers possessed a very charming home on Esplanade Street in New Orleans. It was a large, double cottage, with a broad front veranda, whose round, fluted columns supported the sloping roof. The house was painted a dazzling white; the outside shutters, or jalousies, were green. In the yard, which was kept scrupulously neat, were flowers and plants of every description which flourishes in South Louisiana. Within doors the appointments were perfect after the conventional type. The softest carpets and rugs covered the floors; rich and tasteful draperies hung at doors and windows. There were paintings, selected with judgment and discrimination, upon the walls. The cut glass, the silver, the heavy damask which daily appeared upon the table were the envy of many women whose husbands were less generous than Mr. Pontellier." (Chopin 2000, p. 55)

"Their [the Lebruns'] home from the outside looked like a prison, with iron bars before the door and lower windows. The iron bars were a relic of the old regime, an no one had ever thought of dislodging them. At the side was a high fence enclosing the garden. A gate or door opening upon the street was locked. Edna rang the bell at this side garden gate, and stood upon the banquette, waiting to be admitted." (Chopin 2000, p. 66)

Many locations of importance in the novel are placed in or face towards gardens, setting the scene away from the streets through which the protagonists reach them. In a way, the novel is not situated 'in' New Orleans but rather places itself aside from the city. Contrary to the usual saying, literary texts are not 'set' in a given place; they refer to it, take it as a starting point or create their own setting. It may add to our understanding of a text if we compare it with a real street map. But this does not mean that it always does, nor that it is sensible to map a literary text in the same way. Rather a literary map would have to find a way of rendering this patchy and emblematic 'New Orleans' of streets and rooms, of departures and arrivals. At no point does Kate Chopin's novel elaborate on a description of the city. The characters are described as being on the street, few of which are named.

"It was Arobin who took her home. The car ride was long, and it was late when they reached Esplanade Street. Arobin asked permission to enter for a second to light his cigarette – his match safe was empty." (Chopin 2000, p. 83)

"Madame Ratignolle had not far to go, and Edna stood on the porch a while watching her walk down the street." (Chopin 2000, p. 106)

Likewise, the most detailed views we get are atmospheric rather than specific:

"A profound stillness had fallen upon the broad, beautiful street." (Chopin 2000, p. 101)

> *"Up–away up, over the narrow street between the tall houses, the stars were blazing. The air was mild and caressing, but cool with the breath of spring and the night."* (Chopin 2000, p. 122)

Some difficulties only occur if we take a geographical map as a starting point for literary cartography. In looking for alternatives other difficulties arise, although some seem closer to the sequence of the text and therefore to literary issues: Should we 'fill' the holes of Kate Chopin's literary space, e.g. in Mr Pontellier's walk down the gallery? Should we consider that in this 'New Orleans story' the Mississippi river is barely mentioned? Is it not a significant detail that at the beginning of chapter 17 the narration jumps immediately from Grand Isle into the middle of New Orleans instead of mentioning a return from the holiday home to the 'real home' or describing such a passage in greater detail? What is the perimeter of this New Orleans? The novel mentions Edna's "girlhood home in the old Kentucky blue-grass country" (Chopin 2000, p. 6) and journeys of both her husband and her admirer Robert to New York and Vera Cruz respectively. Is this 'as far as it goes'? Should we convey these places in our map and, if so, in what way? Can we simply put in some dots and lines or shade in some areas on an ordinary map of America (or of a world map for that matter)? Would that really enhance or rather obscure the spatial structure of the text? If maps are about positioning, any accurate positioning requires the idea of a totality in which something is placed as well as a reference point on which to base and calibrate a predefined pattern. Longitude, latitude and altitude, however, are concepts that are alien to literature, as are the zero meridian, the equator and the middle sea level. No one thinks of Greenwich when reading Kate Chopin's novel. Her text sets up its own coordinates: For a start, it has a beginning (and an end). The novel starts where Mr. Pontellier gets up. Thus literary maps have an opening line; they also have their own cardinal points and a line of reference – their 'zero meridian' – which we could identify as the Pontelliers' house, to which Edna has to return from her holidays and from which she yearns to move away.

It is part of the novel's strategy – and a constituent of literature – that its places and their significance only unfold gradually in the course of the story. We are not 'in New Orleans', we do not know what comes next and what is 'out' or 'in there'. As readers of literature we find ourselves in a situation not dissimilar to that of the physicist or astronomer who, relying on commonly accepted laws of physics and new hypotheses, know something about some galaxies and might even attempt a complete picture of the universe, including voids 'all along the way' and at the same time not setting any definitive boundaries. As readers we cannot help but be empiricists. In the beginning, when we open a book, we have some preconceptions but we are outsiders, non-natives to the story. In the course of reading, we increasingly acclimatise to a parallel universe. Instead of concentrating on geographical maps, we might rather draw inspiration from a map of the largest known structures of outer space *(Figure 19.2,* Hartl et al. 1993, p. 14).

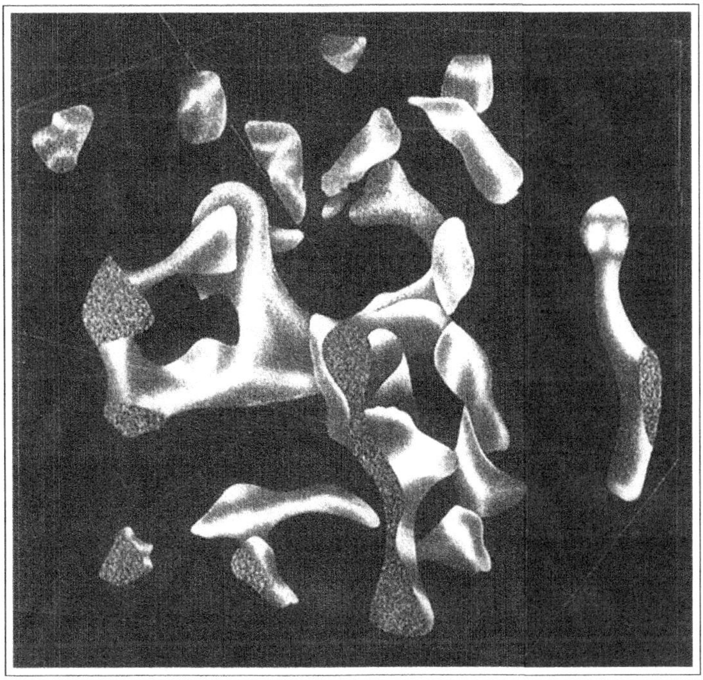

Fig. 19.2. Illustration of largest known structures in outer space

Our universe is said to be expanding and so is the space of any given text in the act of reading. Our world maps too are the result of centuries of exploration. Yet what do we know about the creation of fiction? In reading literature we are constantly as if on an expedition, detecting further details, as it were, on the way. But we are not on any way, we are not free to cut our own path, as an ordinary map would suggest. Instead, as readers, we are taken where the narrative leads us. Nor do we see a continuous space but only various stops that are linked in certain ways, as passengers travelling through a transport network. The map of the London underground, a design classic and a modern version of the traditional itinerary, could serve as another model for maps of literary space, showing points and their connections rather than areas, while neglecting absolute geographic location in favour of the intrinsic relationship between stations, leaving the rest to 'our imaginative faculty'.

19.5 Analyses of Literary Space

The current enthusiasm for maps is encouraged by the expectation that there is nothing that cannot be understood in spatial terms and thus be visualised. Literature and language, however, are quite a different matter and medium. "We experience literary works less as objects than as events" (Attridge 2004, p. 2). Whereas space is characterised by simultaneity, narration follows a sequential order. In the age-old debate about realism, it is prudent to understand "mimesis as recognition" (Compagnon 2004, p. 92) and thus as the result of a constant unfolding of further detail. It has been nicely put that it is not possible to 'narrate' a space but only to convey an idea of it (Schlögel 2003, p. 49); yet in the case of literature, such an idea – a more or less accurate recognition of a given space – is something different from the work of literature from which it derives. We are used to saying that we have 'finished' a book, but this does not mean it was incomplete before. Literature works through a certain order of words (which some might call style). Literary space only exists in the form of constant or irregular extensions, developments and rearrangements. But whereas the cartographic projection of a map suppresses the horizon and does not follow any individual perspective, being instead a view from nowhere (Buci-Glucksmann 1997, p. 40), a text has at the very least the narrative focus of a given sequence. As opposed to physical space, space in literature is not only uneven but accentuated in many ways. A text will stay at or return to its 'locations' with different frequency and duration, will connect them with different personae and moments of varying significance and will use different modes of description, detail, accuracy and emphasis. If the cartogram of a transport network can serve as inspiration for mapping literature, we should draw in the route that a passenger, the reader, actually has to take in order to reach the final destination, the end of a story. On this route we can indicate different values of stops and lines, hint at presuppositions and associations of the implied reader.

An 'atlas of literature' that were to depict the results of such research and prompt further discoveries or understanding should bring together such spatial networks of various genres, plots and motivs, epochs or even regions, as well as show how an individual work makes use of such options. It is important and should be possible to develop cartographic standards that ensure the comparability of such undertakings. Clearly, research on this matter is in its early stages and it is easy to feel overwhelmed by a perhaps infinite number of viewpoints. However, if it is possible to find a notation even for very complicated choreographies – another form of art characterised by a strong temporal and passing rhythm – it is possible to draw some inspiration from any diagram of a simple waltz. Likewise, maps of literature would have to focus on intrinsic relations and vectors rather than 'real' position. They should show structures but not forget chronological aspects of narrative sequence and reflect its rhythms as well as the fragmentary pattern of literary space. In other words, they should highlight what is special about literature: the particular techniques of creating 'space' by means of language. Otherwise maps of literature could just as well show locations of films or paintings.

Any map relies on the principles of selection and reduction for the purpose of rendering or emphasising certain features or contents. A text cannot 'portray' a recognisable appearance, it can only describe it word by word. It can specify – that is, name – a location but not cover an area. It can only tell about it. So instead of concentrating our cartographic efforts on the task of how to shade on a historic street map the streets of New Orleans through which Edna is walking, it seems a better question to ask how Kate Chopin's novel creates a 'New Orleans' in the first place. In short, it does so by linking different elements of different status in different ways but not in a sheer countless amount of literary strategies. Instead they can be narrowed down to a number of narrative mechanisms following both a certain logic of research and empirical observations of what actually happens in texts (von Ungern-Sternberg 2003, pp. 545–716). They constitute features for which future literary maps should develop a pictorial system that could also show varying degrees, as geographic maps use contour lines for altitudes or keys for the size of cites.

The first and perhaps most obvious thing to record on any map is of course MATTER. When New Orleans is mentioned in a text, we imagine houses and streets. But it certainly makes a difference what is mentioned and thus brought to the fore and what is merely implied, remaining in the background. In Kate Chopin's novel we find many stairs and steps, doors, porches and windows but very few walls or fences. It is openings that are closed or barred. Protagonists sail on a boat, take the coach or a cable car, but they hardly ever cross a public square or see a town hall or a market from the corner of their eyes. They meet indoors, not in a restaurant or at a concert hall. Yet a literary map of "The Awakening" should not simply show such different *elements* – whether Mr. Pontellier arrives at a swimming pool or his cottage door – but also the aspect of varying detail, which I would term *density*. While we do not learn very much about the surroundings of Mr. Pontellier's walk or indeed the whole hotel – except that the "bridges" are "narrow" or that the Pontelliers' cottage is "the fourth one from the main building and next to the last" – we become comparatively more familiar with the exterior and several rooms of the Pontelliers' house on Esplanade Street, though it is a smaller property than the Lebrun estate on Grand Isle. In the following example we not only learn that Edna's new home around the block is a small property; it is also packed with detail which we do not find mentioned in other places.

> *"The dining-room was very small. Edna's round mahogany would have almost filled it. As it was there was but a step or two from the little table to the kitchen, to the mantel, the small buffet, and side door that opened out on the narrow brick-paved yard."* (Chopin 2000, p. 111)

But Matter in literary space is not simply 'there'. As we discover it in the form of an ongoing exploration and numerous additions, there are, secondly, different aspects of SEQUENCE which a literary map should record: First, we can distinguish two basic *Directions* of literary space: It can unfold in a *deductive* or *inductive* manner. If "with Balzac to imagine a human being is to imagine a province, a city, a corner of the city, a building at a turn of the street, certain furnished rooms, and finally the man or woman who lives in them" (Lubbock 1968, p. 220), this is clearly a deductive way of building a literary space. A text could also start 'in medias res', zooming away from details to a greater picture and positioning:

> *"Mr. Pontellier found the Doctor reading at the open window of his study. His house stood rather far back from the street, in the center of a delightful garden, so that it was quiet and peaceful at the old gentleman's study window."*
> (Chopin 2000, p. 72)

A literary space is further characterised by something that we could understand in terms of different *Bearings*: When Mr Pontellier arises and walks down the gallery, space is added along his way, what I would call an *accompanying* space. On the other hand space can also remain *disparate* from the narrative sequence, e.g. when Mr. Pontellier observes that "farther down, before one of the cottages, a lady in black was walking demurely up and down, telling her beads" (Chopin 2000, p. 4). Two other forms are an *expanding* or an *approaching* space. The first sentences of the following two examples show features of expanding space:

> *"Edna sat at the window, which looked out over the house-tops and across the river. The window frame was filled with pots of flowers, and she sat and picked the dry leaves from a rose geranium."* (Chopin 2000, p. 107)
>
> *"Two broad dormer windows looked out toward the Gulf, and as far across it as a man's eye might reach. The furnishings of the room were light, cool, and practical."* (Chopin 2000, p. 24)

Neither example is very elaborate, yet they are typical of the rhetoric of space in "The Awakening": In Kate Chopin's novel new segments of space are often added from the position or the perception of a person, yet they remain disparate. Different areas of space tend to be set next to each other – in these examples in separate sentences – without transition or a link between them. This creation of space suits the subject of a novel which is to a large extent about isolation, whether we find Edna "again alone" (Chopin 2000, p. 105), we learn that Edna "though she had married a Creole, was not thoroughly at home in the society of Creoles" (Chopin 2000, p. 12) or we hear about a piece of music that "Edna had entitled 'Solitude'" (Chopin 2000, p. 29). "At a very early period she [Edna] had apprehended instinctively the dual life – that outward existence which conforms, the inward life which questions"(Chopin 2000, p. 16) and in the course of the novel "she [Edna] felt no interest in anything about her. The street, the children, the fruit vender, the flowers growing there under her eyes, were all part and parcel of an alien world which had suddenly become antagonistic" (Chopin 2000, p. 60). The construction of space in Chopin's novel matches other aspects and parts of a larger artistic structure, yet it is not to be confused with them: Space in literature is a separate mechanism quite distinct from other narrative techniques.

Finally, when it comes to Sequence, a text will take its readers on different *Routes* while narrating space. It can extend in a rather *linear* way, Mr. Pontellier's walk being an example, or jump in a *non-linear* fashion, like when we find ourselves suddenly 'transported' from Grand Isle to New Orleans at the beginning of chapter 17. It can constantly expand *into the unknown* or remain *in known* territory, perhaps adding further detail. A text can 'commute' between a limited number of places of equal importance, which become increasingly familiar to the reader, or be centred in one place from where it expands to other destinations on certain singular occasions, thus generating an atmosphere of a well-known base and a diffuse outside world. It can also take us on a long voyage passing through places we never see again or return to some of them or just to the starting point. It might turn out that as the outlines of a literary map such structures are much more appropriate than the coastline of the Gulf of Mexico with New Orleans and Grand Isle.

Thirdly, the space added in the act of reading will be of changing STATUS. It can be of different *Quality:* We can encounter a *static* or a *dynamic* space, as would be the case through upheavals of different nature, be it a revolution or an earthquake. A special form of dynamic force can be changes in the narrative voice adding or correcting further detail and thus perhaps turning out to be deceptive and unreliable (cf. Booth 1961).

We should also distinguish between different *Degrees* of literary space. Space can be added in a *direct* way – e.g. through Mr. Pontellier's walk down the gallery. When we learn, however, that "the Sunday papers had not yet reached Grand Isle" or that "a few days later a box arrived for Mrs. Pontellier from New Orleans. It was from her husband. It was filled with friandises" (Chopin 2000, p. 9) it is a rather *indirect* (implied) extension of an outside world, and a vague one too, for we do not know where the papers are printed or shipped. It is also significant that in this 'New Orleans novel', the world of plantations is only there as a second *grade* space, e.g. in Edna's childhood memories of "her fathers's Mississippi plantation and her girlhood home in the old Kentucky bluegrass country" (Chopin 2000, p. 6) or when "Mr. Pontellier warmed up and grew reminiscent [... and] told some amusing plantation experiences" (Chopin 2000, p. 78) when Edna's father visits them for dinner. There are other third-grade and even fourth-grade areas as well, which are not expressly mentioned, like the distance between Grand Isle and New Orleans, or are not even mentioned at all, like Europe, and are therefore obviously irrelevant for the story and thus for a map of it.

Space is a distinct mechanism of any text, but it closely linked to other structures. There are various means of shading of literary space – which I would term, following an expression of art history, literary *Chiaroscuro* – that contribute to the atmosphere of a text. The following example creates a rather elaborate ambience, but not much space:

> *"Every light in the hall was ablaze; every lamp turned as high as it could be without smoking the chimney or threatening explosion. The lamps were fixed at intervals against the wall, encircling the whole room. Some one had gathered orange and*

lemon branches, and with these fashioned graceful festoons between. The dark green of the branches stood out and glistened against the white muslin curtains which draped the windows, and which puffed, floated, and flapped at the capricious will of a stiff breeze that swept up from the Gulf." (Chopin 2000, p. 26)

While to a large extent this can be understood as *semantic* chiaroscuro, there are other methods of a more *technical* sort, such as age-old mechanisms of contrast and variety or the picturesque or the more recent strategies of ambiguity. The duration a plot remains in certain locations can lend different emphasis as it makes a difference whether a story starts or finishes at a certain place or merely 'passes' through it. Then again, it makes a difference what happens where, be it actions, dreams or an awakening. It is during "that summer at Grand Isle she [Edna] began to loosen a little the mantle of reserve that had always enveloped her" (Chopin 2000, p. 16). It is in their "very charming home on Esplanade Street" (Chopin 2000, p. 55) where "Mr. Pontellier became rude, Edna grew insolent. She had resolved never to take another step backward" (Chopin 2000, p. 63) as it "seemed to her as if life were passing by, leaving its promise broken and unfulfilled" (Chopin 2000, p. 81). It is from the windows of Mademoiselle Reisz' apartment that we have the widest and most specific views across the city: "From her windows could be seen the crescent of the river, the masts of ships and the big chimneys of the Mississippi steamers. A magnificent piano crowded the apartment" (Chopin 2000, pp. 68–69). It is mostly in the open air, on porches or in gardens that people open up to each other. And then again, certain places or structures of places can be associated with class, race and gender.

19.6 Worlds of Literature

The issue of mapping is a new concept to literary criticism. We do not know much about its limits or its potential. It is safe to regard literary maps, like any other map, as a spatial data handling tool. I would demand from them that they show how literature creates and organises 'space'. This is a complicated task if we expect literary maps to address not just questions of subject matter or local flair but characteristics of texts as well as peculiarities of spatiality in literature in contrast to other forms of art or discourse. If we are not content with individual maps drawn up for individual tasks adding up to a mere accumulation of individual pictures that do not say much when put side by side, we have to find and agree on criteria, scales and methods of how to transform the literary rhetoric of space into a pictorial system.

The biggest challenge, however, might be something else: We will have to think in different terms than those of geographical mapping (which is why some cartographers might question the use of the term 'map' for literary space). If we compare "The Awakening" with the beginning of George W. Cable's "Old Creole Days", we are in an entirely different world, of different elements and density, other forms of sequence and status:

> *"A few steps from the St. Charles Hotel, in New Orleans, brings you to and across Canal Street, the central avenue of the city, and to that corner where the flower-women sit at the inner and outer edges of the arcaded sidewalk, and make the air sweet with their fragrant merchandise. The crowd – and if it is near the time of the carnival it will be great – will follow Canal Street.*
>
> *But you turn, instead, into the quiet, narrow way which a lover of Creole antiquity, in fondness for a romantic past, is still prone to call the Rue Royale. You will pass a few restaurants, a few auction-rooms, a few furniture warehouses, and will hardly realize that you have left behind you the activity and clatter of a city of merchants before you find yourself in a region of architectural decrepitude, where an ancient and foreign-seeming domestic life, in second stories, overhangs the ruins of a former commercial prosperity, and upon every thing has settled down a long sabbath of decay. The vehicles in the street are few in number, and are merely passing through; the stores are shrunken into shops; you see here and there, like a patch of bright mould, the stall of that significant fungus, the Chinaman. Many great doors are shut and clamped and grown gray with cobweb; many street windows are nailed up; half the balconies are begrimed and rust-eaten, and many of the humid arches and alleys which characterize the older Franco-Spanish piles of stuccoed brick betray a squalor almost oriental.*
>
> *Yet beauty lingers here."* (Cable 1879, pp. 1–2)

This is what may be aptly though metaphorically called 'local color fiction'. The differences to Kate Chopin's "Awakening" obviously outweigh the vague connection that both mention New Orleans. They share some references, they might work together on the reading public's perception of 'New Orleans', but they do not construct the same literary space. If we were to compile a 'world atlas of literature' it seems of doubtful merit to place them side-by-side or even draw their spaces on the same map, other than to highlight their diversity. As far as the idea of the Old South is a critique of the modern world from the view of an agrarian, patriarchal world, there is a subversive side to the 'setting' of Chopin's novel in which Edna is "demented, dreaming of wild, impossible things, recalling men who had set their wives free" (Chopin 2000, p. 119) while even her own little "'pigeon-house' stood behind a locked gate" (Chopin 2000, p. 102) and the Lebruns' home "from the outside looked like a prison" (Chopin 2000, p. 66). Local color fiction offered "other spaces" for women authors to write themselves "out of the hierarchies of class and gender" (Pryse 1993, pp. 1–15, Toth 1985, McCulloguh 1999, Ewell 1999, May 1970, Ewell and Menke 2002).

Literary space consists of more than references to our world; literary cartography is more complex than giving locations on a map (von Ungern-Sternberg 2003, pp. 875–936). Unsatisfied with what only seems obvious, e.g. what is actually implied in terms like 'regionalism', 'local colour fiction' or 'New Orleans novels', literary cartography is exploring one of the core problems of literary history: the relation between society and rhetoric (Moretti 1999, p. 15). If its maps were to depict what we can know about space in works of literature and then to prompt further discoveries or understanding, it would be more interesting to regroup them according to similarities and correspondences of their rhetoric of space. A world map of literature would have to deal with the literary continents of 'Arcadia', 'Camelot' and 'Anywhere', as well as with locales in outer space or underground on the way to the middle of the earth. The 'Old South' and the 'American West' "belong in large part to the order of social myth" (Tindall 1964, p. 1) as the space of their appearances in literature may stand closer to other groups than that of regional writing. Literary maps including projections of various grades and bearings of literary space may have their very own continental drift of different genres, epochs and styles. It difficult to grasp how multidimensional and warped this space actually is and how unconcerned with our usual perception of the world around us. In some books New Orleans might be just the dream of a dog.

References

Aberley D (1993) Boundaries of Home: Mapping for Local Empowerment, Philadelphia/PA: New Society Publishers

Adorno TW (1970) Ästhetische Theorie, Gesammelte Schriften, vol. 7, Frankfurt a. M.: Suhrkamp

Assert B (1973) Der Raum in der Erzählkunst. Wandlungen der Raumdarstellung in der Dichtung des 20. Jahrhunderts. Tübingen: dissertation manuscript

Attridge D (2004) The Singularity of Literature, London and New York: Routledge

Bal M (1997) Narratology: Introduction to the Theory of Narrative [1985], 2nd edn, Toronto et. al: University of Toronto Press

Baudrillard J (1983) Simulations [frz.], trans. Foss P et al. New York: Semiotext(e)

Blumenberg H (1964) Wirklichkeitsbegriff und Möglichkeit des Romans, in Jauß HR (ed) Poetik und Hermeneutik. vol. 1. Nachahmung und Illusion. München: Fink, pp. 9–27

Booth WC (1961) The Rhetoric of Fiction, Chicago: University of Chicago Press

Broek JOM (1965) Geography: Its Scope and Spirit, Columbus/OH: Merrill CE

Bronfen E (1987) Der literarische Raum. Eine Untersuchung am Beispiel von Dorothy M. Richardsons Romanzyklus „Pilgrimage". Tübingen: Niemeyer

Buci-Glucksmann C (1997) Der kartographische Blick der Kunst [L'oeil cartographique de l'art, 1996], trans. Hiepko A. Berlin: Merve

Cable GW (1879) Old Creole Days: A Story of Creole Life, New York: Scribner

Compagnon A (2004) Literature, Theory and Common Sense. [Le Démon de la théorie: Littérature et sens commun, 1998], trans. Carol Cosman. Princeton and Oxford: Princeton University Press

Carey J (2005) What Good Are the Arts? London: Faber and Faber

Conley T (1996) The Self-Made Map: Cartographic Writing in Early Modern France, Minneapolis: University of Minnesota Press

Chopin K (2000) The Awakening [1899], The Awakening and other Stories, ed. Knights P, Oxford: Oxford University Press

Dorling D, Fairbairn D (1997) Mapping: Ways of representing the world, Harlow: Pearson

Dyer J (1993) The Awakening: A Novel of Beginnings, Twayne's Masterwork Series 130, New York: Twayne

Eco U (1987) Lector in Fabula. Die Mitarbeit der Interpretation in erzählenden Texten [Lector in fabula: La cooperazione interpretativa nei testi narrativi1979] trans. Heinz-Georg Held, München: Hanser

Ewell BC (1999) Unlinking Race and Gender: The Awakening as a Southern Novel, Southern Quaterly. A Journal for the Arts in the South. 37.3–4, pp. 30–37

Ewell BC, Menke PG (eds) (2002) Southern Local Color. Stories of Region, Race and Gender. Athens/GA: University of Georgia Press

Foucault M (1974) Die Ordnung der Dinge. Eine Archäologie der Humanwissenschaften [Les mots et les choses, 1966], trans. Ulrich Köppen, Frankfurt a.M.: Suhrkamp

Flusser V (1992) Das Verschwinden der Ferne, Arch plus 111, pp. 31–32

Frank J (1963) Spatial Form in Modern Literature [1945], The Widening Gyre. Crisis and Mastery in Modern Literature, Bloomington-London: Indiana University Press, pp. 3–62

Gregory D (1994) Geographical Imaginations, Oxford: Blackwell

Hall S (1992) Mapping the Next Millennium, New York: Random House

Hamburger K (1968) Die Logik der Dichtung [1957], 2nd rev. ed. Stuttgart: Klett

Harbison R (1977) Eccentric Spaces, New York: Random House

Harley JB (1989) Deconstructing the Map, Cartographica, 26 (2), pp. 1–20

Hartl G et al (1993) Planeten, Sterne, Welteninseln. Astronomie im Deutschen Museum. München: Deutsches Museum

Harvey PDA (1993) Maps in Tudor England. Chicago: University of Chicago Press

Hearn L (1889) Chita. A Memory of Last Isle, New York: Harper

Hitt J (1995) Atlas shrugged: the new face of maps, Lingua Franca, 5 (5) July–August, pp. 24–33

Holenstein E (2004) Philosophie-Atlas. Orte und Wege des Denkens Zürich: Amman 2004

Iser W (1984) Der Akt des Lesens. Theorie ästhetischer Wirkung [1976], 3rd ed. München: Fink

Kidron M, Smith D (1995) The State of the World Atlas [1981], 5th ed. London: Penguin

Kimerling AJ (1989) Cartography, in Gaile GL and Willmott CJ (eds), Geography in America, Columbus: Merrill, pp. 686–718

King G (1996) Mapping Reality: An Exploration of Cultural Cartographies, London: Macmillan

Lobsien E (1981) Landschaft in Texten. Zu Geschichte und Phänomenologie der literarischen Beschreibung. Stuttgart: Metzler

Lotman JM (1993) Die Struktur literarischer Texte. [Struktura chudozestvennogo teksta, 1970], trans. Rolf-Dietrich Keil, 4th ed. München: Fink

Lubbock P (1968) The Craft of Fiction. [1921]. London: Jonathan Cape

May JR (1970) Local Color in the Awakening, Southern Review 6, pp. 1031–1040

McCormick BH et al. (eds) (1987) Visualisation in Scientific Computing, Special Issue of Computer Graphics, 21 (6), New York: ACM Siggraph

McCulloguh K (1999) Regions of Identity. The Construction of America in Women's Fiction, 1885–1914, Stanford/CA: Stanford University Press

Marcotte E (1974) The Space of the Novel, Partisan Review 41, pp. 263–272

Monmonier M (1991) How to Lie with Maps, Chicago: University of Chicago Press

Moretti F (1999) Atlas des europäischen Romans. Wo die Literatur spielte, [Atlante del romanzo europeo 1800–1900, 1997], trans. Daniele Dell'Agli, Köln: DuMont

Mukařovský J (1974) Studien zur strukturalistischen Ästhetik und Poetik, München: Hanser

Munro T (1970) Form and Style in the Arts. An Introduction to Aesthetic Morphology, Cleveland and London: Case Western Reserve University Press

Paulston RG (1996) Social Geography: Mapping Ways of Seeing Social and Environmental Change, New York and London: Garland Publishing

Pellegrino P (1994) Sémiologie générale et sémiotique de l'espace, Fictures architecturales, Formes urbaines. Actes du Congrès de Genève de l'Association internationale de sémiotique de l'espace, Genève

Pickles JA (2006) History of Spaces: Cartographic Reason, mapping and the geo-coded world [2004], London and New York: Routledge

Pile S, Thrift N (eds) (1995) Mapping the Subject: Geographies of Cultural Transformation, London: Routledge

Pryse M (1993) Distilling Essences: Regionalism and Women's Culture, American Literary Realism 25.2 (Winter 1993), pp. 1–15

Rose G (1993) Feminism and Geography: The Limits of Geographical Knowledge, Minneapolis: University of Minnesota Press

Schlögel K (2003) Im Raume lesen wir die Zeit: Über Zivilisationsgeschichte und Geopolitik, München and Wien: Hanser

Spencer M (1994) In the Canon's Mouth' Being Lucid about the Local, in Lawall S (ed). Reading World Literature. Theory, History, Practice, Austin/TX: University of Texas Press, pp. 328–350

Stockhammer R (2005) Verortung. Die Macht der Kartographie und die Literatur, in R Stockhammer (ed), Topographien der Moderne: Medien zur Repräsentation und Konstruktion von Räumen, München: Fink, pp. 319–340

Tindall GB (1964) Mythology: A New Frontier in Southern History, in Vandiver FE (ed) The Idea of the South. Pursuit of a central theme, Chicago: University of Chicago Press, pp. 1–15

Toth E (1985) Regionalism and the Female Imagination. A Collection of Essays, New York: Human Sciences Press

Tyner JA (1982) Persuasive cartography, Journal of Geography, July-August, pp. 140–144

von Ungern-Sternberg A (2003) Erzählregionen: Überlegungen zu literarischen Räumen mit Blick auf die deutsche Literatur des Baltikums, das Baltikum und die deutsche Literatur, Bielefeld: Aisthesis

Werlen B (2004) Sozialgeographie: Eine Einführung [2000], 2nd rev. ed. Bern et al: Haupt

Wood D (1992) The Power of Maps, New York: Guildford Press

Wright JK (1966) Human Nature in Geography: Fourteen Papers, 1925–1965, Cambridge, MA: Harvard University Press

Zipfel F (2001) Fiktion, Fiktivität, Fiktionalität. Analysen zur Fiktion in der Literatur und zum Fiktionsbegriff in der Literaturwissenschaft, Berlin: Erich Schmidt.

20 Robert Smithson's Pursuit: Mapping some Traces of Remnants

Nils Plath

School of Design, University of Applied Sciences, Münster, Germany

... all the maps you have are useless, all this work of discovery and surveying; you have to start off at random, like the first man on earth; you risk dying of hunger a few miles from the richest stores. ...

Butor 1996

Abstract

In drawing close attention to selected works by artist Robert Smithson and film-maker James Benning, this essay deals with questions of remaining and recollection in landscape, art works, and literary texts. In the view expressed here, mainly with reference to Smithson's lesser known map works and to Bennning's film on his art work Spiral Jetty in the Great Salt Lake in Utha, the perception of space and time – that is, landscape as we see it – always begins and ends with a map. Meaning, with a representation in mind. This is so because we learned to read maps in some other place, somewhere else than the place we are when we find ourselves consulting them at a later point in time. Thus, the use of maps can be viewed and intepretated as a practice of re-reading.

20.1 Remains to begin with

What is left to say, when confronted with the unattainable goal of marking certitudes on maps, of labeling what is inscribed in them as new or old world orders, destined to fail over the course of centuries? After all, time makes all of the age-old results of discovery and surveying look like forms of fiction in the end.

Fictions do not transmute that easily into history, of course, even if everlasting boundaries between the two realms cannot be drawn. Is the reverse also true for history that claims not to be fictional?

Some reassuring assertions will serve as a point of departure for the following remarks, which aim to address attempts at mapping and cartography:

Wherever maps are displayed, histories are unfolded. Maps tell numerous, most often conflicting, stories of appropriation and exploitation. Maps are signs of multiple relationships. As time pieces, maps furnish particulars about time and its observations; they show how these correlate with the process of establishing locations, through which time itself is framed in turn. Maps make it possible to see how connections are established between the presumptions behind the process of representation and the things represented by the historical modes of representation that they embody. Although legible through cultural mediation and for the purposes of orientation, maps as works in progress cannot be totally permeable in any given context; neither can they be exhaustively interpreted. Nor can they remain decorative artifacts in the display cases of new practices of historical interpretation. These assumptions require that some readings be continued.

The key task of mapmaking has long been regarded as the delivery of knowledge for the purpose of orientation. As numerous examples from throughout the course of history prove, maps may be imprecise due to incorrect measurements or formats, or the deliberate or visually necessary falsification of perspectives. Again and again, the ability of maps to reproduce reality is revealed as temptingly truthful, yet this claim to be truthful is but a mere pretense. Still, regardless of whether a map is read or interpreted as an image or as an instrument, as a narrative or a document, it continues to convey a measure of knowledge concerning cultural orientation. The type of orientation it illustrates affects both space and time equally and simultaneously: both are inseparably related, and no so-called spatial turns in cultural discourses can dissolve the reciprocal relationship between them. Maps testify to the materiality of what they claim to depict; as images of places they claim to represent, they literally put these places in the hands of the reader.

Since they are put forth as narratives, the mediatized landscape depicted – although only captured in a formal sense – appears to be a section of the world at large, more than just a means of transport or a vehicle for representation, exchange, and communication: it historicizes not only the ways that landscape is used, but also the ways it is read. A map, when it is viewed, tells of this, right in front of (and in) the eye of the reader. Throughout history, maps have been archives of legible time, and in the history of art, they are assured a prominent position in more than one spot along the timeline.

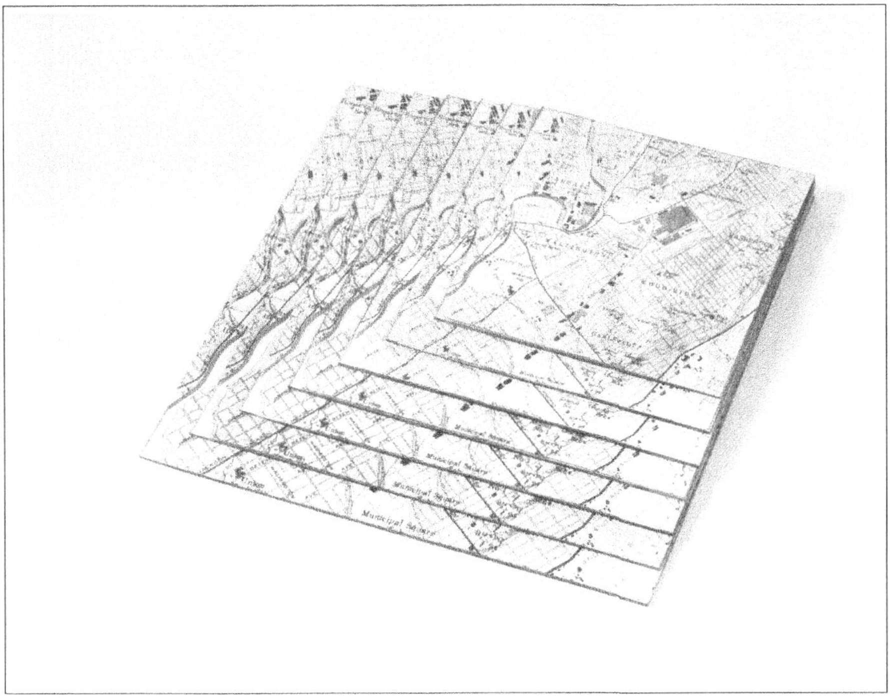

Fig. 20.1. Robert Smithson: Untitled (Map on Mirror – Passaic, New Jersey), 1967

20.2 Mediations and Materialities revisited

Following the introductory citation here, which is taken from the novel Degrees by French writer Michel Butor, there is a section titled "Mapscapes or Cartographic Sites." It starts with one of the many essays by the late American artist Robert Smithson: "A Museum of Language in the Vicinity of Art," originally published in the March 1968 issue of Art International. A glance at a quote in this text takes us back to the nineteen-sixties, to the heart of the discussion concerning two topics in particular: one, how to determine if an artist's work really can be classified as a work of art; and two, the issue of reading and interpreting the phenomenal world. In all areas of life, the latter was usually first considered a secondary, mediated experience, a substitute for a general encounter with the void. At the same time, Smithson's words describe the conditions under which, as he saw it, the art of his day was produced. He characterizes the artist as a "site-seer" faced with an omnipresent, "absolute void in time," with the overall disintegration of nature into "an infinite series of movie 'stills'." The artist is looking for a fiction "that reality will sooner or later imitate." (Smithson 1967)

Smithson writes:

From Theatrum Orbis Terrarum of Orrelius (1570) to the '[p]aint'-clogged maps of Jasper Johns, the map has exercised a fascination over the minds of artists. A cartography of uninhabitable places seems to be developing – complete with decoy diagrams, abstract grid systems made of stone and tape (Carl Andre and Sol Le Witt), and electronic 'Mosaic' photomaps from NASA. Gallery floors are being turned into collections of parallels and meridians. (Smithson 1967, p. 91)

The obvious density of these few lines – in which Smithson links the situation in his day to history and manifest promises of a future to come – remains challenging for those currently attempting to understand maps and art. Smithson mentions the fascination that maps and technology alike have held for artists in every era. And as he knows, maps do appear on the pages of literature, too. As Smithson reminds us in his essay, a map makes a prominent appearance in Lewis Caroll's famous poem "The Hunting of the Snark." The map depicted in Caroll's piece shows "nothing," and yet, because it has a frame, this map of "nothing" actually makes the "nothing" visible, for it illustrates the object recalling the written framework that determines every map. However, in the last of Carroll's novels, the reverse is true. According to Smithson, Sylvie and Bruno talks about a map that contains "everything":

In Chapter 11, a German Professor tells how his country's cartographers experimented with larger and larger maps until they finally made one with a scale of a mile to a mile. One could very well see the Professor's explanation as a parable on the fate of painting since the 50s. Perhaps museums and galleries should start planning square mile interiors. The Professor said, 'It has never been spread out, yet. The farmers objected: they said it would cover the whole country, and shut out the sunlight! So now we use the country itself, as its own map, and I assure you it does nearly as well. (Smithson 1967, p. 93)

Obviously, Smithson applies the interest in representation and its imaginary effect on the order of things to the way the viewer sees himself with regard to maps as illustrative references in texts. He insists that the ultimate determinant is the missing center, for a missing center implies the lack of a gravitational center, as well as points of reference that can be used for orientation: "Where is the central point, axis, pole, dominant interest, fixed position, absolute structure, or decided goal? The mind is always being hurled towards the outer edge into intractable trajectories that lead to vertigo." (Smithson 1967, p. 95)

An imaginative notion like Carroll's – of the map as a palimpsest of the world, as a reproduction of empirical space that could potentially cover an entire country, the unfolding and spreading of which provokes the inhabitants of the country involved in farming and ranching to object to this kind of academic production of reality – must have pleased Smithson the artist. After all, his work was about unfolding and spreading out standards; it was about constructing perspective, and allowing the critical eye to examine given formats and claims to property; he made unexpected analogies and created references in order to both expand and destabilize these things, too: on paper as well as in the so-called outdoor space, the gallery, or film. To do this, he shifted around sand, dirt, stones, trees, mirrors, words.

Smithson's works and words implement and reveal the frames of narrative, giving them their designated space by avoiding a central perspective and employing (sometimes speculatively or satirically) different modes of discursive language to articulate key questions about ways of mapping descriptions and the use of directional aids – in this case, art and the map. Because of this, Smithson's works are simultaneously articulations and disarticulations of fixed points of orientation and reference in reading, which – because scholars perceive and thus appropriate a work in different ways – normally assign points of view to those reading maps and their legends. This is the case, not only when a work is first viewed, but also later, when art historians position and classify a work in the context of art history. When one returns to it, in other words. In Smithson's works, maps are references to a world that is abstractly illustrated and defined by those who have a stake in evaluating it. Cut up and mounted, and yet more than just simple references to William Burroughs's cut-up collages for instance, folded, in one case at least, to make an allusion to contemporary artistic styles of the time, Smithson's maps can be viewed as complex reflections upon the usability and the usage value of representations at large.

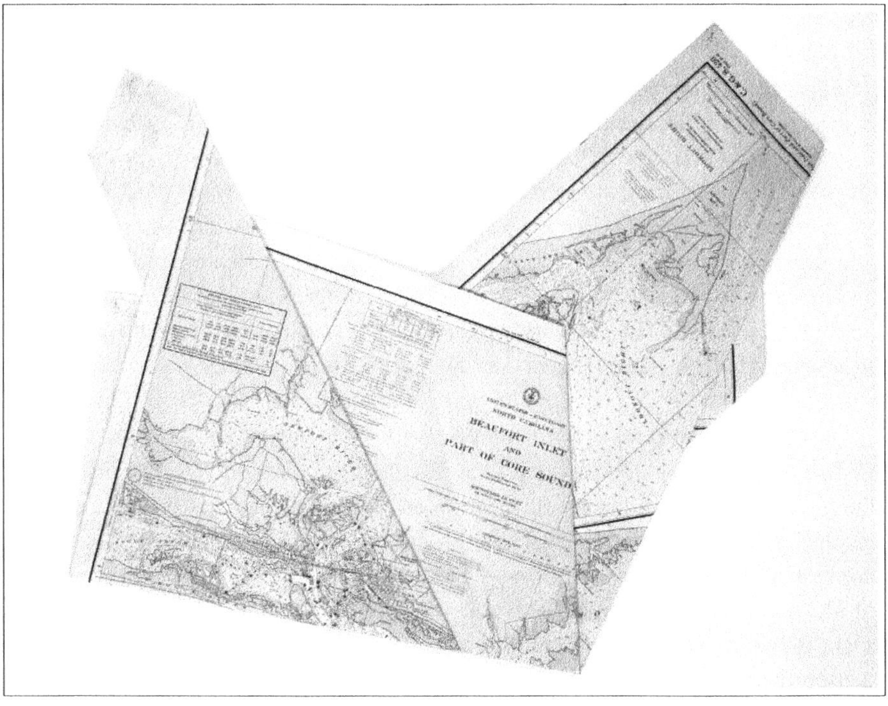

Fig. 20.2. Robert Smithson: Untitled (Folded Map of Beaufort Islet, Map 16), ca. 1967

Instrumentalized as material that will contain or support content, as a subordinate visual element, as a supplement to something else, or as eye candy, maps seem to identify a kind of viewpoint of a world marked by usage. This perspective is clearly fascinated with the prospect of inventorying these phenomena, including the conditions required for their creation, in abstract ways and in anticipation of the Structuralist analysis and controversy that broke ground globally a short while after Smithson's demise. This process is one of appropriation, accomplished by creating catalogues of patterns, like the ones made as an industrial product (see Dan Graham's typology of suburban row houses in New Jersey, to which Smithson refers several times in his writings, and Ed Ruscha's serial photographs of buildings along Sunset Strip in Los Angeles, also cited by Smithson). However, this process of classification and categorization is answered in turn by an original impulse to create disorder.

It can be asserted that Smithson's fundamental belief – that objects or phenomena of the material world and the notions they came to represent over time in various discourses – are only of temporal value. Thus, all of the historical order of what was called cultural history will be subject to decay and evanescence. This explains Smithson's lifelong fascination with traces of primeval history, its sorting, classification, and presentation in museums of natural history and printed works.

In an essay on cinema Smithson explains why he mistrusts an enduring assumption on which every order relies – especially the assumption that locations and their representations manifest something already represented in a different medium at a different time some place else:

> *There is nothing more tentative than an established order. What we take to be the most concrete or solid often turns into a concatenation of the unexpected. Any order can be reordered. What seems to be without order often turns out to be highly ordered. By isolating the unstable thing, we can arrive at some kind of coherent, at least for a while. The simple rectangle of the movie screen contains the flux, no matter how many different orders one presents. But no sooner have we fixed the order in our mind that it dissolves into limbo. Tangled jungles, blind paths, secret passages, lost cities invade our perception. The sites in films are not to be located or trusted. All is out of proportion. Scale inflates or deflates into uneasy dimensions. We wonder between the towering and the bottomless. We are lost between the abyss within us and the boundless horizons outside us.* (Smithson 1971)

Drawn to the ambivalence of paradoxes, Smithson's words and works mirror something Deleuze and Guattari would have called re-territorialization. With the expected results: Although they engage in the process of destroying illusions, Smithson's works also create their own illusions. They are entirely products of their time, critical of ideologies and institutions, artistic in their critique of artistic concepts – and as such, they have become models for later generations of artists. Smithson's works, densely interconnected and rich with references, are transformed into re-mythologized surfaces, upon which further appropriations in altered contexts are projected.

Just as they destroyed myths, these works, with their long half-lives, created one of the late twentieth century's most successful myths about the process of creating art.

A brief hunt for traces and remnants in written and visual images shed light on some modes of perception that need to be recounted in an act of interpretation or translation. This is a way to examine references in Smithson's work that go beyond what is visible; a way to begin telling a tale that can only be briefly alluded to here by presenting what can be seen at first and second glance.

Grain upon grain, one by one, and one day, suddenly,
there's a heap, a little heap, the impossible heap.
Beckett 1974

20.3 Landscape and Language

This narrative will have to be condensed to just a few aspects. Just as all maps testify to limitations and selectivity, as well as to the relationship between universalism and particularism in general, these aspects are intended to do nothing more than shape a framework for reflection, where we undertake to situate everything: a process of translation. In looking at the maps, one notices the materiality of the words and works, and this materiality indicates the fact that the human being, who always appears to be absent in Smithson's works, consistently leaves behind a variety of traces, both in language and in his environment. And this attests to the fact that he has continually attempted to recognize and interpret linguistic similarities in the wilderness, in the starry sky. Smithson's exploration of an environment determined by entropic processes can probably be called obsessive, and his exploration of things within contexts existing outside the temporality of the individual was more than just a superficial critique of the ideological gestures of his day. In his own words, they were "drillings" or "borings" that produced ground samples and had their own esthetic value.

The wide-ranging repercussions of this are described by Smithson in an essay written for the magazine "Landscape Architecture" and published in April 1968. He sketches out what he calls an "esthetic method," which brings together anthropology and linguistics with regard to the "building":

Boring, if seen as a discrete step in the development of an entire site, has an esthetic value. It is an invisible hole. [...] All language becomes an alphabet of sites. The boring, like other works, is becoming more and more important to artists. Pavements, holes, trenches, mounds, heaps, paths, ditches, roads, terraces, etc., all have an esthetic potential. [...] By extracting from a site certain associations that have remained invisible within the old framework of rational language, by dealing directly with the appearance of what Roland Barthes calls 'the simulacrum of the object,' the aim is to reconstruct a new type of 'Building' into a whole that engenders new meanings. From the linguistic point of view, one establishes rules of structure based on a change in the semantics of building. (Smithson 1968a, pp. 95 and 96)

As well as in the semantics of mapping, one might add. The plan for the landscape has to be rewritten – according to Smithson's statement, which was informed by early Structuralist writings and later considered premature post-structuralism – by means of establishing a fundamental change in perception.

Accordingly, Smithson's land inscriptions – his earth works such as the Spiral Jetty in Utah, the Amarillo Ramp in Texas, or the Broken Circle in Holland – should not be regarded as monuments intended to remain static, but as objects whose expiration dates and decay are inherent. This is demonstrated, for instance, in Partially Buried Woodshed (1970), a work installed on the campus of Kent State University, which actually did collapse (and became a landmark, commemorating the four students killed by the National Guard in an anti-war demonstration on May 4, 1970). Instead, these works on mapping are allegories written into a world that is perceived as a map. These repeated inscriptions of written language into the environment are primarily remnants of acts of orientation, performed as acts of writing oneself into a language meant to be a testimony to the will to create, the will to leave behind traces as documents of the finality and transitory nature of everything in the world. Even the most impressive building projects represent more than mere functionality when they are regarded and interpreted not only as inscriptions of certain languages (the language of architecture being only one of many other languages), but also as inscriptions of the will to communicate over the course of time.

How, then, can landscape and maps be perceived, interpreted, read? When and where? As work or writing? Since we are confronted with these questions whenever we look at Smithson's pieces, one can see them as indeterminate leftovers, remnants, and reminders of the plurality of perspectives that make a work and give

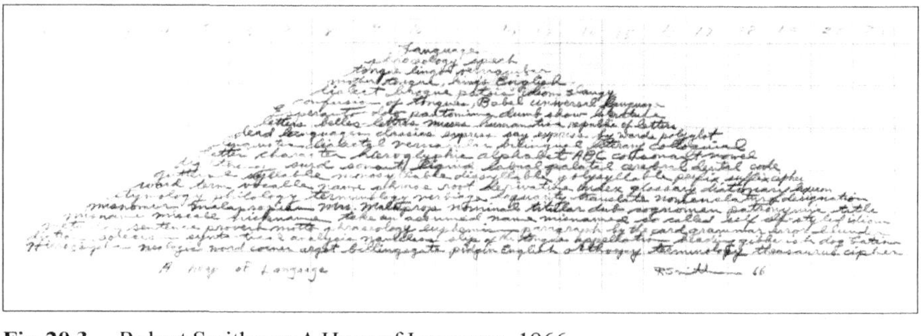

Fig. 20.3. Robert Smithson: A Heap of Language, 1966

it its multiple identities. Yet Smithson himself – who, after his premature death in 1973, was remembered mainly for his Land Art – supplies us with an answer to all of these questions, in the form of a work that is rather fragile in comparison to the massive sites where he left his handwriting on the map. The work is a delicate pencil drawing, entitled Heap of Language. It depicts a pyramid made up of linguistic terms: words that are key terms or that form an instruction manual for texts and literature. One can decipher words and terms such as "confusion of tongues," "Babel," "belles lettres," "bilingual," "name phrase root," "hieroglyphic," and so on. "Like so many of Smithson's pieces," says a well-informed commentary on the work by Gary Shapiro, "it forces us to attend to the way in which it transgresses the presumed boundary between work and text. It asks whether all work consists of writing and whether all writing is matter to be arranged." (Shapiro 1995, p. 157) If language can be built up into a mute heap or pyramid, says the commentary, then "…the geological strata can be read as lines or paragraphs that constitute a text." (Shapiro 1995, p. 160)

To come to a quote from Smithson:

The strata of the Earth is a jumbled museum. Embedded in the sediment is a text that contains limits and boundaries which evade the rational order, and social structures which confine art. In order to read the rocks we must become conscious of geologic time, and of the layers of prehistoric material that are entombed in the earth's crust. When one scans the ruined sites of prehistory one sees a heap of wrecked maps that upsets our present art historical limits. (Smithson 1968a p. 89, quoted in Shapiro, p. 160)

This is precisely what is shown in the maps incorporated into Smithson's works on paper. They mean for us to take away ideas that affect both our shared and our uniquely individual ways of seeing and orienting things in space and time. For Smithson, the cartography of space and time is a consequence of travel, of the process of locating particular places, the result of scanning the landscape, comprehending it through the use of maps and by actually being on location. The definition of outside and inside thus becomes a process of delimitation through the perception of time and space – two phenomena that can only be perceived as representations, anyway. An interviewer once asked Smithson: "Do you actually denote the boundaries? Like you draw a map of a particular square and constrain yourself from traveling other routes?" Smithson answered: "No, first I find the sites. Then I draw the squares. I'm scanning the physical material before I start to set up the plan; in other words, the map layout follows the scanning. It's not reconceived, so that it's discovered rather than pinpointed." (Smithson 1968b)

In one of his many articles that complement – or rather, should not be separated from the works, since both testify to his enduring interest in question about time, entropy, and finitude – Smithson registers the materiality of words that manifests in their fragmentary character:

> *The names of minerals and the minerals themselves do not differ from each other, because at the bottom of both the material and the print is the beginning of an abysmal number of fissures. Words and rocks contain a language that follows a syntax of splits and rupture. Look at any word long enough and you will see it open up into a series of faults, into a terrain of particles each containing its own void. This discomforting language of fragmentation offers no easy gestalt solution; the certainties of didactic discourse are hurled into the erosion of the poetic principle. Poetry being forever lost must submit to its own vacuity; it is somehow a product of exhaustion rather than creation. Poetry is always a dying language but never a dead language.* (Smithson 1968c, p. 107)

Smithson seems affected by the fact that one can encounter a form of language in any landscape and its elements. The construction of landscape – through the succession of the landscape painters who first used this term, adopted into the English language from the Dutch – has to assume the existence of a language, if the aim is to connect oneself and landscape's various elements in order to make a whole. Getting down to the essence of denoting things, one has to come to a realization: that things contain, in Smithson's words, their "own void." No "gestalt solution" can make this uneasiness and discomfort go away, or master the "discomforting language of fragmentation."

20.4 Moving Images of Time

Maps and depictions of landscape incorporated by the viewer can be recovered in narratives that tell the story of their historical production, which bears witness to the loss and gain of orientation in and through language. James Benning is an American filmmaker well-known to European cineastes who are interested in avant-garde film beyond Structuralism. His landscape films can be described as cartography in moving images, so we might look to his work to deliver this type of narrative. Benning remembered and described his first trip, made in 1989, to Robert Smithson's Spiral Jetty, located in Utah's Salt Lake. While being on the road with Dick Hebdige, who kept a travel log of their trip, Benning tells of his first encounter with the work, which inspired some of his works, including his penultimate film to date, Casting A Glance. The story is a two-fold reading of a map, one might say. On the one hand, it was the reading of maps – or non-existent maps – that led Benning to Smithson's enormous Land Art piece, or rather, led him astray on his way to the work. On the other hand, the narrative is a reading of the landscape itself, interpreted as a map: as a man-made perspective of space, centuries in the making, and infinite, while Smithson's work is an inscription on this perspective. Taking into consideration Benning's film on the Spiral Jetty, Casting a Glance (2007), one can even hear three different readings of maps when listening to Benning's narrative.

Fig. 20.4. Film still from James Benning: Casting a Glance (2007), filmed at the Spiral Jetty on Rozel Point in the Great Salt Lake, Utah

Traveling through the country in the late nineteen-eighties, Benning decided to pay a visit to Smithson's work. At the time, the Spiral Jetty was under water, largely forgotten by the art world. Benning did not have a map to find it, so he had to rely upon his memory of the description Smithson himself had given in order to find the jetty. Without a map, he recounted to himself Smithson's notes on how to find the path. The park service did not give out maps of the location at the time, and there were certainly no art tourists making pilgrimages to the site, so Benning could not find the original jetty, which he considers to be a main influence on his own films, or "in any case, for Deseret, in which the Jetty appears three times." He captured the wrong site with his camera, and even used the footage later to document his trip in one of his films. (Hebdige 2007, p. 144) In his conversation with Hebdige, Benning talks about how he missed Smithson's Spiral the first time: "And when I came back to film North on Evers it was still under water, not exposed like it was today ... When I came back I realized straightaway it wasn't even the Spiral Jetty I'd walked along the first time but the commercial jetty ... you saw how it goes way, way out into the lake and then kind of curves a little at the end." All of these things – the way he missed it, his capturing the wrong site, the void, the disorientation – became constitutive elements of the narrative in which, once again, depictions of landscape are told as stories about claims to possession and use. They make us see landscape as an on-going process of re-telling; they document landscape as something that will never be fully or exhaustively mapped, as something that leaves behind its remains, as a leftover, a void, a trace, a reminder of the passage of time, whose task it is to mark a certain spot in a very idealized space.

One can assume that this is the point that interests those who author the history of cartography with their arsenals of moving and still images of landscape. Smithson himself claimed that his film on the Spiral Jetty was part of the work he placed into the landscape – the Salt Lake. This means that it is difficult to say where the work as a work should be positioned. If, in the case of the Spiral Jetty, the work is a film, an object in a specific location, or an essay on the pages of a magazine. This uncertainty, as to how to view the "works," as well as the fact that it is impossible for the works to be a totality (and thus be perceived as such), are also what make the works so appealing to those who reflect upon the frames and framework in which narratives of space and time unreel. There seems no place imaginable where it would be possible to collectively grasp and experience landscape without a narratological consciousness, meaning, the consciousness that stories are told when we look at maps. In the view expressed here, the perception of space and time – that is, landscape as we see it – always begins and ends with a map. Meaning, with a representation in mind. This is so because we learned to read maps in some other place, somewhere else than the place we are when we find ourselves consulting them at a later point in time. When reading a map, our stories and histories lead us back and forth, toward and around places we are not, right at the moment in which we are engaged in the process of orienting ourselves in the place we actually are: movie theaters, studies, galleries, libraries, lectures halls, or the pages of a book.

References

Beckett S (1974) Endspiel, Fin de Partie, Endgame, Frankfurt/Main: Suhrkamp 1974 (1957), p. 10.

Butor M (1996) Degrees, quoted in Robert Smithson, A Museum of Language in the Vicinity of Art, in: Robert Smithson, The Collected Writings, ed. Jack Flam, Berkeley, Los Angeles, London: California UP 1996, pp. 78–94, originally published in Art International, March 1968

Hebdige D (2007) Reeling in Utah: The Travel Log Trilogy, James Benning, edited by Barbara Pichler and Claudia Slanar, Vienna: Synema 2007, p. 144

Shapiro G (1995) Earthwards: Robert Smithson and Art after Babel. Berkeley, Los Angeles, London: California UP 1995

Smithson R (1967) A Museum of Language in the Vicinity of Art, in: Smithson, Writings, p. 78–94 ; Smithson illustrated his piece with a number images of maps and a copy of Jasper John's Map (1967, based on Buckminster Fuller's Dymaxion Airocean World, 1967)

Smithson R (1968a) A Thing is a Hole in a Thing It is Not, in: Smithson: Writings, pp. 95 and 96, originally published in Landscape Architecture, April 1968

Smithson R (1968b) Interview with Smithson, Four Conversations Between Dennis Wheeler and Robert Smithson (1969–1970), in: Smithson, Writings, p. 217

Smithson R (1968c) A Sedimentation of the Mind: Earth Projects 1968, in Smithson, Writings, p.107 (originally published in Artforum, September 1968)

Smithson R (1971) A Cinematic Atopia, Smithson: Writings, pp. 140–141, originally published in Artforum, September 1971

Smithson R (1995) Writings, p. 89, quoted in Shapiro (1995, p. 160)

Picture Credits

1. Robert Smithson: Untitled (Map on Mirror – Passaic, New Jersey), 1967, map on mirror, 14" x 14" x 7/8", © Estate of Robert Smithson / licensed by VAGA, New York, NY, VG Wort, Bonn; Image courtesy James Cohan Gallery, New York
2. Robert Smithson: Untitled (Folded Map of Beaufort Islet, Map 16), ca. 1967, folded map, 36" X 31", © Estate of Robert Smithson / licensed by VAGA, New York, NY, VG Wort, Bonn; Image courtesy James Cohan Gallery, New York
3. Robert Smithson: A Heap of Language, 1966, pencil on paper, 6 ½" X 22", © Estate of Robert Smithson / licensed by VAGA, New York, NY, VG Wort, Bonn; Image courtesy James Cohan Gallery, New York; Collection: Museum Overholland, Nieuwersluis
4. Film still from James Benning: Casting a Glance (2007), 80 min., color, sound

21 Towards a Literary Geography of the Interwar Australian Novel

Harriet Edquist

School of Architecture and Design, RMIT University, Melbourne, Australia

Abstract

This paper discusses the possibility of constructing a literary geography of twenty Australian novels published by writers from New South Wales between the Depression and the end of World War II. Using Franco Moretti's approach that examines where action takes place in novels (Atlas of the European Novel 1800–1900, Verso, London and New York, 1999) it argues that in the Australian novels, action is confined to relatively few places. These are the city of Sydney itself, a few harbourside suburbs and the rural hinterland in the south and west of New South Wales. The paper further argues that each locale gives rise to its own characteristic action: the city is the space of political struggle, the harbour suburbs are places for reflection on the European colonisation of the land, and the hinterland is the place where white Australians wrestle with issues of national identity. It concludes by observing that the most striking quality of this geography is its partial nature in that it ignores suburbia, the urban condition that has defined Australian settlement since the nineteenth century.

21.1 Introduction

This paper examines the ways in which some of the propositions put forward in Franco Moretti's *Atlas of the European Novel 1800–1900* may uncover new readings of the inter-war Australian novel.

> *An atlas of the novel. Behind these words, lies a very simple idea: that geography is not an inert container, is not a box where cultural history 'happens', but an active force, that pervades the literary field and shapes it in depth. Making the connection between geography and literature explicit, then mapping it: because the map is precisely that, a connection made visible – will allow us to see some significant relationships that have so far escaped us.* (Moretti 1999, Preface)

Moretti plotted Jane Austen's novels on a map of England to show where her stories, and her heroines' journeys, begin and end. From this study a particular image of England emerges: for example there is no 'Celtic fringe' in the geography of the novels. 'No Ireland; no Scotland; no Wales; no Cornwall ... only England: a much smaller place than the United Kingdom as a whole. And not even all of England: Lancashire, the North, the industrial revolution – all missing. Instead, we have here the much older England celebrated by the 'estate poems' of topographical poetry: hills, parks, country houses' (Moretti 1999, p.13). Over against this 'small England' however, Moretti also shows how Austen sets out an understanding of the formation of England as a developing nation-state, by taking her tradition-bound 'local gentry' and attaching them to the rising, wealthy 'national elite'.

Moretti's technique of opening up the novel to this sort of cartographic scrutiny, and Barbara Piatti's development of his research, seemed like useful approaches to the Australian novel (Piatti et al. 2008). As a way of testing their methods I assembled a sample of novels, written not by one author but by several authors over a limited period of time, the interwar years. The choice was not difficult to make. The 1930s saw an explosion in the popularity of the novel as a literary form in Australia. According to Drusilla Modjeska 'the ten years between 1917 and 1927 saw the publication of only twenty-seven novels as against eighty-seven volumes of verse, whereas for the years 1928 to 1939, there were 106 novels and only fifty-seven volumes of verse'(Modjeska 1981, p. 5). The novel was, as Marjorie Barnard (1897–1987) and Flora Eldershaw (1897–1956), writing under the M Barnard Eldershaw pseudonym, observed, 'the typical form of the period; large, rich, confused, intricate'; it was 'the organ of becoming, the voice of the world in flux'; it was 'an attempt to get the chaos of circumstances into some sort of shape, using every method of attack, every ingenuity. A natural organic reaction' (Barnard Eldershaw 1983, pp. 78, 80). Its dramatic rise in popularity in the late 1920s and 1930s, sustained until the 1940s, filled in important cultural gaps as Geoffrey Serle observed:

> *when there were still few Australian historians and almost no political scientists or sociologists, and when newspaper comment was of an abysmal quality, these able novelists were almost alone in attempting to define, analyse and comment on the nature of Australian society. At the very least, the novels of the 30s, in the comparative absence of normal sources, provide valuable historical evidence for what Australia was like.* (Serle 1974, p. 123)

Serle also pointed out that a 'feature of the period was the prominence of women-novelists' and that 'most of the best novelists were women' (Serle 1974, p. 123). In 1938, when publishing *Essays in Australian Fiction* M Barnard Eldershaw noted that 'women had won their spurs [as novelists] before anyone thought of telling them they were incapable of doing so' (Chisholm 1987, p. vii). The majority of this group of leading novelists who were producing their best work at this time came from New South Wales and it was from amongst their work that the sample discussed below was chosen.

Having plotted the geography of the twenty selected Australian novels published between the Depression and the end of World War II, I found that, like Austen's novels, their primary action occupied quite a narrow geography. While this geography is somewhat untidier than the elegant country/littoral axis of Austen's novels which guides her heroines' movements from family to marriage; from country estates to the marriage market at Bath or Brighton; from the ancient ties of land, to the lure of new money, it is similarly organised around oppositional spaces. If Austen's 'two Englands' which according to Moretti 'form a field of narrative forces, whose reiterated interplay defines the nation *as the sum of all its possible* stories' (Moretti 1999, p. 20) much the same can be said of the Australian set although there is the qualifying intermediary space of Sydney Harbour, as a third geography to contend with. What is similar to Austen is the radical containment of all possible Australian geographies into a very small New South Wales. If one drew a map of this geography it would exclude vast tracts of the middle and outer suburbs of Sydney, most of the suburbs on the north shore and west of the central city, nearly all the northern and southern coastal regions, and the industrial heartlands of Newcastle and Wollongong. It would include the Monaro region in southern New South Wales and the north western hinterland over the Great Dividing Range. It would also have to take into account a few excursions to and from Victoria, a trip to Canberra and voyages both from London, a lengthy description of which opens Miles Franklin's novel *Back to Bool Bool* and back to London which occupies the second half of Christina Stead's *For Love Alone*. The Australian geography of these novels can thus broadly be divided into three: the city, the hinterland, and the harbour.

21.2 The City

Of the twenty novels in the sample, thirteen are set wholly or partly in Sydney. All cover much the same terrain so a recognisable, if broadly conceived, geography of the city emerges (Edquist 2008). M Barnard Eldershaw's two early novels, *A House is Built* (1929) and *Green Memory* (1933) are historical novels set in old Sydney town, parts of which still remained in the late 1920s and 30s. Their last novel, the futuristic *Tomorrow and Tomorrow and Tomorrow,* published in censored form in 1947 and not in its entirety until 1983, is set 400 years in the future after Sydney has been destroyed in the ongoing conflict following World War II. What was once the twentieth century city is now part of a new technologically advanced but politically disenfranchised global society. However, the narrative is structured as a novel-within-a-novel, and the interior story of Harry Munster is set in pre-war and war-time Sydney, primarily amongst the crumbling slums of Darlinghurst. Between these two points of their career M Barnard Eldershaw published two historical works on early Sydney, *Phillip of Australia* (1938) and *The Life and Times of Captain John Piper* (1939) which can be understood as the mirror of the novels in the imaginative recreation of a lost city. *Tomorrow and Tomorrow and Tomorrow* however takes history into a dystopic future. In all these works the setting for the action remains the historical heart of Sydney.

The first novel of Christina Stead (1902–1983) *Seven Poor Men of Sydney* (1934) written in Europe but set in her home town, established her as a writer of the first rank; its action begins in the fishing village of Watson's Bay (renamed Fisherman's Bay) east of the city just below South Head but gravitates to the central city where the seven poor men of the title work; the action then moves back and forth between these settings, on occasion migrating to the North Shore. In the first half of Stead's semi-autobiographical *For Love Alone* (1945) Teresa Hawkins navigates between her stifling family life in Fisherman's Bay and her ambitions fuelled by love and intellectual ambition in Sydney, the implications of which are played out in London in the second half of the novel.

The geography of the novels of of Eleanor Dark (1901–1985) is often subservient to their narrative, particularly in psychologically driven plots like *Prelude to Christopher* (1934). However, both *Waterway* (1938) and *The Little Company* (1945) develop their action within recognisable locales of Sydney, the latter setting up a contrapuntal movement between the isolated and somewhat precious North Shore and the vibrant and engaged modern metropolis. The social realist Kylie Tennant (1912–1988) was an astute recorder of the city and in *Foveaux* (1939) the suburb itself assumes the role of the novel's central character. The human figures are never as vividly evoked as the vibrant, dilapidated streets and houses of the working-class suburb of Surry Hills on which Foveaux is based. In Tennant's *Ride on Stranger* (1943) Shannon Hicks arrives in Sydney by stages: from rural Victoria to a semi-suburban Victorian town; from marginal and alternative communities hovering on the outskirts of Sydney to the centre of the city itself. *Time Enough Later* (1945) reverses this narrative movement as Bessie Drew escapes from the slums of Redfern firstly into the city's bohemian life and then into rural suburbia. In *Jungfrau* (1936) Dymphna Cusack (1902–1981) depicts Sydney through the eyes of her protagonists, three young women working in the professions of medicine, teaching and social work, while *Pioneers on Parade,* a satirical novel set in the 1938 sesquicentennial celebrations in Sydney combines Cusack's urbane and sophisticated view of the city with Miles Franklin's jocular but earnest valorisation of life in the rural hinterland.

The Macquarie Street precinct, including the Domain, Botanical Gardens, Mitchell Library and a few adjacent streets, the most beautiful part of the city and most redolent of its colonial past, is the privileged geography of central Sydney for all the novelists. Macquarie Street provides the urban boundary to the Domain and along its length are the jewels of Sydney's colonial city: Hyde Park Barracks, the

Mint, Sydney Hospital, Houses of Parliament, Mitchell Library. Over against these architectural statements of political and cultural authority is the Domain itself, the symbolic heart of popular politics where every Sunday people voiced their opinions from their soapboxes. This literary city of Sydney is a city moulded by and unable to escape its past. The geography of Sydney mapped in these works – a geography that drives people from Circular Quay up the hill to Macquarie Street and the Domain – is the geography of the colonial city centred on Sydney Cove. The streets are redolent of the past. Governor Lachlan Macquarie's name is everywhere, as is the evocation of the old city, its plan activated in the present, its unresolved legacies of illegal settlement infecting its modernity (Dark 1938, p. 229). What we have in the Sydney novels is an understanding of a city, of its history and of the way one is dependent on, indeed arises from, the other.

Hard up against the city's Central Station are the three slum areas that provide the setting for most of the domestic action covered in these urban novels: Surry Hills, Redfern and Darlinghurst. The degraded urban environment, its overcrowded boarding houses, brothels and pubs, is the geography of *Foveaux*, *Tomorrow and Tomorrow and Tomorrow* and *Time Enough Later* and to a lesser extent of *Seven Poor Men of Sydney*, *For Love Alone*, *Jungfrau* and *The Little Company*. *Foveaux* is set almost entirely in Surry Hills, a suburb that grew out of large land holdings of Joseph Foveaux, a soldier and early colonial administrator and much of its action revolves around political issues to do with housing and slum clearance. Kylie Tennant builds a symbolic geography of Foveaux around hilly Lennox Street at the top of which the affluent old families live while at the base of the hill the slums, pubs and factories fester. Mary Street is midway between the two, the carefully observed point below which, if one descended, one literally, 'went downhill'. Much of the action of the novel centres on movements up and down the hill following the fortunes of Foveaux's citizens during WWI, the nineteen-twenties and the Depression. Adjacent Redfern 'which just missed being a slum by a narrow margin' (Tennant 1945, p. 6) is the home of Bessie Drew before she escapes to the city; Harry Munster never escapes the slums of Darlinghurst, although he tries (Barnard Eldershaw 1983). One of the outcomes of the picture of Sydney projected by these novels is of an old city – even an old-world city, rather than a city of the new world; as one character observed, King's Cross is 'more like Montparnasse than ever with the foreigners about' (Barnard Eldershaw 1983, p. 180)

21.3 The Hinterland

The mirror of Austen's rural England is, according to Moretti, the marriage market which takes place on the littoral – Bath, Ramsgate, Brighton (Moretti 1999, pp. 18–21). The mirror of the harbourside city of Sydney is the hinterland, the rural districts beyond the Great Dividing Range to the south and west of the city. The tensions between these two places are best seen in the works of the oldest of the writers, Miles Franklin (1879–1954). Franklin's six-volume romantic family saga written under the pseudonym Brent of Bin Bin, as well as *All that Swagger* (1936) published under her own name are set in the south of New South Wales where her family were pioneering settlers. It is from the lives of these people that Franklin attempts to construct an Australian cultural identity. Contrasted with the city, rural Australia is for Franklin the geography of authentic nationhood, an idea she shares with numbers of nineteenth-century painters and poets. Miles Franklin spent her early years on the family home station in the Monaro region of southern New South Wales; as circumstances changed the family gradually moved closer and closer to the urban centre where they finally arrived in 1914. The family's migration from the rural hinterland, land of wealth and independence, to the inner suburbs of Sydney, mirrored the movement of the nation as a whole as it became increasingly urbanised, indeed one of the most urbanised countries on earth. In *Back to Bool Bool* (1931) a novel with an overtly nationalistic flavour, Franklin gathers together several artistic members of her intertwined fictional families from England and Sydney, herds them over the mountains to Bool Bool where, exhilarated by experience of station life, they collaborate (under the baton of a British actress) in a pageant celebrating their creation of a new Australian culture. It is a culture which celebrates the land, its formation and vegetation, the honest toil of the family, and, very much in tune with the contemporary European affirmations of fast disappearing 'folk' cultures, a subtle xenophobia.

Eleanor Dark (1901–1985), like Franklin, also uses a road trip from Sydney to the rural hinterland to explore ideas to do with nationhood and identity. *Return to Coolami* (1936) is a 'road' novel beginning in Sydney and ending at Coolami, a station in the north west of NSW somewhere past Mudgee. Tom Drew, a wealthy, self-made man and his wife Millicent offer to drive their daughter and son-in-law back to their property Coolami; their marriage is in trouble and the daughter has spent time in Sydney with her parents. Millicent's family also comes from the area around Coolami, but Tom has never been there and has always felt a stranger, excluded from his wife's early rural life. Ensconsed in their expensive new tourer this uncomfortable family set off on what unfolds as a literal and metaphoric journey home. As they travel away from Sydney their relationship to each other, and to the land, changes. Home takes on the character of both the physical place and a reconciliation with history. Tom is the key figure here and his interrogation of the road map the metaphor by which Dark documents his change of spirit. As they set out Tom explodes in exasperation at the indigenous place names: 'Parramatta. It had a silly sound, a jabbering sound, the kind of sound that a child might make experimenting with vocal noises! And over there to his left still another – Kirribilli! Well, they sounded exactly what they were – the language of savages! (Dark 1936, p. 16).

Gradually, as the trip unfolds and they drive further and further into the hinterland, Tom undergoes a change:

> *And Good God, what an unholily seductive thing a map could be! Black magic, no more, no less. For after you'd seen it the road you drove was no longer just a road, but a valiant little black line, weaving and threading and picking its way across plains and through vast tracts of bush and over wild blue mountains. ...*
>
> *Names which you must have seen last night in small black print were like voices now, calling out of some primeval past; Yarragrin, Cobborah. They were like something you had forgotten a thousand years ago and to which you were returning now, not only in miles along a road but in spirit through a dissolving barrier of time...*
>
> *Time! Again he was aware, uneasily, of the difference even, sometimes the antagonism, between what one knows and what one feels. And he found himself suddenly abashed because he realised that until now he had always felt that this land of his had been born out of the womb of Mother England in the year 1770 with Captain James Cook for midwife.* (Dark 1936, pp. 295–296)

While Dark situates her characters' questioning of their identity in rural New South Wales, she understands it to be an engagement with history and a coming to terms with the implications of European conquest, even if that proves illusory. This becomes a major theme in her later work, explored in *Waterway* (1936) which prefaces each chapter with journal entries from the First Fleet, and the historical novel *A Timeless Land* (1941) a reconstruction of the earliest days of European settlement and the first book in a trilogy of historical novels that included *Storm of Time* (1948) and *No Barrier* (1953).

For Kylie Tennant the reality of the rural hinterland in the 1930s was unremittingly harsh. Altered irrevocably and despoiled by poor farming practices and drought it is cast in altogether a different light from that of Miles Franklin. Tennant set *The Battlers* (1941) like her earlier novel *Tiburon* (1935) in the state's north-west but in her second novel of the Depression the hinterland is a geography of dispossession and suffering. Her family is an unlikely group thrown together by chance and necessity. Travellers, their landscape is the road, temporary camps and country towns where they are allowed to pick up the dole but not congregate. Itinerant, their home is compressed into their cart and what they carry with them. The idea that rural life expresses some notion of nationhood such as that put forward by Franklin and developed by Dark is severely challenged by Tennant.

21.4 The Harbour

Sydney is a harbour city but in most of the novels the harbour itself plays little part in the action; it is, rather, a part of the general backdrop for action. Only in the Watson

Bay novels is the ferry ride, for example, of major dramatic importance; overall the novels are curiously land-bound, particularly when compared with today's image of Sydney, a city almost entirely defined by the harbour, the Bridge and the Opera House. Even the construction of the Sydney Harbour Bridge which opened in 1932 had little impact on these writers. Indeed the bridge is mentioned, fleetingly, only three or four times, in spite of its radical effect on the geography of the city, its visual imposition on the harbour and the fact that it became the prime symbol of progress for a number of modernist painters at the time. The one exception to this lack of interest is Dark's hostile treatment of the Bridge in her war-time novel *The Little Company* where a character observes its destructive, rather than constructive legacy to the people of Sydney:

> *The obliteration of the old Milson's Point wharf and railway station, to make room for the sweeping concrete curve of the bridge, was one of those facts which still occasionally gave her a slight shock... Going to town had meant for her, in her childhood, crossing the harbour in a ferry, embarking and disembarking at a wharf where an old man with a white beard say year in and year out, playing on a concertina.*
>
> *...Now the electric train, tearing across the bridge, rattled up the gently climbing hills where long ago the steam train had puffed and laboured. One got home from town much faster. Did that matter?* (Dark 1945, pp. 96–97)

Insofar as the action in our group of novels is set in suburbs other than the slums of Sydney, it occupies only the harbour villages tucked in the bays to the east of the city – Darling Point, Rose Bay, Vaucluse and Watson's Bay, and, in *Sun Across the Sky,* a fishing village on an unspecified northern beach. The affluent harbour suburbs of Darling Point, Rose Bay and Vaucluse appear as minor sites for action but Watson's Bay which lies just behind South Head at the entrance to Port Jackson is a major site in three novels.

Guarding the entrance to Sydney, South Head had a military encampment and opposite, North Head, the quarantine station. Stead spent part of her youth in Watson's Bay and her descriptions of it (renamed Fisherman's Bay) are vivid:

> *The hideous low scarred yellow horny and barren headland lies like a curled scorpion in a blinding sea and sky. At night, house-lamps and ships' lanterns burn with a rousing shine, and the headlights of cars swing over Fisherman's Bay. In the day, the traffic of the village crawls along the skyline, past the light-house and signal station, and drops by cleft and volcanic gully to the old village that has a bare footing on the edge of the bay.* (Stead 1999, p. 2).

Fisherman's Bay and the Baguenault family who 'had settled in the bay directly after its arrival from Ireland some thirty years before, and had its roots growing down into the soil and rocky substratum' open and close the narrative of *Seven Poor Men of Sydney,* giving the otherwise complex and many-stranded novel a symmetrical narrative structure. We first explore the bay with the young, illegitimate, Michael Baguenault:

> *Early in the morning, through the open window, the people hear the clatter of anchors falling into the bay, and the little boys run out to name the liners waiting there for the port doctor, liners from Singapore, Shanghai, Nagasaki, Wellington, Hawaii, San Francisco, Naples, Brindisi, Dunkirk and London, in the face of all these old stone houses, decayed weatherboard cottages, ruinous fences, boat-houses and fishermen's shanties.* (Stead 1999, p. 2)

Throughout the novel Michael returns to the bay, attracted as much to its vertiginous cliffs as to family ties and it is from the Gap that he finally and inevitably leaps to his death. The novel ends with the journey home of his young cousin Jo Baguenault:

> *Roar on, O Pacific sea, and blow the golden weather in with morning; in the dark bad deeds are done, pillows are wet, hearts despair; in the morning everything is resolved. Underneath sleep and snore in their beds the hutted people of the Bay, like dry leaves fallen from the heavenly tree.* (Stead 1999, p. 318)

As the novel is encircled by the ocean and its storms, its characters, particularly the Baguenault family, and their stories take on an elemental and ancient patina; they appear wedded to place: 'There is no place in the estuary, though, so suited for an old tale as this fish-smelling bay, first in the port', announces Stead in the opening pages.

For Love Alone is a novel set in an epic mould that again invokes a seafaring culture at the outset, perhaps deliberately picking up the theme that closes *Seven Poor Men of Sydney*. Divided into two parts – 'The Island Continent' and 'Port of Registry: London' – which mirror each other, Stead prefaces the novel with a Homeric invocation:

> *Sea People*
>
> *This island continent lies in the water hemisphere. On the eastern coast, the neighbouring nation is Chile, though it is far, far east, Valparaiso being more than six thousand miles away in a straight line; her northern neighbours are those of the Timor Sea, the Yellow Sea; to the south is that cold, stormy sea full of earth-wide rollers, which stretches from there without land, south to the Pole.*
>
> *It is a fruitful island of the sea-world, a great Ithaca, there parched and stony and here trodden by flocks and curly-headed bulls and heavy with thick-set grain. To this race can be put the famous question: "Oh Australian, have you come from the harbour? Is your ship in the roadstead? Men of what nation put you down – for I am sure you did not get here on foot?"* (Stead 1982, pp. 1–2)

This is the question Telemachus asks Athena when she visits Ithaca disguised as Mentes to set the ball rolling for Odysseus' return home. The *Odyssey* is both a quest and a homecoming and the invocation of the poem at the beginning of her novel allows Stead to pose universal questions about the nature of home and of love –

the driving force of Odysseus's quest, as it is of Teresa Hawkins. Geographically, this literary conceit also entwines Australia in larger world narratives, its mythic structures, which the other novelists do not do. Where their characters are aligned to world issues, it is through the day to day realities of political action or war. And they seem, if anything, more isolated than Stead's characters who commune with the larger world by sea.

Stead sees Australia as a cartographer might: she maps her action across vast terrain. Her Australian novels are set on the rim of the continent, looking outwards, as we saw in the opening lines of *Seven Poor Men of Sydney*. *For Love Alone,* written after years of separation from Australia, positions Australia against its hemispherical opposite:

The other world – the old world – the land hemisphere – is far above her as it is shown on maps drawn upside-down by old-world cartographers. From that world and particularly from a scarcely noticeable island up toward the North Pole the people came, all by steam; or their parents, all by sail. And they live round the many thousand miles of seaboard, hugging the water and the coastal rim. Inside, over the Blue Mountains, are the plains heavy with wheat, then the endless dust, and after outcrops of silver, opal, and gold, Sahara, the salt-crusted bed of a prehistoric sea, and leafless mountain ranges. There is nothing in the interior; so people look toward the Water, and above to the fixed stars and constellations which first guided men there (Stead 1982).

It was probably in reference to Stead that Eleanor Dark set her early novel *Waterway* (1938) in Watson's Bay. If Stead is the Homeric bard, Dark is the Greek tragedian. Like Stead in *Seven Poor Men of Sydney,* Dark lists her characters at the front of the book, as in a play, but unlike Stead the action takes place over one day marked by the rising and setting sun – a unity of place, time and action one finds in classical drama. While Stead emphasises the ancient geography of the bay, Dark investigates its not-so-ancient historical geography, interleaving the chapters of her book with excerpts from the 1798 diary of David Collins and the founding of white settlement. For Dark the historical resonances of Watson's Bay are central to the drama as one character, Ian Harnett, who works on the Harbour, makes clear:

He realised that as he walked along the path beside the Bay, thinking with one part of his mind that it must have been a much prettier beach, prettier Bay altogether, in the days of that Robert Watson after whom it had been named. ... He glanced again at the memorial and ... began to speculate about Robert Watson, and to wonder if that Quartermaster of the Sirius, Pilot and Harbourmaster, and Keeper of the Macquarie Light had come at last to love the harbour which had been his charge, or whether he had felt himself, to the end, an exile in a barbarous and inhospitable land. (Dark 1938, pp. 38–39)

This relationship between white settlement and the ancient land, between two cultural geographies, one superimposed on the other, is a major point of tension in

Dark's fiction; in *Waterway* we see this in the struggle of writer Lesley Channon to find an authentic voice in her work while in the historical novels such as *A Timeless Land* Dark returns to the origins of European settlement in an attempt to clarify the issues of the present, 'both to explain the contemporary brutality and to lay claim to a radical tradition that bore the possibility of change. *The Timeless Land* identifies a heritage of humanism and anti-fascism in Australian history and holds out the possibility that it could be continued, despite the weight against it' (Modjeska 1981, p. 240).

The Watson's Bay setting for Stead and Dark is therefore a potent geography of settlement, its site at the entrance of Port Jackson a powerful image of both white colonisation and the fraught contemporary connections with Europe. Australia's geographical situation in the 'antipodes', at the bottom of the world, is one that clearly helps to shape these novels. Geography is the metaphor by which these authors grapple with issues of national identity and Australia's relationship to the world.

21.5 Conclusions

The literary geography of this selection of interwar Australian novels can be broadly defined in this way: the city is the space of political struggle, and cultural and social change. The harbour is primarily the site of white occupation and reflection on its present ramifications. The hinterland is the place where white Australians wrestle with the issues of national identity. It is a remarkably narrow range of possible geographies from which to fashion a national literature.

The most striking thing about this geography is that there is no suburbia – that urban condition that has defined Australian settlement since the nineteenth century. The Australian Dream (of owning one's own house) is part of the mythology that has shaped the twentieth century in this country, and low-density suburban development with a high level of owner-occupied houses is the preferred model of living; indeed Australia was the first suburban nation. As architect and critic Robin Boyd famously noted in 1952 'Australia is the small house. Ownership of one in a fenced allotment is as inevitable and unquestionable a goal of the average Australian as marriage. Australia has more per head of population than other nations and has maintained a higher standard of living in a greater number of them. Collectively, they are an achievement' (Boyd 1952, preface). Where Austen protected her small, historical, England from the Celtic fringe, the writers discussed here protected the small historical settlement of old Sydney from the 'fringe' – the actuality of suburbia. We can posit various reasons for this. By restricting the geographical spread of their action they are able to investigate the conditions of modern life as though in a controlled environment divided into distinct locales each of which has characteristic inhabitants – the slums dwellers; the bohemians; the battlers; the landed gentry; the struggling lower middle class. In literature these are clearly defined groups with long lineages.

Suburbia's middle classes didn't have this literary lineage; the vast tracts of suburban development that encircle every major Australian city and town had to wait until after World War II to be widely represented in fiction. What could be said about suburbia? For Miles Franklin it is a place to be repudiated. In *Back to Bool Bool* suburbia is depicted as both physically and intellectually barren and repulsive: 'the August sun shine blazed upon miles of bungalows like rabbit hutches separated by wide dusty roads... Ashville... was now a populous and execrably ugly city of well-to-do working people, retired or active', and so on (Franklin 1956, pp. 32–33). Or it is a space in the landscape to be traversed quickly, eyes averted:

> *The train ran through Suburbia. The pattern sorted itself out, took form and coherence, villas in gardens, tree-lined streets, garages with red and yellow bowsers, cinemas bedizened with light, shopping centres, churches and little parks, schools and fire-stations, clean and separate like the grains of well cooked rice.* (Eldershaw 1983, p. 52)

The twin geographical points of these interwar Sydney novels – the city and the hinterland – squeeze out the in-between, the suburban reality that had not as yet found its voice. Like Austen, our novelists carefully delimit the literary geography in which their novels' action will unfold and also like Austen they do so in order to reflect on the political and social upheavals of a fast-changing nation state under the extreme pressures of change.

References

Boyd R (1952) Australia's Home. Its Origins, Builders and Occupiers, Melbourne: Melbourne University Press

Brent of Bin Bin (1956) Back to Bool Bool, Sydney: Angus & Robertson

Chishome A (1983) Introduction in Eldershaw MB Tomorrow and Tomorrow and Tomorrow, London: Virago

Cusack D (1936) Jungfrau, Sydney: The Bulletin,

Dark E (1937) Prelude to Christopher, London: Collins

Dark E (1936) Return to Coolami, New York: The Macmillan Company

Dark E (1937) Sun Across the Sky, London: Collins

Dark E (1938) Waterway London: Collins

Dark E (1958) The Timeless Land, London: Collins

Dark E (1945) The Little Company, Sydney: Collins

Edquist H (2008) 'Reading Modernity. Architecture and Urbanism in Sydney in the Inter-War Australian Novel', Fabrications

Eldershaw MB (1947) A House in Built, London: George G Harrap & Co
Eldershaw MB (1931) Green Memory, London: George G Harrap & Co
Eldershaw MB (1983) Tomorrow and Tomorrow and Tomorrow, London: Virago
Franklin M (1986) All That Swagger, Sydney: Angus & Robertson
Franklin M, Cusack D (1988) Pioneers on Parade, Sydney: Angus & Robertson,
Moretti F (1999) Atlas of the European Novel 1800–1900, Verso, London
Modjeska D (1981) Exiles at Home. Australian Women Writers 1925–1945, Sydney: Sirius Books
Piatti B (2008) 'Mapping Literature: Towards a Geography of Fiction', Proceedings of the Cartography and Art – Art and Cartography Conference: Vienna
Serle G (1974) From Deserts the Prophets Come: The Creative Spirit in Australia 1788–1972, Melbourne: William Heinemann
Stead C (1999) Seven Poor Men of Sydney Sydney: ETT Imprint
Stead C (1982) For Love Alone, London: Angus & Robertson
Tennant K (1936) Tiburon, Sydney: The Bulletin
Tennant K (1946) Foveaux, Sydney: Syrius Publishing Company
Tennant K (1954) The Battlers, London: Macmillan
Tennant K (1994) Ride on Stranger, Sydney: Angus & Robertson
Tennant K (1945) Time Enough Later, London: Macmillan

22 Mapping Informal Geographies

Michaela Kinberger

Department of Geography and Regional Research, University of Vienna, Austria

Abstract

In traditional mapping we usually visualise data with defined relationships to geographic space and time. Informal geographies are about non-physical/human geographic information, where, usually, the content cannot be 'geotagged'. Before these topics can be mapped the information has to be spatialised in some way. Due to the fact that the resulting depictions lack 'traditional' geographic information they are mostly not considered to be 'maps' in a common sense. These maps do not represent geographies that map users are accustomed to, but maps of 'other' geographies. This paper will introduce informal geographies, their spaces and visualizations. It will also report on the CHIS (Cultural History Information System) project located at the University of Vienna and sponsored by the Austrian Science Found, where informal geographies are being mapped using contemporary techniques.

22.1 Introduction

In January 2007 an interdisciplinary project called 'The Cultural History of the Western Himalaya from the 8th century' started in Vienna. This National Research Network (NRN), funded by the Austrian Science Fund, includes cartographers, art historians, numismatists, Buddhist philosophers, and Tibetan and Sanskrit philologists. The main objectives are to intensify research on the cultural history of the Western Himalayas as well as to develop a map-based Cultural History Information System (CHIS) for sharing the outcomes with other experts and the interested public.

After intensive discussions with each of the project partners it is clear now, that not every project works with information related to geographic space. There is a lot of important non-physical/human geographic information, which will be referred to in the following text as informal geographies. As one of the research outputs is an information system based on a mapping environment, the informal geographies have to be integrated in a visual way. One problem is to combine this other geography and associated mapping techniques to a meaningful depiction. A common method in information visualisation to get a 'map-like' representation of non spatial information is the application of spatial metaphors also known as 'spatialization' (Skupin & Buttenfield 1997, p. 117). Due to the fact that 'mapping informal geographies' is one of the main research topics within the CHIS project, this chapter will also give a brief summary of the ongoing work.

In the following text the most important properties used to distinguish different geographies and to visualise their content will be described. Due to the fact that the CHIS is a map-based information system, this is a necessary process for being able to select the level of integration of the information (geographies) collected by the different sub-projects. The differences will be explained on the basis of the type of information and the different spaces that form these geographies. According to this division formal and informal geographies included in the CHIS-project are established. Following this, the different types of maps and different ways of mapping informal geographies are presented, including map examples found on the Internet and in books. If information is available, their production processes are also described. The chapter ends with some concluding words and recommendations for further research.

22.2 Formal and Informal Geographies

Multidisciplinary research projects lead to a huge amount of non-physical/human geographic information with undefined geo-spatial relation. As one of the research outputs is an information system based on a mapping environment, the informal geographies have to be integrated in a visual way. The best way to distinguish the information is to define the different geographies by describing their main properties; containing information, reference system and space.

22.2.1 Formal Geographies

When defining informal geographies it is easier to define the formal ones first. Formal geographies are about physical/human geographic information. These are the geographies traditionally mapped by cartographers.

The term ' formal information' refers to any information, which is either representing phenomena of the real world, well defined in space and time (geodata), or information directly linked to the geodata (thematic data). Geodata are the references for thematic data and are presented in a spatial model like a geodatabase as points, lines or areas. Each dataset consisting of different features (point, line, area) needs a defined spatial reference system. Geodata used in the mapping system can either be vector data (mostly Shape-files) or raster data (satellite images, elevation models, maps). Usually a special set of geodata is compiled and visualized to present the base map. The base map is important for orientation within an area and as a visual reference for thematic data.

Thematic data can be any information which is directly related to one or more features of the geodata. It is important for the correct integration of all data, that the person gathering information for a topic is linking the information clearly to space and time. One possibility to connect thematic data to a spatial reference is a gazetteer, presenting a list of already known geographic locations. Another way to add spatial location to thematic data is the application of a GPS-device.

Spatial information is always related to geographic space. Geographic space is presenting the spatial framework used for geographic information. According to Fabrikant & Buttenfield (2001, p. 271) geographic space represents a phenomenon's semantic structure with a metaphor, which is based on a locational ordering principle (Cartesian coordinate system). This includes concepts of distance, direction, magnitude (height), etc.

22.2.2 Informal Geographies

Informal geographies are about non-physical/human geographic information, where, usually, the content cannot be geotagged. This non-geographic information cannot be related to any of the mentioned geodata like other thematic data. Traditionally these geographies are visualized in forms of diagrams, float charts, or other representations not based on a spatial reference system. Before these topics can be represented in a 'map-like' representation the information has to be spatially referenced in some way.

One possibility is spatialization, a technique to add space to information based on Waldo Toblers 'First Law of Geography'. Spatialization according to Skupin and Fabrikant (2003) extends geographic principles and cartographic techniques to the visualization of non-geographic information, and is designed to summarize large data repositories and provide opportunities for visual query and sense-making of large data collections.

In context with the concepts and development of the CHIS any non-geographic information collected within the process of academic research is referred to as informal information. This information can currently be stored in any digital or analogue kind of way.

For the integration and visualisation within the CHIS analogue information has to be digitised. At the current state of the project it is not clear, how this data should be stored. A lot of 'map-like' visualisations described in literature (Old 2002, Skupin and Fabrikant 2003, Dodge and Kitchin 2001) are based on text documents translated to a XML document. The XML Format could also be a solution to store the information of manuscripts or other text within this project.

To be able to apply traditional cartographic means a new, so called 'information space', is created by spatialization. Considering the Topic of the sub-project this could be for example a "philospace" for the Philosophy project. Using the spatial metaphor of distance this means that philosophical ideas which are more similar are nearer and ideas which have no or little alikeness are further away. In a third dimension (philosophic elevation model) this could mean that ideas which are stronger or more distributed are at a higher level (visualised as a peak). The result will be a philosophical landscape.

22.2.3 Geographies Within the CHIS-Project

Next to the coordination project and the subproject developing the CHIS there are five subprojects collecting information which will be included in the information system. These are Art History, Tibetan Manuscripts, Tibetan Inscriptions, Numismatics and Philosophy. According to the kind of geography they are dealing with, their results are integrated in the application. If it is formal, all objects are georeferenced and can be mapped. If the content is informal, a kind of 'map-like' representation has to be applied.

Due to the distributed information storage, the visualizations will also be dependent on the data access and structure given by the different subprojects. For the ongoing work the subprojects of the NRN have been divided into projects mainly structuring their information in formal or informal geographies:

- Formal geographies: Numismatics, Art History, Tibetan Inscriptions; and
- Informal geographies: Numismatics, Tibetan Manuscripts, Philosophy.

The border between formal an informal geographies is not a hard line. Considering the numismatic research for example, there are the coins, physical objects which carry nearly all the information important in the research field. Their current location or the location they were found is a direct link to geographic space. But this location is irrelevant to the research, because there are only a few different locations, which can be linked to thousands of ancient coins. More important for the research is the connection between the different coins like the chronological order. Hence in this case there will be more than one kind of visualization.

22.3 Types of Maps of Informal Geographies

The concept of mapping informal geographies comprises map-like representations of non geographic spaces. Visualising this kind of information in a map is a way of simplifying the communication of schemes and coherencies of complex data. In this context the term "spatialization" is used by Skupin & Fabrikant (2003, p. 95), which describes the extension of geographic principles and cartographic methods to visualisations of non geographic information.

The distances between spatialized phenomena on maps of informal geographies cannot be measured in metric units (e.g. kilometres), but in their degree of similarity. Fabrikant & Buttenfield (2001, p. 266) distinguish two forms: semantic and geometrical spatialization. With semantic spatialization the distance between objects is computed by analyzing the textual content of objects. This data will later on be transformed into numerical values. The geometrical spatialization calculates the distances between the objects directly from numerical data (Jongh and Ormeling 2003).

In the following some of the major fields of application for informal geographies and examples of maps created through spatialization of non-geographic information are presented. Furthermore if information is available the processes of data preparation and data visualization are identified schematically. The chapter represents in no way a complete list of applications/maps of informal geographies, but has to be seen as an attempt of showing descriptive examples which could help understanding the goals, ways and methods of the rather new research field.

The resulting maps are not a mere cartographic product, but information graphics involving different disciplines. There are only few other areas in geographic/cartographic research, where the borders between science and art are that fuzzy.

At the Indiana University in Bloomington (USA), Katy Börner directs the Information Visualization Laboratory, where research on various visualizations of informal geographies is conducted. Some of their projects are introduced below. Additionally the concept of moral fantasy maps is described as a prime example for earlier maps of informal geographies.

22.3.1 Knowledge Domain Visualizations

Mapping knowledge domains means the process of "charting, mining analyzing, sorting, enabling navigation of and displaying knowledge" (Shiffrin and Börner 2004, p. 5183) with the goal of easing access to information and making evident the structure of knowledge. Semantic spaces identifying the correlations of researchers, publications, funding and other facts are visualized in a way that major research areas become apparent.

"By studying maps of non-spatial phenomena, we are able to distil the contributions of maps (i.e. their fundamental geo-spatial connotations) in analysing 2D or 3D space, especially data space, without being sidetracked by the actual geographical information on the real world, be it geometrical, topological or semantical, provided by the maps." (Jongh and Ormeling 2003)

Various projects on mapping knowledge domains have been undertaken at the InfoVis Lab in Indiana including maps on the research on visualising knowledge domains itself, maps on research fields, on topics and topic bursts. Below two examples are described; the Emergent Mosaic of Wikipedian Activity and Topic and topic burst maps of PNAS.

The Emergent Mosaic of Wikipedian Activity is a map created in 2007 by fellows of the InfoVis Lab visualizing the activity of the online encyclopedia Wikipedia and its authors. The main intentions were to analyse the basic statistics, to identify semantic structures and age of the categories and the content coverage of the Wikipedia authors as well as to visualise all of that, using the map as a metaphor.

Analysis required the comparison of different versions of the online encyclopedia. For that reason complete database dumps of the English Wikipedia using eXtensible Mark-up Language (XML) format were downloaded and imported to an own database. Not only the articles but also the category and linkage data were used for further examination.

The creation of a base map was done by comparing the co-occurrence of categories in articles. Two categories were assumed to be similar to each other and were connected by a link if they were used in the same articles.

After weighting the linkages for the category network, the data were loaded into *VxInsight,* a program that allows interactive exploration of vast datasets. The output of the program was a pair of coordinates for every category. Now the data could be analyzed and clusters could be shown comparing different attributes (e.g. author, last revision). A map (see *Figure 22.1)* was created were pictures from Wikipedia are shown in the background from the most prominent article in the respective tile of the map.

Topic and topic burst maps of PNAS are very similar to the mosaic of Wikipedian activity. It is a mapping project which was done by the InvoVis Lab in 2004. The goal was to show the dynamics of scientific knowledge via analyzing all papers that were published in the "Proceedings of the National Academy of Sciences" (PNAS) of the United States in the years 1982–2001. This way the importance of topics and the introduction of new topics could be visualized on a map. Visualisation processes were very similar to those of the map above.

The map of topics and topic bursts (see *Figure 22.2)* visualizes with different colours, when a topic was first introduced to PNAS. A network shows the linkage and therefore the similarity of the various topics. The sizes of the coloured circles show how many articles were written about the topic. Altogether the map is a clear and relatively easy to understand depiction of the structure, importance and time of appearance of topics and topic networks.

Fig. 22.1. Mosaic of Wikipedian Activity (Holloway et al. 2007, http://scimaps.org/maps/wikipedia/)

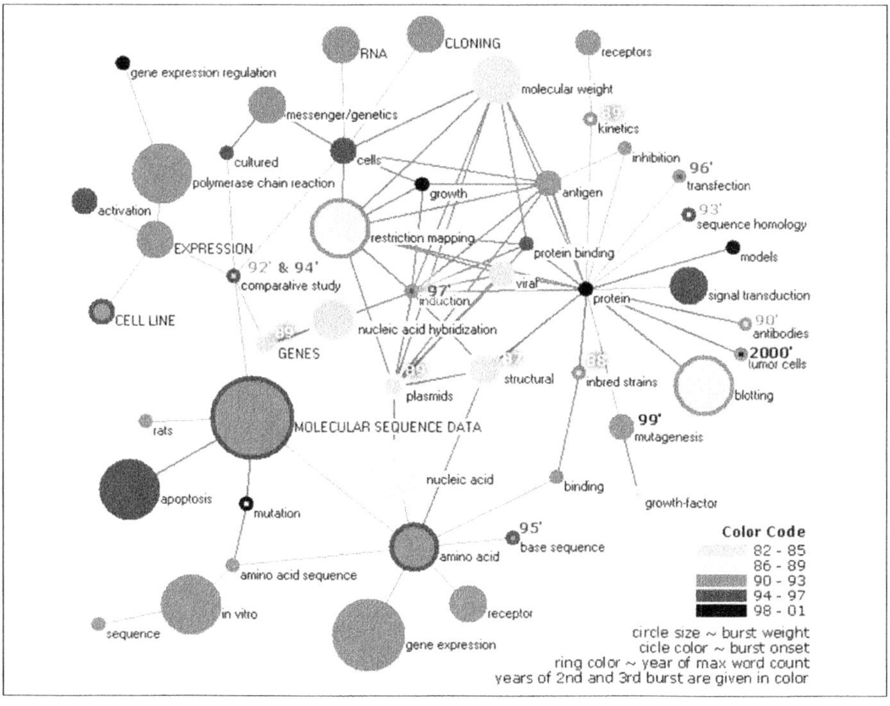

Fig. 22.2. Map of topics and topic bursts in PNAS (http://ella.slis.indiana.edu/~katy/gallery/pnas-large.gif)

22.3.2 Maps of Virtual Worlds and their Users

Interactive virtual 3D worlds allow users to walk around and have interactions with objects or other users of that virtual world. Those diffusion and activity patterns can be mapped in combination with the objects of the virtual world to get an insight on how people react to the virtual environment and to other users.

This kind of information visualization is used to disclose "the evolution of three-dimensional (3D) virtual environments, the distribution of their virtual inhabitants over time and space, the influence of group leaders and environmental and social diffusion patterns" The emerging maps help to evaluate and improve the design of the virtual worlds, to ease social navigation and to provide means to study the communities evolving in virtual environments. (Börner and Penumarthy 2003, p. 182).

Figure 22.3 shows a map of objects and user diffusion patterns by time from the 3D virtual world Quest Atlantis created by the InfoVis Lab and featured in the *Atlas of Cyberspace*.

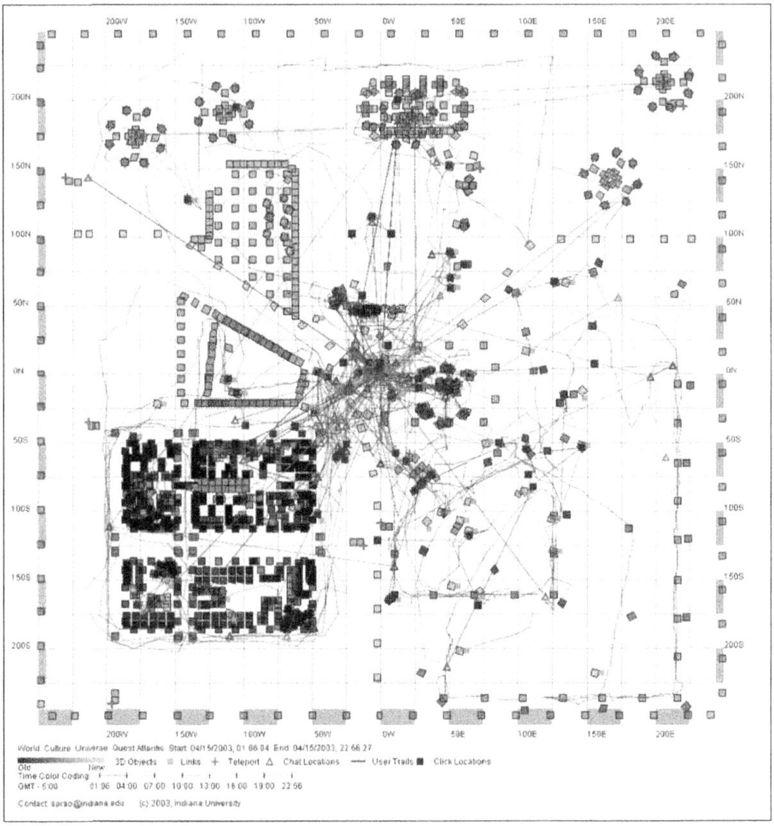

Fig. 22.3. Map of users and diffusion patterns from the virtual online world Quest Atlantis (http://ella.slis.indiana.edu/~katy/gallery/qa-culture-large.jpg)

22.3.3 Moral Fantasy Maps

Moral fantasy maps, which were very popular in from the 16th to the 19th century, visualise the things that would happen in your life (e.g. falling in love, getting to heaven) when you moved to a certain direction from a certain station in life. Those fantasy maps are not showing spatial reality, but options you can choose from, when living your life. They let you know what to expect and what the consequences of your choices will be (Jongh and Ormeling 2003).

The figure below shows a typical example of a moral fantasy map, which depicts the journey of life. Everyone starts in the middle and by the course of action someone takes he either is moving towards hell or towards heaven.

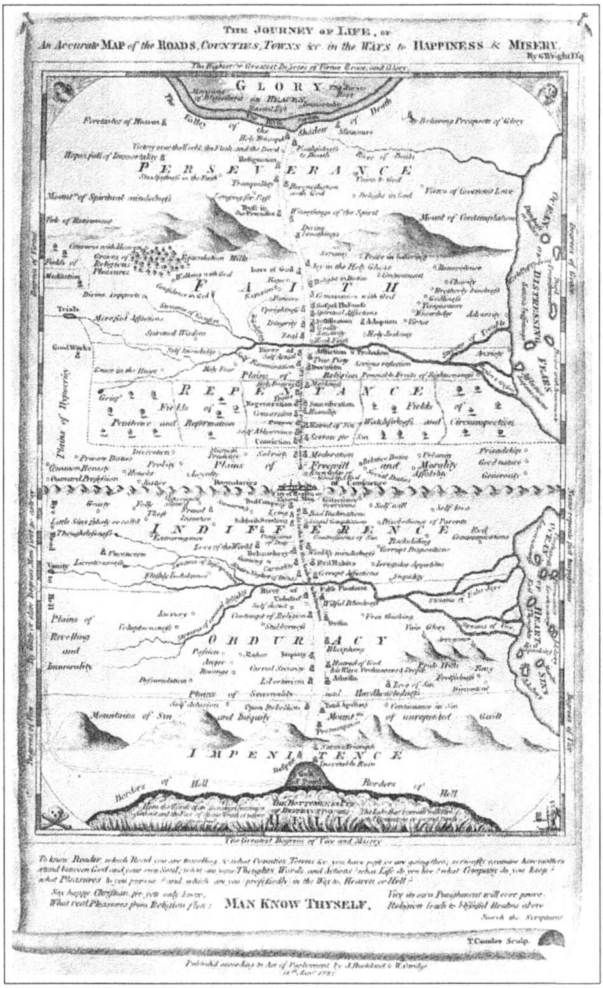

Fig. 22.4. New Map of the Journey of Life (Harmon 2003, p. 50)

22.4 Conclusions

As the examples in this chapter show, mapping of informal geographies is a vast research field. The overall goal is to visualise information in a way that the user can read it most efficiently. Not only maps of knowledge domains with hundreds of thousands of texts as a basis for data, or moral fantasy maps, or maps of virtual worlds but also any map-like representation that visualizes information not georeferenced, can be called a map of informal geographies.

The underlying database structures for producing a map/application of informal geographies are not very well delineated in the available references. XML is mentioned very often as a key technology when it comes to methods of standardized data exchange. Further research on the databases of informal geographies might probably depend on discussions and communication with experts in this field.

Concerning the CHIS-project this means, that after the division of the research information in formal and informal geographies, the storing and mapping process has to be considered now. The participating subprojects of the NRN and their content of research were divided into formal geographies (Numismatic, Art History, and Tibetan Inscriptions) and informal geographies (Numismatic, Tibetan Manuscripts, and Philosophy). This has not to be seen as a differentiation in two different qualities of subprojects, instead it is a necessary process to find the most suitable kind of visualization.

Considering only the topic of the subprojects and their general way of working they described in the subproject interviews, it is not easy to select the correct kind of visualisation. For all the formal geographies it is clear, that they will be implemented in the CHIS as in other map-based information systems. Traditional cartography is all about mapping geographic information; as long as the relation between geodata and thematic data is clearly defined, the results of the research process can be visualised in the CHIS.

In a next step the best way to spatialize the information of each of the informal geographies within the project has to be found. This means establishing rules for the transformation of non spatial data to spatial data, finding new ways of storing informal geographies, and selecting sound cartographic depictions for the new spaces.

References

Börner K, Penumarthy S (2003) Social diffusion patterns in three-dimensional virtual worlds. In: Information Visualization (2003) 2, pp 182–198. http://ella.slis.indiana.edu/~katy/paper/03-ivj-diffusion.pdf [12.12.2007]

Card SK, Mackinlay JD, Shneiderman B (eds) (1999) Readings in Information Visualization, Morgan Kaufmann Publishers, San Francisco, CA

Dodge M, Kitchin M (2001) Mapping Cyberspace. Routledge, London

Harmon K (2003) You are here – Personal geographies and other maps of the imagination. New York

Herr BW, Holloway TM, Börner K (2007) Emergent Mosaic of Wikipedian Activity. Google Map at http://scimaps.org/maps/wikipedia/ [15.12.2007]

Holloway TM, Božičević M, Börner K (2007) Analyzing and Visualizing the Semantic Coverage of Wikipedia and Its Authors. Complexity, Special Issue on Understanding Complex Systems. Vol. 12(3), pp 30–40. http://ivl.slis.indiana.edu/km/pub/2007-holloway-vizwiki.pdf

Jongh C, Ormeling F (2003) Mapping Non-Spatial Phenomena. In: Proceedings of the Seminars on Developing the ICA-CET Internet Cartography Course held at Beijing (China), August 9, 2001 and Helsinki (Finland), May 28, 2002. http://lazarus.elte.hu/cet/publications/proc13-ormeling.htm

Fabrikant SI, Buttenfield BP (2001) Formalizing semantic spaces for Information access. In: Annals of the American Geographers, 91, pp. 263–280

Old LJ (2002) Information Cartography: Using GIS for visualizing non-spatial data. Proceedings, ESRI International Users' Conference, San Diego, CA, July 2002. http://gis.esri.com/library/userconf/proc02/pap0239/p0239.htm

Shiffrin RM, Börner K (2004) Introduction – Mapping knowledge domains. Proceedings of the National Academy of Sciences Vol. 101. April 2004. The National Academy of Sciences of the USA. http://ella.slis.indiana.edu/~katy/paper/04pnas-intro.pdf

Skupin A, Buttenfield BP (1997) Spatial Metaphors for Display of Information

Spaces. Proceedings, AUTO-CARTO 13, Seattle, WA, Apr. 7–10, pp. 116–125

Skupin A, Fabrikant SI (2003) Spatialization Methods: A Cartographic Research Agenda for Non-Geographic Information Visualization. Cartography and Geographic Information Science. 30 (2) pp. 99–119

Tufte ER (1983) The Visual Display of Quantitative Information. Graphic Press, Cheshire, Connecticut

23 Mapping the Sacred Landscape of Lahaul – The Karzha Khandroling Mandala

Verena Widorn[1] and Michaela Kinberger[2]

[1] Department of Art History, University of Vienna, Austria
[2] Department of Geography and Regional Research, University of Vienna, Austria

Map me no maps, sir, my head is a map, a map of the whole world.
Henry Fielding (English dramatist and novelist, 1707–1754)

Abstract

The following chapter discusses the mapping of sacred landscape and ritual space by the example of a pilgrimage route in Lahaul, India. This area is a place of symbolic and ritual significance for Tibetan Buddhists. The popular pilgrimage route around the peak of the holy mountain Drilbu ri transforms the mundane geography of the region into a *manṇḍala,* a cosmic diagram and paradigm of the spiritual structure of the universe.

23.1 Introduction

In various spiritual traditions, the term *maṇḍala* (Tibetan: *dkyil 'khor*) is used for establishing a sacred space, symbolising the universe. Originating in Hindu tradition, the *maṇḍala* has become one of the most popular representations of the micro- and macrocosms, metaphysically or symbolically, in the art of the Tibetan Buddhism. In its figural representation, the Buddha or the principle deity is worshipped in the centre [dkyil], while the retinue, minor deities or emanations of the main god are placed in the outer circles, the surrounding area [khor]. But as a cosmic diagram the *maṇḍala* depicts any plan, chart, geometric pattern or landscape as 'a model or image for the transcendent world existing within the mind of every being' (Snodgrass 1985, p. 107).

This is the case of Lahaul, a district in Himachal Pradesh, North- India, where the landscape is regarded by the local people as such a *maṇḍala*, formed by two rivers and with the holy "bell-shaped" mountain Drilbu ri in the centre *(Figure 23.1).* But

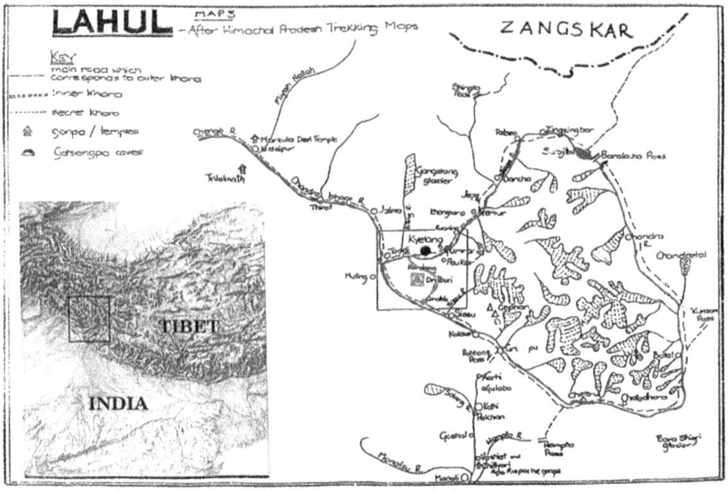

Fig. 23.1. Map of Lahaul, Himachal Pradesh. After Stutchbury 1991 (modified)

can one map a sacred landscape that has mainly been defined through the meditational powers of the saints and the yogis, who spent some time in this area and marked the landscape by the imprints of their bodies and their miraculous acting? Is it possible to draw a map of a region whose spiritual power associated with several places and monuments causes a sacredness that exceeds the pure geographical and natural landscape?

This chapter is intended to combine the characteristics of the geographical and religious landscape of Lahaul into an informal map that shows how these different features are connected to the idea of landscape as a *maṇḍala,* as it is regarded and used by the local people. This idea will be represented by the example of the day-long ritual circulation around the Drilbu ri (based on the experiences of the authors)[1] and the sites associated to this pilgrimage route.

23.2 The Geographical and Religious Setting of Lahaul

The district of Lahaul is situated at 32° latitude and 77° longitude.[2] The area is surrounded by the high peaks of the Great Himalayan Divide to the south and the

[1] The joint research trip to Himachal Pradesh and Lahaul in August/ September 2007 was kindly funded by the National Research Network "The Cultural History of the Western Himalaya from the 8th century" directed by Deborah Klimburg-Salter and financed by the Austrian Science Fund (http://www.univie.ac.at/chwh). We would also like to express our thanks to Anna Filigenzi who has supported this paper with her precious comments and suggestions.

[2] Most of these statistics, including the elevation of the peaks and passes, have been taken from Balokhra (1998). These figures, which vary in every publication on Lahaul, are just approximate, since the valley district has not been thoroughly surveyed.

Zanskar Himalayan range to the north. Only three passes[3] permit vehicular travel, enabling economic exchange with the neighbouring regions. A few other passes that vehicles cannot cross are used mainly by the tribal people. The average elevation of the peaks and glaciers in Lahaul lies between 5,480 meters and 6,400 meters. Originating from opposite corners of Chandratal Lake near the Baralacha Pass and flowing in different directions, the rivers Chandra and Bhaga join again at the base of the sacred mountain Drilbu ri[4] at Tandi and flow as the Chandrabhaga or Chenab River towards Jammu and Kashmir and further on to Pakistan. The rivers' courses run through the three sparsely populated valleys of Lahaul – the Chandra Valley, the Bhaga Valley and the Chandrabhaga Valley.

Lahaul is a cultural borderland, situated at the social, religious and artistic crossroads between India, Tibet and Central Asia. In the valleys of Lahaul we are confronted with a mixture of Hindu, Buddhist and indigenous worship *(lung pa'i chos)*[5], the latter sometimes identified with Bon. Nature worship is represented in Lahaul by tree and mountain worship.[6]

The emergence of the different religions and their development in this area are difficult to trace back to certain events or personalities. Popular legends and the local traditions provide only vague information without any historical evidence. Especially Tantrism plays a pivotal role in the religious setting of Lahaul. The area with the establishment of pilgrimage sites for both, Buddhist and Hindus, and the fluid nature between the ritual practices of these two religious groups, supports Fritz Staal's opinion that 'at present, Hindus and Buddhists in these areas may look very different from each other. In reality, even a cursory inspection of past events demonstrates that the transitions between Hinduism and Buddhism have generally been fluid and uncertain' (Staal 1982, p. 50). The early development of the Tantric or *siddha* tradition of the Vajrayāna[7] in this region is particularly connected with three important personalities, who are closely associated with the pilgrimage route around the Drilbu ri, namely Padmasambhava (Guru Rinpoche), Siddha Ghāṇṭapa, and the pilgrim and saint rGod ts'an pa.

Elizabeth Stutchbury (1991) was the first who has particularly emphasised the sacredness of the landscape in Lahaul. She has published a map and diagram of the *Karzha Khandroling,* the central part of this sacred geography with the circulation

3 The main route through the narrow valleys connects Lahaul to the South via the Rothang Pass (3,978 m), to the East via the Kunzum Pass (4,520 m) and the North via the Baralacha Pass (4,890m).

4 The Tibetan word "ri" means mountain.

5 This term was first used by Harcourt (1871 reprint 1972, p. 72) to characterise the popular belief of the local people.

6 In addition to the Drilbu ri the high peak in the Chandra Valley at Sissu is worshipped as the abode of Gephang, the mountain protector of Lahaul.

7 The Vajrayāna or Tantric Buddhism is considered as one main phase in the development and evolution of Buddhist thought and religion.

Fig. 23.2. The pilgrimage route around the peak of the Drilbu ri. M. Kinberger 2007

route around the peak of the holy mountain, as it is seen and understood by the local people (Stutchbury 1991, p. 69).

23.3 The Karzha Kandhroling

The sacred geography in Lahaul has a strong ritual reflection in the *Karzha Khandroling*[8], the popular name of the clockwise circuit around the Drilbu ri *(Figure 23.2)*. It starts from Kardang in the Bhaga Valley and ends either in the village of Gondhla in the Chandra Valley or at the Guru Ghantal monastery above

[8] Karzha is the local name for Lahaul. The genesis and meaning of the name Lahaul is uncertain. Cunningham has proposed a Tibetan origin, from the terms *lho-yul* "southern land" (in comparison to Ladakh, which is therefore situated in the north) and *lha-yul* "land of the gods" (Hutchinson and Vogel 1933, p. 479). Another possibility is *la-yul* "country of passes", as suggested by Tobdan (Tobdan 1984, p. 7). The Tibetans and inhabitants of the region prefer the name Gar zha (Hutchison and Vogel 1933, p. 476), also written Garzha or Garsha (Sahni, 1994, p. 21), Gar źa, dKar źva (Klimburg-Salter 1994, p. 46) or Karzha (Stutchbury 1991, p. xii). Tobdan refutes Francke's and Gergan's proposed relationship of the term "Garzha" to an ancient tribe and suggests instead that the word refers to the name "Ga-sha" mentioned in the biography of Padmasambhava, which has been identified as Lahaul (Tobdan 1984, p. 10).

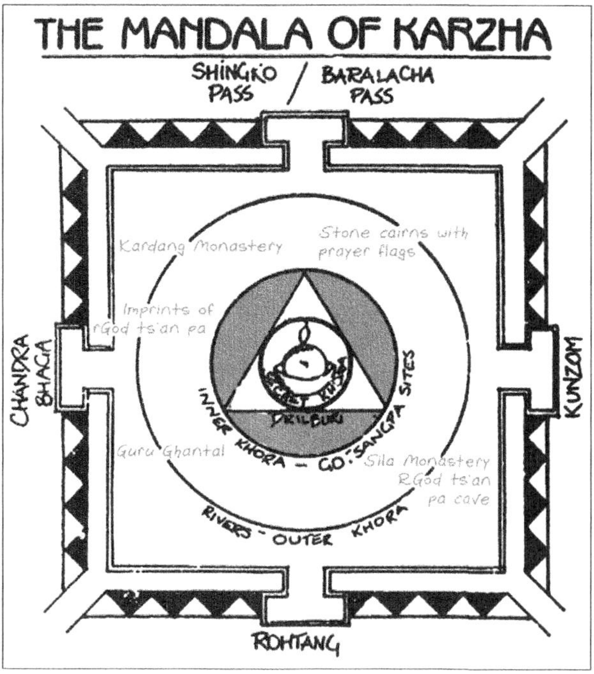

Fig. 23.3. Diagram of the sacred landscape in Lahaul, showing the Karzha Khandroling with the sites of the inner *khora*. After Stuchbury 1991, p. 69 (modified)

the confluence of Chandra and Bhaga. The pilgrimage route follows the physical and spiritual configuration of the landscape, which is composed, like the *maṇḍala* usually is, by concentric circles within a square having as a reference point the triangle of the Drilbu ri.

Based on her observation of the local opinion, Stutchbury has drawn a diagram of the sacred geography that transforms the existing landscape into a geometric pattern *(Figure 23.3)*. The centre of the cosmic diagram is a triangle that can be identified as the peak of Drilbu ri, the abode of Cakrasaṃvara.[9] The triangle contains a circle symbolising the secret *khora*, the closest circumambulation around the summit that is usually reserved to high-ranking clerics or Tantric practitioners. A second circle touches the three corners of the triangle and emblematises the inner *khora,* the day long circuit around the Drilbu ri that is conducted by all the other pilgrims and lay-devotees. The courses of the two rivers Bhaga and Chandra correspond to the third circle, the outer *khora* that encircles all the sacred sites in the area. The square frame

[9] Cakrasṃavara (Tibetan: bde mchog) is literally translated as 'the binding of the wheel' and is one of the most prominent Yidam (deity for meditation) of the Tibetan Buddhism. He is normally depicted as a male, wrathful god with blue body-colour, four heads and twelve arms. He is usually standing on a red Kālarātrī (a manifestation of the destructive Kali) and a black Bhairava (a god from the retinue of the Siva). Cakrasaṃvara's consort is Vajravārāhī, whom he embraces in sexual unification.

of the *maṇḍala* with entrances on all four sides marks the border of Lahaul with the four passes that lead into the region from each cardinal direction.

Especially the Cakrasaṃvara *maṇḍala* has a territorial component. According to the myth, Cakrasaṃvara took over Siva's position at the top of Mount Meru (or Kalaish) and became manifest in twenty-four sacred (Saiva) sites scattered all over India, transforming the landscape of India into a *maṇḍala.*[10] The places not only are correlated to the ten classes of pilgrimage places[11], but they are also mapped on the human body. The twenty-four sacred sites and the associated deities therefore correspond to various parts of the microcosm of the human body. The primary support for the ritual practice is the visualisation of the *maṇḍala* and the deities within the human mind and the adept's body. But each *maṇḍala* do not only depict the inner but also the outer world and has in principle two architectonic/ geographical forms, the *maṇḍala* palace like the abode of the deity on the top of the mountain, and the offering shrine, which constitute a hierarchic ordering of the spatial and physical environment.[12]

One of the many legends to the sacred sites of Lahaul refers to Ghāṇṭapa, one of the eighty-four *mahāsiddhas*[13], who is supposed to have lived in the ninth century.[14] The oral tradition reports that Ghāṇṭapa whose name means "the one with the bell" (Drilbupa in Tibetan) lived together with a beautiful young girl, a *ḍākinī* (a "celestial woman")[15] in the Chandrabhaga Valley. Since the local people started gossiping about this relationship, the siddha became angry and threw his wife into the air. The *ḍākinī* turned into the bell-shaped mountain above the confluence of the Chandra and Bhaga River. Ghāṇṭapa himself turned into Cakrasaṃvara, flew after her and lives there with his wife until today. He converted the peak of the Drilbu ri into a *maṇḍala* of Cakrasaṃvara, by performing yab-yum[16] with his consort (Stutchbury 1991, p. 64).[17]

[10] The oral tradition of Chamba, however, reports that Śiva being disturbed by some yogis left the Drilbu ri and flew to the Manimahesh Kailash in Chamba.

[11] David B. Gray lists the ten different types of pilgrimage sites, which seems to derive form Hindu scriptual sources (Gray 2006, p. 295). Among this list we also find the sites Maru and Kulutā that might be identified as Udaipur and the confluence of Bhaga and Chandra. Both are classified as "Subsidiary Charnel Grounds", which especially for the latter is definitely true.

[12] As Toni Huber has pointed out 'one can identify several levels of reality and fundamental orders of spatial reference as the Buddhist cosmic space of the world system, the Tibetan geographic space of the trans-Himalayan zone, and the local topographic space with its specific architectonics of mandala landscape' (Huber 1999, p. 40)

[13] In the Indian and Tibetan tradition, a *siddha* or a *mahāsiddha* is a person, who by means of meditational practice has achieved great miraculous powers.

[14] Although the activities of the siddha are normally placed in the area of Nalanda and Orissa, Dowman K (1985, p. 274–275) claims that the saint was instructed to go to Uddiyana, the mystic pilgrimage centre, which is said to be the place where Padmasambhava was born.

[15] Lahaul is also called the land of *ḍākinīs.*

[16] In Tibetan yab-yum means literally "father-mother" and refers normally to the representation of a male deity in sexual union with his female consort, symbolising the union of opposites.

[17] The dualism of female and male power is particularly distinctive in Lahaul. The Chandra

The pilgrimage route around the Drilbu ri is not only very popular among the inhabitants of the Bhaga and Chandra Valleys. The crowd of worshippers can in principle by divided into three groups: the villagers of the surrounding settings (particularly from Kyelong on the opposite of the Drilbu ri), who perform their daily or weekly rites; practitioners following the regional pilgrimage circuit; and pilgrims who come to Lahaul on their trans-regional journeys.[18] The latter group consists to a large part of Buddhist pilgrims set off mainly from Tibet or Ladakh to the Swat Valley and ancient Uḍḍiyana, traversing Lahaul on their journey. The second and third group circle the peak of the Drilbu ri in connection with their visits of the sanctuaries of the small villages Triloknath and Udaipur (both in the Chandrabhaga Valley). These three sites form a trinity with the Drilbu ri in the center. The regional pilgrimage route is described in the *māhātmya* (guide book)[19] of Triloknath, collected by Francke and translated by Johannes Schubert (Schubert 1935, pp. 76–78, 127–136).[20]

23.4 The Peak Circuit Itinerary[21]

The circumambulation around the Drilbu ri normally takes place on the day before full moon. The local people affirm that one can do the *khora* in four to six hours although the duration depends upon the route and the sites visited by the pilgrim.

('moon') is normally regarded as the male, while the Bhaga represents the female power. In the legends of Lahaul this polarity is switched: Chandra is considered to be the daughter of the moon and the Bhaga the son of the Sun-god (Gill 1972, p. 15). The legend reports how the two rivers place a trail through the mountainous landscape of Lahaul to be united at the foot of the Drilbu ri. The confluence of the rivers symbolises the union of the female and the male 'which geographically represent the yogic or Tantric pratices which achieve that union within the adept's mind-body' (Stutchbury 1991, p. 60). In order to achieve this union the *yogi* is not only supposed to meditate and locate the *mandala* of twenty-four places within his own body, but also to go on pilgrimage to visit these holy places associated with the *vajrakāya* of Buddha.

18 While the regional- and sub regional-level pilgrimage seem to be specific-purpose oriented, the high-level sacred places of the supra regional character are visited largely for general purification.

19 According to the colophon, which contains the names of the author, the carver and the printer, the block print can be dated to 1905 (Schubert 1935, p. 76).

20 In the book it is explained that Drilbu ri is the place of the Buddha's Body associated with Cakrasavara, while Triloknath is the place of Speech associated with Avalokiteśvara, and Markula (the Mirkulā Devī Temple in Udaipur) is the place of Mind/ Heart of Vajravārāhī or Dorje Phagmo (Stutchbury 1991, p. 64). The three sites form a trinity with the peak of the Drilbu ri in the centre.

21 We are especially grateful to Amar Singh (Brokpa Tour and Travel in Kyelong) from the Thakur family of Barbog, who accept the burden to guide our group from Vienna (the authors together with Ewa Lewandowska) along the pilgrimage route and who showed great patience and optimism in a successful accomplishment of the circulation around the Drilbu ri. We also like to thank Tashi Karpa from Kyelong, who made our stay in the Tashi Deleg Hotel as most comfortable as possible.

Fig. 23.4. Kardang Monastery, Lahaul. V. Widorn 2007, Western Himalaya Archive Vienna, Department of History of Art, Univ. Vienna.

There is no written guidebook that describes the circuit and maps the path; the pilgrim visitor depends on the description based on oral maps or on a local guide as we did. In making the circumambulations, the adept follows the ancient traces of famous saints and *siddhas* as Ghāṇṭapa and rGod ts'an pa, who performed their Tantric rituals and miracles in this area, sanctifying it and defining the internal circles and external boundaries of the *maṇḍala* within the local landscape.

The pilgrims start their journey at the small village temple of Kardang, the ancient capital of Lahaul on the south side of the Bhaga River. There, the believer prays for an auspicious travel around the peak and offers small alms to Buddha and the deities worshipped in the small sanctuary. While turning the prayer wheels that are fixed along the outer walls of the temple, the pilgrim further visits the preserved imprints of the pilgrim rGod ts'an pa, whose historical presence in Lahaul in the thirteenth century is verifiable from his *namthar*, his travelogue (Tucci 1971, pp. 377–378). The oral tradition has it that the famous yogi travelled through the air and that he left his imprints while pursuing a *ḍākinī* in the shape of a dove.[22]

[22] The former abbot of Kardang monastery Kun-dga rinpoche (died in 1967) was highly honored as the incarnation of rGod ts'an pa. The person and reputation of Kun-dga rinpoche is not undisputed. The literature draws a quite different image of the lama, which is, however, not accepted by the local people, who highly respected him (Stutchbury 1991, pp. 83–85). However, he reported that there were many more stories and legends about rGod ts'an pa with regards to Lahaul, as well as with regard to his teachings and Tantric practices. His favorite meditation site was a small cave

Fig. 23.5. Amal Singh fixing prayer flags to a stone pile. V. Widorn 2007, Western Himalaya Archive Vienna, Department of History of Art, Univ. Vienna

The second stop is the large Kardang monastery *(Figure 23.4),* a little bit outside of the actual village, and one of the oldest and most important foundations in Lahaul. The old temple, which is affiliated with the Brug pa bKa' brgyud sect, was recently renovated and enlarged; old parts were integrated in the new structure and are hardly distinguishable. The walls also depict an image of Cakrasaṃvara.

A huge old deodar tree marks the point, from which the pilgrim – invigorated with the blessing of the worshipped deities and confidently aiming for merit and empowerment – starts climbing up the steep slope of the range. A small path leads uphill to a narrow passage that crosses the ridge at a height of 4,400 metres. At the top some stone cairn carved with mantras are piled up high with prayer-flags flapping in the wind *(Figure 23.5).* It is the first time one faces the peak of Drilbu ri and the pilgrims perform their ritual practises and prayers. People normally fix freshly printed prayer-flags, make some offerings, or burn incense at this point.

This is also the place where the inner *khora* separates from the secret *khora,* the circuit circles closest to the actual mountain top, which is usually reserved only for high-ranking monks and lamas. The latter route allows for the most intimate direct

near Sila monastery that one passes on the way down from the steep slope of the Drilbu ri. The yogi is also considered to be the founder of the rGod st'an monastery, between Kardang and Guru Ghantal on the left bank of the Bhaga River, and also of the Yoldong monastery, just on the opposite side of the river. The oral tradition claims that he used to fly between these two monasteries.

Fig. 23.6. Wooden ceiling with the Cakrasaṃvara maṇḍala of the Guru Ghantal monastery, Lahaul. V. Widorn 2004, Western Himalaya Archive Vienna, Department of History of Art, Univ. Vienna

devotion and encounter with the spiritual power, which is considered as a powerful and possibly hazardous spiritual experience.[23]

The secret *khora* leads to the small monastery of Guru Ghantal or Gandhola high above the confluence of Bhaga and Chandra. The monastery is said to go back to the times of Padmasambhava and is considered to be the oldest Buddhist foundation in this area. However, the monument has since then undergone heavy restoration efforts and structural changes. The building is now protected by a tin roof that has displaced the old pyramidal roof with wooden shingles. The wooden ceiling inside the main hall is covered with textiles that are painted with a depiction of the Cakrasaṃvara *maṇḍala (Figure 23.6).* The lantern ceiling itself is divided into several square and triangular shaped sectors, each decorated with non-figurative, geometrical representation of small *maṇḍalas,* representing the deities of the Cakrasaṃvara cosmos. The equivalents of the abstract patterns, the figurative depiction of the gods and goddesses, are shown on rectangular panels set along the wall below the ceiling. The colourful image of Cakrasaṃvara is twice as large as the other panels and placed on the east wall, corresponding to the direction of the peak of the Drilbu ri. A careful identification of the iconography of the other sixteen

[23] Elizabeth Stutchbury was invited to continue this secret circuit with the lamas accompanying her, although she was alerted that this route is not only physically dangerous because of the strenuous and stony path (Stutchbury 1991, p. 71).

panels on the south, west and north side as well the exact correlation of each deity to the symbolised *maṇḍalas* of the ceiling are still missing.

The inner *khora* for the other pilgrims and mere mortals leads either to the rGod ts'an pa cave near Sila monastery, or directly to the village of Gondhla, where the large tower of the Thakur refers to the former political power and importance of this small princedom. We had chosen the direct very steep descent to Gondhla looking at rGod ts'an pa's meditation cave only from afar. A jeep finally brought us back from the village to our starting point in Kyelong.

23.5 Conclusions

David Gray has defined that 'the maṇḍala is a complex symbol, subject to multiple deployments in discourse and practice, by which it engenders specific subjectivities within a definite social space' (Gray 2006, p. 301). We have travelled a *maṇḍala* as big as an entire region. The projection of ritual practice and worship on a local landscape sanctified by the miraculous acting and meditation of yogis and *siddhas*, has reshaped the physical space. The believers and worshippers keep this tradition alive and animate the sacred environment by walking the pilgrimage route. The Karzha *maṇḍala* in Lahaul shows the tight conflation of geographical elements (like rivers, mountain tops and passes) and spiritual sites (like the imprints of saints, meditation caves, art and artefacts or monasteries) with ritual practices, transforming a mundane landscape into a sacred one.

References

Balokhra JM (1998) The Wonderland Himachal Pradesh. New Delhi

Dowman K (1985) Masters of Mahamudra. Songs and Histories of the Eighty-Four Buddhist Siddahs. New York

Gray DB (2006) Mandala of the Self: Embodiment, Practice, and Identity Construction in the Cakrasmavara Tradition. The Journal of Religious History. Vol.30, No.3, October

Harcourt AFP (1972) The Himalayan Districts of Kooloo, Lahoul and Spiti. New Delhi

Huber T (1999) The Cult of Pure Crystal Mountain: Popular Pilgrimage and Visionary Landscape in Southeast Tibet. Oxford

Hutchinson J, Vogel JPh (1933) reprint 1994,. History of the Panjab Hill States. Vol 1,2. Lahore

Klimburg-Salter D (1994) Tucci Himalayan Archives Report, 2: The 1991 Expedition to Himachal Pradesh. East and West New Series Vol. 44 (1), 13–82

Sahni RN (1994) Lahoul, the Mystery Land in the Himalayas. New Delhi

Schubert J (1935) Der tibetische Māhātmya des Wallfahrtsplatzes Trilokināth. Artibus Asiae Vol. 5 (2–4), 76–78, 127–136

Snodgrass A (1985) The Symbolizing of the Stupa. Ithaca

Staal F (1982) The Himalayas and the Fall of Religion. In: D. E. Klimburg – Salter, The Silk Route and the Diamond Path. Esoteric Buddhist Art on the Trans-Himalayan Trade Routes. Los Angeles, 38–51

Stutchbury E (1991) Rediscovering Western Tibet. Gonpa, Chorten and the Continuity of Practice with a Tibetan Community in the Indian Himalaya. PhD Thesis, Canberra

Tobdan (1984) History & Religions of Lahul. New Delhi

Section VI: Art and Media Design

24 Sculpting Place Through Ceramic Maps

Edward L. Kinman

Longwood University, Farmville, VA, USA

Abstract

Place is a fundamental theme of geographic inquiry that has recently seen a renewed interest. As a multifarious concept, place can be a challenge to portray through maps. The influence of empirical science and new technologies on cartography can obscure meaning attached to place that is not observable. To better portray place, the role of art needs to be reconsidered in making maps. *Domain,* the collaborative artwork of a geographer and an artist, depicts the diverse meaning of place that continues to evolve in south central Virginia of the United States. Constructed of stoneware, glazes, and steel, the piece consists of panels portraying the contentious relationship between an ethnic neighborhood and a state-supported university. Collectively, the work presents multiple temporal and spatial perspectives showing property acquisition adjacent to the university.

24.1 Place

The concept of place is considered one of the fundamental themes of geographic inquiry. In the United States, understanding place was written into the National Geography Standards Project for primary and secondary education in part to provide a unifying rubric for the discipline (Boehm and Bednarz 1994, Smith 1996). At the college level, most introductory geography textbooks emphasize that all locales have certain characteristics that make them unique from any other (De Blij et al. 2007, Fellmann et al. 2007, Jordan-Bychkov and Domosh 2006, Rubenstein 2008).

While importance is attached to place, nomothetic approaches currently dominate the discipline. Technological innovations, like those associated with geographic information systems and global positioning systems (GIS/GPS), have profoundly impacted geography and have made it an integrating discipline as never before (Richardson 2004). Despite the expanding and exciting research frontiers afforded by technology, recent years have seen a resurgence of humanistic modes of inquiry within geography. The launching of the journal *Health and Place* in 1995 and subsequent publication of *Putting Health Into Place* (Kearns and Gesler 1998) are two examples of the utility of this humanistic approach to studying issues that were once only viewed as belonging to the realm of empirical science. Place, as a concept, has endured the test of time and is deeply embedded within the discourses of human geography (Earle et al. 1996).

Understanding place, however, can be a challenge. As Franke (in Jordan-Bychkov and Domosh 2003, p. 25) wrote over seventy years ago, "The essence of place derives from a creative force at work over millennia; to comprehend and capture in prose this spirit of place is worth the sweat of the noblest among us." More recently, Tuan (1974) notes the complexity of place as it involves items such as symbolism, sensation, values, and morality. He states that there is not one single type of place. One category is that of public symbol, which yields its intended meaning to those who see it because it commands attention and is meant to impress. Conversely, a field of care is a type of place requiring the inside knowledge of those who inhabit the area. Because fields of care are not easily identifiable visually, lengthy narratives are typically used to demonstrate this inside view.

Place is complicated as it is constantly changing. Sack (1997) developed a dynamic relational framework comprised of meaning, social relations, and nature to examine place. A place's character depends on the mixture it draws from each of the three realms. Since meaning, social relations, and nature rarely go unchallenged at any locale, the mixing of elements is constantly changing, which in turn, changes a place's character.

Both Tuan and Sack are geographers, but scholars outside the field have also thought about place. Smith (1996) provides an insightful review of how place has been conceptualized by diverse people throughout history and numerous disciplines with varying levels of theoretical sophistication. He concludes that these multifarious conceptions have created an ambiguity that is both useful and attractive when examining place and which rewards the reflective geographer.

24.2 Depicting Place Through Maps

Maps have the ability to communicate complex geographic information visually. Any introductory cartographic or GIS textbook will proclaim the utility of maps and spatial analysis for a variety of functions. However, communicating place through maps can be difficult; as Silberman (1999, p. 37) notes, "geographical maps are primarily about the external, physical world". Such works, while useful in many environmental and land-use applications, can mislead in their appearance of accuracy. A topographic map, for example, does not convey aspects of social relations and meaning that together combine with the landscape to constitute place. But even those objects captured on topographic maps are problematic. Wood (2006) contends maps contain signs of detachment that create a sense of objectivity, which masks the social construction of maps. A mapmaker, after all, chooses what aspects of the world to include as well as those being excluded, along with the manner in which the selected objects are illustrated. Further, maps are not representations of the milieu; rather they are propositions that actively assert what is found at a locale by the cartographer (Wood and Krygier 2007).

For maps to better depict place, a renewed emphasis on the artistic aspect of the map-making process is needed. Cartography is defined by the International Cartographic Association (ICA) as "the art, science, and technology of making such maps, together with their study as scientific documents and works of art" (Meynen 1973, p. 1). Although this definition emphasizes both science and art, contemporary geographers are arguably more likely to produce maps influenced by scientific advances in remote sensing and GIS. Harley (1989, p. 5) goes so far to say that art "is being edged off the map" and that its role is only cosmetic rather than constituting a central role in cartographic communication. Such a role relegates art to adorning or beautifying maps. Art in cartography involves much more; according to Dent (1999, p. 17), it "is the cartographer's ability to synthesize the various ingredients involved in the abstraction process into an organized whole that facilitates the communication of ideas". While I don't disagree with Dent's application of art as a skill, it misses aspects of the creativity that is better captured by Tuan (1999, p. 23) when he states "and so, if the depiction of reality is the challenge [of a map] – and by reality, I mean not only what has been or is but also what can plausibly be – then the person best equipped to meet it is one who combines the analytical intelligence of the cartographer with the detailed, multihued imagination of the artist".

24.3 Domain

With Tuan's admonition in mind, I set out to collaborate with ceramic artist John Williams to create *Domain*, which depicts the diverse meaning of place that continues to evolve in south central Virginia of the United States (Kinman and Williams 2007). Constructed of stoneware, glazes, and steel, the piece consists of four panels depicting the contentious relationship between an ethnic neighborhood and a state-supported university *(Figure 24.1)*. Collectively, the work presents multiple temporal and spatial perspectives related to property acquisition adjacent to the university. Each panel consists of four to twelve tiles, varying in size from 18 by 21 centimeters to 31 by 31 centimeters and 10 centimeters thick. We began collaborating in 2005, where over eight months we conducted numerous interviews and consulted records to learn about Longwood University's expansion. Producing

Fig. 24.1. Domain 2006, glazed stoneware with inlay and incised design; steel

the artwork required long hours pounding out clay tiles, scaling, and drawing graphic elements with exacto knives. Once dried, the tiles were bisque fired. Upon cooling, colors were inlayed into cut lines, multiple glazes were applied, and then the tiles were fired again. The finished tiles were mounted to steel panels and installed in the Chichester Science Center at Longwood University and are part of the Longwood Center for the Visual Arts permanent collection.

The center panel of this work maps in 20-year increments how Longwood expanded from its tiny niche on High Street to occupy an area 20 times larger than in 1900. The outer panels of *Domain* encode place from diverse perspectives. In the mid-1960s a decision was made to expand the college grounds. Land south and west of campus, occupied predominantly by African Americans, was targeted for acquisition. These residents, trying to preserve place, contested Longwood's plans. As mapped on the right panel, place experienced by this community was rooted in their homes. The left panel depicts a different view, that of the institution's, portraying the same area as a patchwork of properties devoid of people. Here the task of purchasing so many parcels is viewed in economic terms; the different colors represent the affordability of the properties based on assessed values.

To better appreciate *Domain,* additional background information is needed to understand the complex nature of place we wanted to capture. When I moved to the area in 2002, Longwood's landscape was going through considerable change. A devastating fire had destroyed two campus buildings, one clearly representing place as a public symbol. Upon reading postings on a university web site written by alumni, I quickly realized the fire resulted in a loss of place typified as a field of care. Entries recalled how Ruffner Hall, Longwood's oldest academic building, had caused many students to realize their connection to the school's past every time they stepped onto the worn front step. By helping to continually wear away the slate step, students were made to feel part of the fabric of the institution by thinking of those who had preceded them. So powerful was this field of care that when Ruffner was rebuilt, the institution not only followed the original architectural plans, the front step was honed to replicate the worn slate that was destroyed in the fire to reconstruct place.

24.3.1 But Whose Place Is it?

Most locales have numerous parties who interact with the same landscape. Area residents who never took classes in Ruffner do not have the same interactions as Longwood students. Their sense of place is different, which became very apparent at a community meeting to discuss the town of Farmville's offer to give a parcel of land to build a new library. The African-American community protested the proposed site as it was in front of a predominantly African-American cemetery, a sacred space. The well-attended public hearing resulted in a passionate defense of their place, to preserve the area adjacent to the cemetery and their community in general. Many references were made to white insensitivity by the town council and library board. Some even stated this proposal was part of a long history of discrimination rooted in the area.

It should be noted that Farmville's race relations, and to an extent how they have contributed to place, have been examined periodically by a number of people. William E. B. Du Bois in 1898, after becoming the first African American to graduate from Harvard University with a doctorate, conducted a study on the social conditions of African Americans in the area for the U.S. Department of Labor. Du Bois (1898, p.21) documented the dissatisfaction of African-American women in domestic service for the white community, stating that the women "are coming to regard the work as a relic of slavery and as degrading, and only enter it from sheer necessity". There was one exception to the misery just described, a small community one mile west of Farmville, called Israel Hill. Racial antagonism was at a minimum as there were no other whites living in the area, which was originally populated in 1810 by emancipated slaves on land that was willed to them by their deceased master (Ely 2004). Du Bois's work develops fields of care for the various black communities in the Farmville area, and goes into considerable detail describing the role churches and secret benevolent societies played in creating place, one of which was responsible for creating and maintaining the cemetery where the town was offering the adjacent land to build the new public library.

One additional example of race relations in the area should be mentioned. In April 1951, a student-led strike at the Moton High School in Farmville set in motion events that Don Baker (2001, p. 10) of the *Washington Post* claimed "forever changed the landscape of American education, and arguably marked the start of the modern civil rights movement in the United States". The National Association for the Advancement of Colored People learned of the student actions and brought legal suit against the county school board. Subsequently this case was successfully argued before the United States Supreme Court with four others under the title of Brown versus Board of Education of Topeka. Despite the ruling, massive resistance to court-ordered desegregation continued in the county, resulting in the closing of public schools from 1959 to 1964.

So the pleas to preserve place regarding the plot of land in front of the African-American cemetery were part of something much larger. Perhaps not unexpectedly, further discussions with members of the African-American community uncovered distrust with our affiliated institution, as the university has acquired over 100 properties of the adjoining ethnic neighborhood over a forty-year period, a practice that continues to this day. In the process, most of the houses were torn down and land cleared for new buildings and parking lots. How these properties were purchased troubled some informants, while all lamented the loss of their homes that represented a topophilia for the neighborhood they once knew. The lower right panel portrays a community being pulled apart *(Figure 24.2).* Tiles recording the footprints of homes are separated. The red inlaid lines lack precision compared to those on the other panels, to not only convey the vernacular nature of the architecture but also suggest vulnerability and insecurity. The green glaze further reinforces the neighborhood's precarious situation with the thousands of cracks that develop during firing. By contrast, the top right panel contains campus buildings placed next to each other. The composite image, while smaller than the lower panel, suggests the unity, order, and power of an institution in control of its destiny.

Interviews of current and former administrators, by contrast, revealed different opinions about the land acquisition and campus expansion. Words such as "vision," "strategic directives," and "growth" reflected the views of university officials. These individuals, while not oblivious to the plight of the ethnic neighborhood, viewed the benefits associated with the campus expansion as outweighing the costs to local residents. They contested the idea that this area was suitable for housing, pointing out how heavy traffic going through the neighborhood was dangerous for children and that most of the buildings were substandard. Officials always stated the money being offered for the properties was more than their appraised market value and the right of eminent domain was not being used. To emphasize the disconnection with residents' sense of belonging, the left panel is devoid of elements suggesting fields of care by illustrating only streets and land parcels *(Figure 24.3).* As such it facilitates rational economic decisions based solely on property values. This representation is akin to notions of place offered by the noted economist Adam Smith, whose conceptualization of place is reduced to marketplaces best illustrated as land consisting of "terrain of fluctuating prices for labor, stock (capital), and rent" (Smith 1996, p. 198). After all, how can one place a price on community and sense of belonging? Detachment was further achieved by classifying property values with the Fisher-Jenks algorithm with the intention to place parcels in the same bin based on the position of property values along a number line (Slocum et al. 2009).

In addition to the four panels, the artists' statement plays an integral part of *Domain*, acting as a legend for the entire work. It begins by informing how landscape is transient, always changing. Multiple examples of former uses of the land now part of Longwood are described. The statement then explains how over time

Fig. 24.2. Domain right panels, lower panel portraying homes in the adjoining ethnic neighborhood targeted for acquisitions by Longwood. The upper panel illustrates Longwood's campus in the mid-1960s.

people interact with a landscape and place develops. As places evolve, they become unique and distinct. They also bear diverse meanings depending on the type of interactions people have with the same landscape, as exemplified by the diverse perspectives of the two outer panels. Finally, the statement speaks of the collaboration process of a geographer and artist working together to produce *Domain,* including a reference to the definition of cartography as a product of art, science, and technology.

Since its installation in 2006, *Domain* has created diverse responses. Before the title and artists' statement were attached, colleagues in biology and chemistry speculated that the center panel represented the development of a microbial organism.

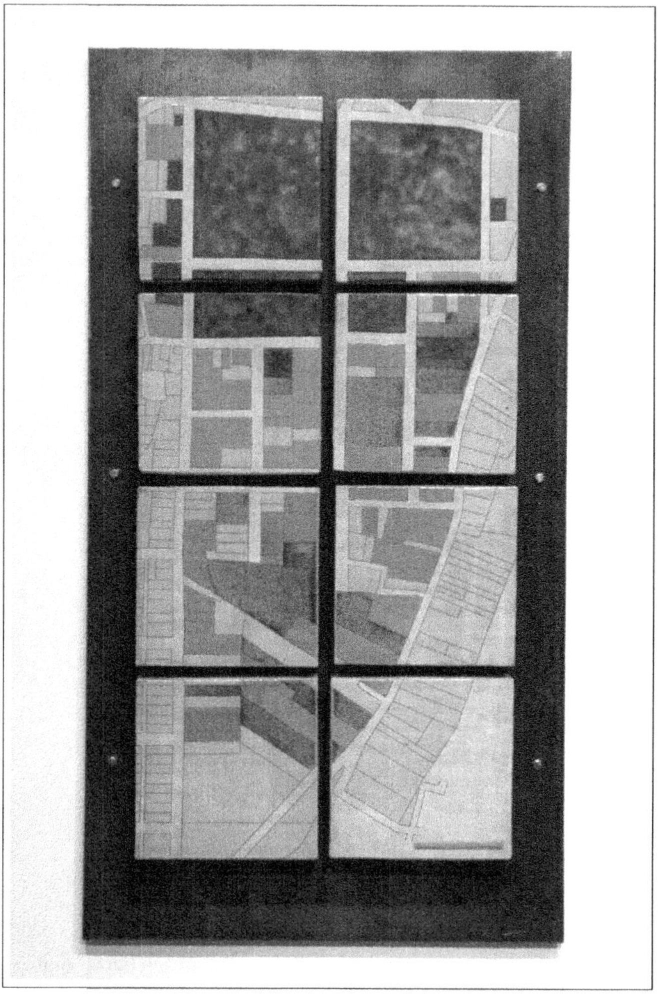

Fig. 24.3. Domain left panel contains land parcel valuations for adjoining properties

The outer panels, by contrast, were difficult for them to interpret. Although familiar landmarks are depicted, the work lacked the expected interpretive guides in the form of labels and legends they expect and are familiar with. Even with an explanation, many were uncomfortable with a work presenting multiple realities when they are used to pursuing and teaching absolutes.

Of course, *Domain* was designed for a broader audience. It is intended to make people reflect on the landscape in which it is embedded, the Longwood campus and the town of Farmville. Students, as they have encountered *Domain,* have found the work makes them reflect on what they thought was familiar. Many say they had no idea of the diversity of prior land uses for a landscape that is obviously designed for their use now. Some express anger over the perceived arrogance of those in power to grab land for their intended purposes. They have expressed an interesting form of appreciation for opening their eyes to what already existed, but was hidden from view. Most often, students speak of empathy for those who lost their homes.

A number of community residents thanked us for speaking loudly about the eradication of place. Some former residents, who used to live in houses that have now been razed, have sought us out to recount stories about their former neighborhood, sometimes with teary eyes. They have taken pictures of the artwork and a few have sent prints to family and friends who have moved away. While *Domain* was never intended as a memorial, certain community members view it that way and it has become a public symbol. As such, place is again being constructed as *Domain* organizes space into a center of meaning (Tuan 1974). More importantly, this reaction by those most affected by Longwood's expansion demonstrates that map art can be a medium to communicate place.

24.4 Conclusion

In summary, the geographic concept of place is challenging to capture and portray. While science and technology have assisted in creating maps of the external, observable world, they lack the artistic creativity to portray the complex social relations and meanings associated with place. *Domain,* the product of a collaborative experience of a cartographer and artist, took the form of multiple panels to help disentangle the complex social relations and means that are part of place for land now called Longwood. Our proposition is that both of these realities are true to different groups of people based on their sense of place. While some may argue that polarizing positions are being anchored on either side of a work that distorts historic events, our role is to actively assert what is there (Wood and Krygier 2007). We sought to give equal weight to both perspectives, insuring the softer voice would not go unheard.

References

Baker D (2001) Closed. Washington Post, 4 March, 10

Boehm R, Bednarz S (1994) Geography for life: National geography standards. Washington, DC: National Council for Geographic Education

Dent B (1999) Cartography: Thematic map design, 5th ed. New York: McGraw Hill

De Blij H, Murphy A, Fouberg E (2007) Human geography: People, place and culture. Hoboken, NJ: John Wiley and Sons

Du Bois WEB (1898) The negroes of Farmville, Virginia: A social study. Bulletin of the Department of Labor, 14, 1–38

Earle C, Mathewson K, Kenzer M (1996) Back to basics: The virtues of key concepts in human geography. In Earle C, Mathewson K, Kenzer M (eds), Concepts in Human Geography, pp. xi-xxv). Lanham, MD: Rowman and Littlefield

Ely M (2004) Israel on the Appomattox: A southern experiment in black freedom from the 1790s through the Civil War. New York: Knopf Publishing

Fellmann J, Getis A, Getis J (2007) Human geography: Landscapes of human activities. New York: McGraw Hill

Harley J (1989) Deconstructing the map. Cartographica, 26(2), 1–20

Jordan-Bychkov J, Domosh M (2003) The human mosaic: A thematic introduction to cultural geography, 9th ed. New York: W.H. Freeman and Company

Kearns R, Gesler W (eds) (1998) Putting health into place: Landscape, identity, and well-being. Syracuse: Syracuse University Press

Kinman E, Williams J (2007) Domain: collaborating with clay and cartography. Cultural Geographies, 14(3), 433–444

Meynen E (ed) (1973) Multilingual dictionary of technical terms in cartography. International Cartographic Association, Commission II. Wiesbaden: Franz Steiner Verlag

Richardson D (2004) Forward. In Brunn S, Cutter S, Harrington J (eds), Geography and technology (pp. xi-xii). Dordrecht: Kluwer Academic Publishers

Rubenstein J (2008) An introduction to human geography: The cultural landscape. Upper Saddle River, NJ: Pearson Prentice Hall

Sack R (1997) Homo geographicus. Baltimore: Johns Hopkins University Press

Silberman R (ed) (1999) World views: Maps and art. Minneapolis: University of Minnesota Press

Slocum T, McMaster R, Kessler F, Howard H (2009) Thematic cartography and visualization, 3rd ed. Upper Saddle River NJ: Prentice Hall

Smith J (1996) Ramifications of region and senses of place. In Earle C, Mathewson K, Kenzer M (eds), Concepts in Human Geography (pp. 189–211). Lanham, MD: Rowman and Littlefield

Tuan Y (1974) Space and place: Humanistic perspective. Progress in Human Geography, 6, 233–246

Tuan Y (1999) Maps and art: Identity and utopia. In Silberman R (ed), World views: Maps and art. Minneapolis: University of Minnesota Press

Wood D (2006) Map art. Cartographic Perspectives, 53, 5–14

Wood D, Krygier J (2007) Ce n'est pas le monde. Paper given at Geography and the Humanities Symposium, Charlottesville, VA, 23 June

25 The Map of Fashion

Laurene Vaughan

School of Applied Communication, RMIT University, Melbourne, Australia

Abstract

The relationship between the body, fashion, the map and the making of place is a rich area of exploration by a variety of creative practitioners, anthropologists and theorists. This essay discusses the proposition that fashion, or the clothed body, is the 'stuff' of the inhabited landscape. Conceived of a map of fashion, this map documents and communicates an embodied landscape through a topography that lives, breathes, pulsates and evolves through the everyday actions of those that are there. The 'place' of this landscape is both here, and there. It is a landscape that transitions as the individual moves through the various locations of their lives. It is a landscape of the collective and the singular. My clothes mark and identify me, our clothes create an individual collective; a landscape of colour, shape and form, that is positioned within the constructed and natural world. Through the *Map of Fashion* this paper and the creative artefact that it discusses, explores how every garment is an intimate map of the place making of lives lived.

25.1 Introduction

The fundamental proposition of this essay is that fashion, and in particular the clothed body is the 'stuff' of the inhabited landscape; it is the *Map of Fashion*. This map documents and communicates an embodied landscape through a topography that lives, breathes, pulsates and evolves through the everyday actions of those that inhabit it. This map presents a landscape that transitions and evolves as the individual moves through the various locations and events of their lives. The form of this map cannot be found as an image on paper, in a book or as a digital representation. The *Map of Fashion* exists in the clothes of our wardrobes or the shirts on our backs. The 'place' that the landscape of this map represents is both here, and there. It is a landscape of the collective and the singular. My clothes mark and identify me; our clothes create a collective of individuals. Together we create a landscape of colour, shape and form that is positioned within the constructed and natural world.

The relationship between the body, fashion, the map and the making of place is a rich area of exploration by a variety of creative practitioners, anthropologists and theorists. There is an intimacy in the construction and understanding of place, an inherent connection between the individual and the location. Locating and communicating this connection is more than a science of markers on a grid; it is an art or aesthetic practice that calls for the poetic in the known and the unknown. Many references to the relationship between the map and the body focus on the tattooed body; a body which is intimately and painfully marked with records of travel or other life experiences (e.g. sailors, soldiers and college pranks). In other texts we see the evolution of fashion and textiles as a means for representing the passages of time, trade and technology, or specific socio-cultural groupings: for example there are the histories of the silk-road, the evolution of tartan, or the migrations of people from one land to another. This is a particular way to understand and use the *Map of Fashion;* this essay and the creative object that it discusses, moves beyond this and explores how every garment is an intimate map of the place making of lives we live.

25.2 Site for Transformation

The *Map of Fashion* refers to a collection of work that has been evolving since 2003. One object that has been created within this collection and which facilitates this discussion about fashion's relationship to cartography is entitled, *Site for Transformation.* This is an object with two forms, it is both a hand felted, embroidered wrap for the body, and a map of memories, place and associations of a specific period in my life – Japan 1986–1989. *Site for Transformation* provides a subtle and multi-layered narrative that tells tales and hides details depending on where you are and what you know. It is a map of time, as well as being a map of meaning and of associations. Know me, and you potentially will know more about the map.

Site for Transformation is, what we might call, a hyper-garment. It is a functional garment, made using some of the oldest and most traditional techniques for garment construction and decoration (felt, appliqué and embroidery). If worn, it would function like any other body wrap; it would keep you warm, it is light and comfortable when placed on the body and, it is aesthetically pleasing even though it was not designed for comfort or beauty. Although all these characteristics are true, this wrap was not created with this end use in mind. *Site for Transformation* is a hyper-garment that was created as an extreme example of the meaning of fashion, and the subtle nuances of the discourses of ownership and wear. It is a collective rather than a singular item. It is both itself and it represents the many items of our wardrobes.

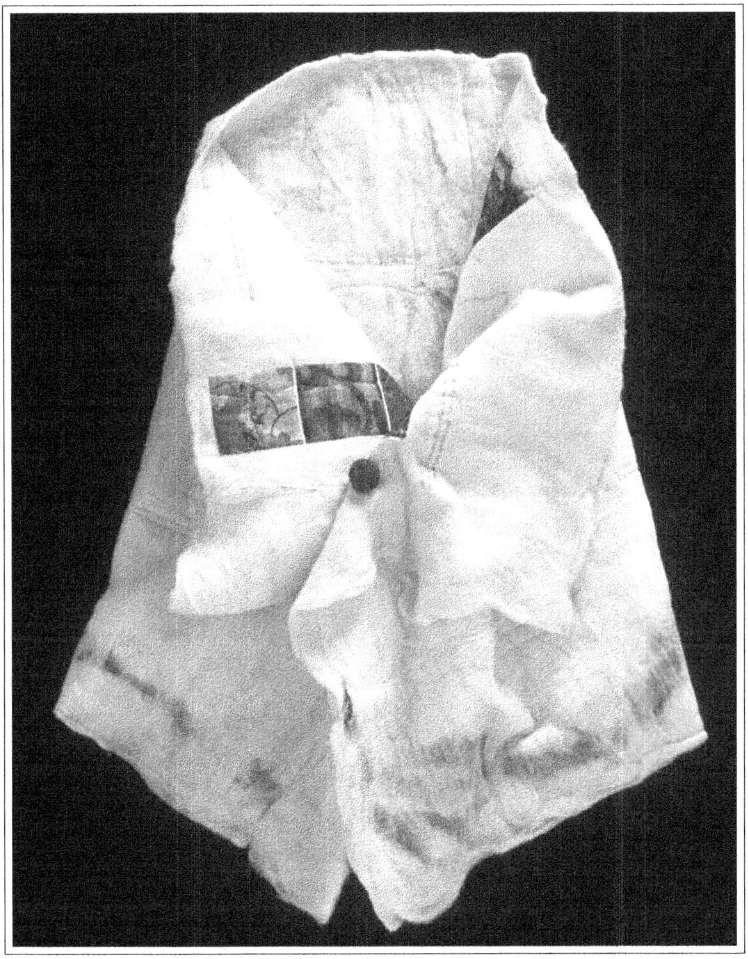

Fig. 25.1. Site for Transformation

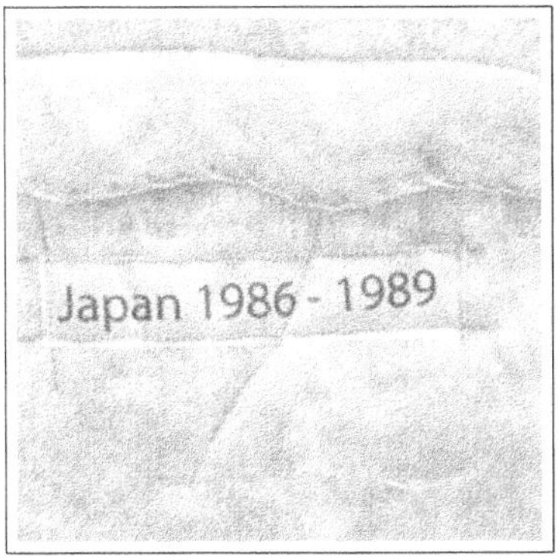

Fig. 25.2. Detail

Site for Transformation has two faces: one public and the other private. Simplistically these are what are conventionally known as the inner and outer sides of the garment. Within the terminology of the making of garments these are referred to as the right and the wrong sides of the item. Although the garment is usually understood as a whole (e.g. my jacket, sweater or wrap) it is also accepted that this whole has parts or sides (e.g. sleeve, pocket or lining). This multiplicity of understanding is integral to the ways in which we engage with the *Map of Fashion* and *Site for Transformation* in particular.

The public outer face of this hyper-garment is cream dusted with a red on the border. At a distance this is all you can see. As you move closer you begin to notice faint lines and subtle stitch marks. These marks are the visible traces of a secret

Fig. 25.3. Public Face

Fig. 25.4. Public Inner

inner world, one that viewer may not know unless they know the garment or the wearer. In order to see what lies beneath you would have to ask them to show you, and tell you, the garments tale.

In contrast, the private inner side of the garment is marked with images, text and stitches; each a reference and a placeholder of relationships, experiences and memory. Their meaning is unique to the owner/wearer of the garment. This is not a garment that was mass-produced; its production is within the tradition of couture and bespoke production. The artisans handiwork is visible in the final form. This hyper-garment was made for someone, but not as a garment to be worn but as an account of what has been, somewhere at sometime. This is a personal narrative of place and travel captured in cloth.

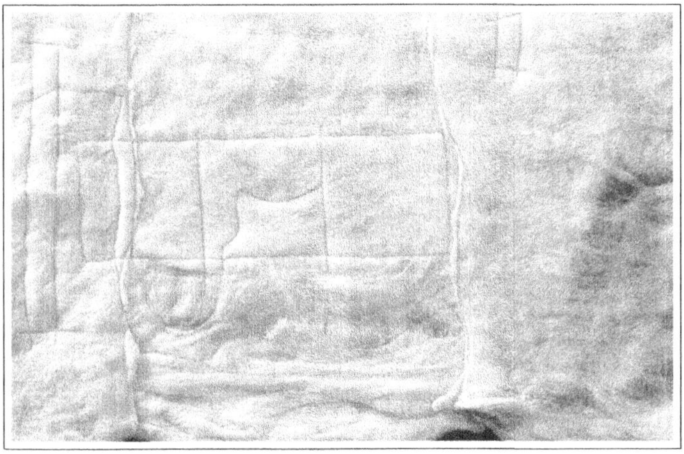

Fig. 25.5. Site for Transformation, detail

Fig. 25.6. Site for Transformation, detail

Fig. 25.7. Site for Transformation, detail

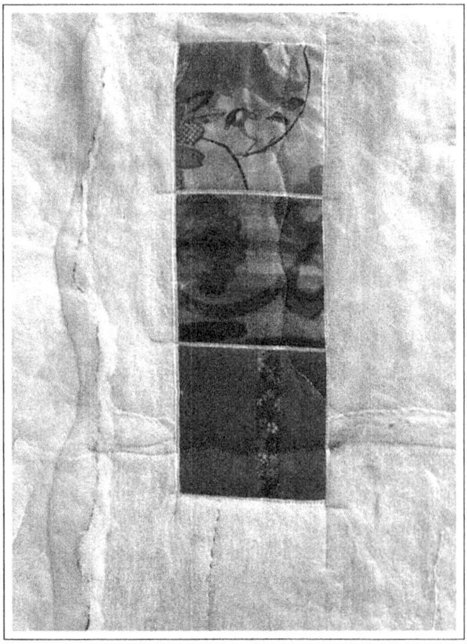

Fig. 25.8. Site for Transformation, detail

Fig. 25.9. Site for Transformation, detail

25.3 Illusion, Perception, Subjectivity: Narrative

There is a commonly held belief or assumption made by map-readers that a map represents the 'truth' about something and somewhere; there is a belief that, what is represented on a map is real or actual and objective. However as Turchi observes, '(c)artographers must continually confront the fact that there is no such thing as objective representation' (2004, p. 74). The very fact that a map is a representation of something or somewhere opens the possibility of interpretation and modification. The representation of the three- dimensional world on a two-dimensional plane necessitates a 'certain adjustment' (Ptolemy in Turchi p. 75) if it is to occur. We (the map-readers) accept and most often are ignorant of this adjustment to the 'truth,' in fact this adjustment may be necessary for us to be able to engage with the map at all. In this way the cartographer like the author, artist or designer creates for us a 'world' in which to live and we take on this world as our own. Their truth becomes our truth, and with time we come to own it for ourselves and create our own interpretation of it.

Fashion has many definitions; fashion is transition, fashion it is temporal, and to fashion something is to make or create. Fashion is the narrative of life, style and evolution; it reflects political, technological and cultural development. Fashion in the form of garments is intimate and public, individual and collective. In its various forms it is owned and worn by each of us, placed against our skin, it is a marker and a maker of our identity. As a collective we create fashion and styles, creating a communal, even global outer appearance, yet for every individual within the collective, the meaning and experience of each item is unique. The proposition of *Site for Transformation* is one of mapping and map creation and map-reading. Yet this is not a map of propositional directions, one that shows us where to go or how to get there. *Site for Transformation* is a map of knowing about ourselves, and others, as deeply as we may be able to know through fashion or any representation. The direction of this map is inwards, it may build a bridge between owner and viewrer, but the real objective is to capture the narrative of what is and the owner's experience of getting and being there. As such it is reminiscent of an explorers map, marked with discoveries, realisations and places of significance.

In some ways *Site for Transformation* connects to the *Memory Maps* of Jake Barton and Nancy Nowacek (Krygier 2006). Barton and Nowacek created these maps as a means to engage people in develop a collective narrative of place. These were first developed as part of the annual 'Folklife Festival' in Washington DC in 2001. At this event, people were invited to post notes on a large map of New York, recounting their experiences and locations of meaning within that city. In this case the individual contributes to the shared narrative of the collective, their experiences and the associated meaning of places becomes embedded with the experiences of others. The individual is acknowledged but it is the collective meaning of a city that is being sought. The device for doing this is the map of the city. The tale of the collective is laid onto the predetermined boundaries of the set location as marked by the lines on the map. These are memory maps of community and diversity. *Site for Transformation* is a far more personal map of memory. There are no city or place limits that determine the shape or form of the location being recounted. The locations, places and streets that are being captured within this map are free to associations and subjectivity just as the connections of memories are. On its private/inner side this is not a map of public declaration available for all to see, that is, unless I choose to turn my personal representation (the garments of my identity) inside out and declare it to the world.

Site for Transformation is an historical phenomenological map of actions, experience and embodied knowing. It is a fashion object, which is by its very nature temporal. Like many maps it is a representation of the placed narrative of time and experience.

As the philospher Jeff Malpas argues, '… just as time and space are properly understood only together as interconnected forms of dimensionality, so, too, is narrative properly understood as connected, not to time alone, but also to space. Just as narrative provides a means to grasp and to articulate the unity of space or region, so, too, is narrative of itself necessarily worked out in relation to such spatial and topographic structures' (1999, p. 87). This is the challenge of this object – to explore through the narrative of image, texture and form, the connection between individual, time and place.

There are many ways to experience and capture the narratives of humanity. There are conversations, novels, films, telephone texts, discussion boards, diagrams, images and email. The form, temporality and connectedness of these conversations varies; some are synchronous, others historic, some may be face to face, one to one, others may be between a group (a discussion), or one way in the form of the telling of a tale or giving a lecture. *Site for Transformation* represents an historic narrative that is constantly in the making. It is a conversation between the object and myself (the wearer). Sometimes this conversation opens up to include another should I invite them in.

Site for Transformation is a fashion map that captures the narrative of individual experience. In this way, just as the novel (one of the ultimate forms of narrative) has its origins in the travelogue then the garment too, can be seen to be a travelogue of a journey or the many journeys of a life. It is a travelogue that is an articulation of spatial and topographic experiences of life.

There are many ways to capture and articulate the maps of experience. One-way is through the marking of trajectories of travel on maps: for example, marking pages in street directories or tracing paths on a tourist map. This image of a trace of one days walking in the streets of Tokyo on a tourist map is one way for me to record and share my travels in a foreign yet familiar city. So what makes it different to *Site*

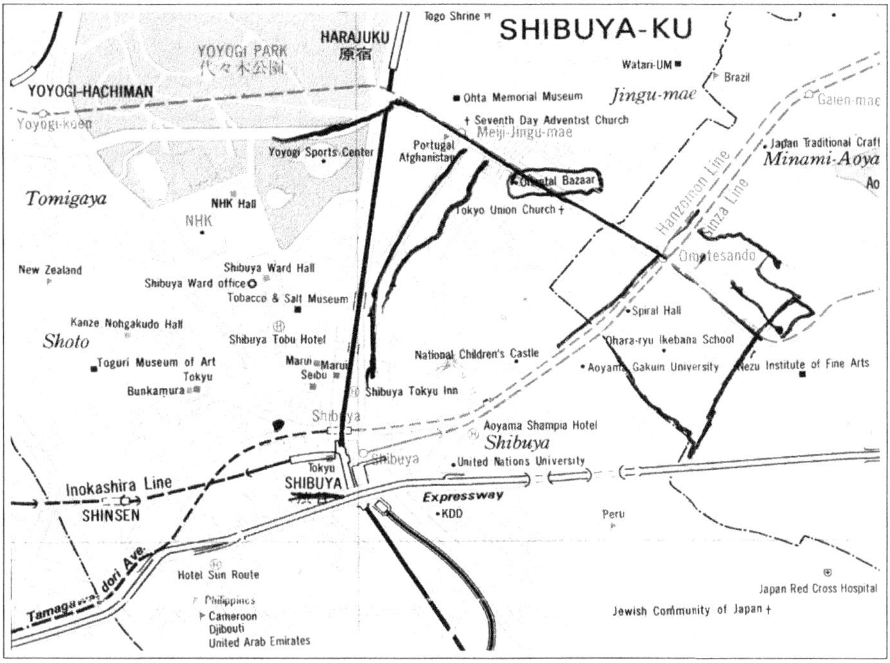

Fig. 25.10. Map of Tokyo

for Transformation? Surely it is an easier way to create a record of a city, and both are accounts of my time in Japan. I could add images and annotate the map. I could create a personal travelogue version of a *Memory Map.* It could be argued that such a map has the same meaning as *Site for Transformation* but just in a different form. And it is the difference of form, which is fundamental. As a body wrap, *Site for Transformation* relates directly to the body of being. The historic narrative with its layers of personal interpretation presses against the body. As a hyper-garment, this wrap is a model and metaphor for the meaning that gets embedded in the everyday actions, thoughts and other musings of our beings. Fashion, more than any other designed object not attached to the skin, has the greatest connection to the nature of our being. The map of *Site for Transformation* is as much in the object as it is in the owner; it is the true map of memory captured in cloth. The public face is like the tourist map that lives on the shelf, a little tattered and used, it suggests by its presence and its signs of use, that the map represents a place that the owner has been too. The body wrap is not about the external map that others can see; it is about the narrative and the influence on the owner as a result of having been. It is a map of transience and influence. Maps don't just tell us things or direct us to places; they enable experiences, which are ongoing in our lives even when we leave. Made from cloth and placed against the body, *Site for Transformation* acknowledges the fragility of our connections to places and our need to capture or record them.

> *My spaces are fragile: time is gong to wear them away, to destroy them. Nothing will any longer resemble what was, my memories will betray me, oblivion will infiltrate my memory, I shall look at old yellowing photographs with broken edges without recognising them.*
>
> *Space melts like sand running through one's fingers. Time bears it away and leaves me only shapeless shreds.*
>
> *To write: to try meticulously to retain something, to cause something to survive, to wrest a few precise scraps from the void as it grows, to leave somewhere a furrow, a trace, a mark or a few signs.* (Perec 1997, p. 91)

The cloth and thread details of *Site for Transformation* should not be seen as decoration or ornament. They are not a way of embellishing an object in order to create a notion of beauty or to fill in gaps for unknown places. Rather, they are, as George Perec proposes, aesthetic marks of time and life, points of time captured before they become 'shapeless threads'.

In this instance this map has been deliberately created, the wrap is an historical record. But what does this mean beyond the hyper-garment *Site for Transformation?*

Designers and academics have for some time, explored the idea that every garment has semiotic meaning. That each item is a representation of time, place and social standing. Each garment in our wardrobes (when worn on the street, when it sides up beside another), creates a mobile contemporary living, breathing map of

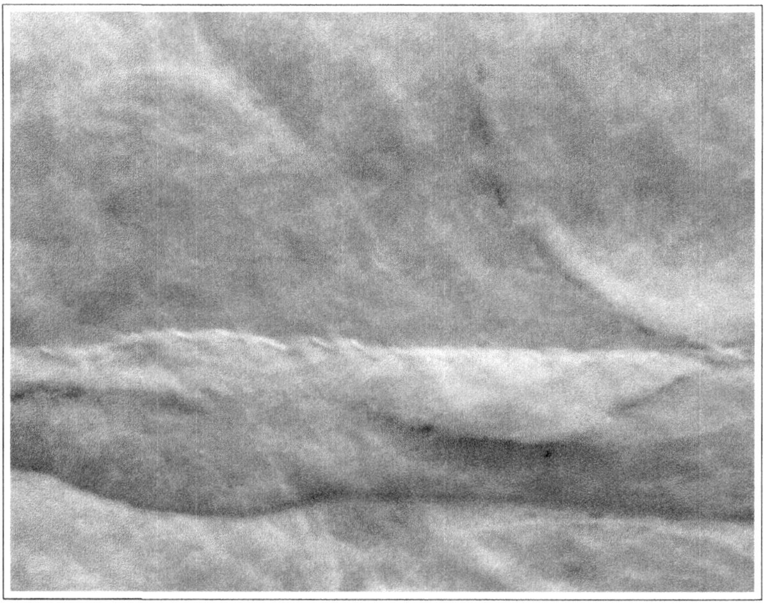

Fig. 25.11. Site for Transformation, detail

social identity. This perspective is important for two different aspects of the *Site for Transformation,* the individual and the collective.

Site for Transformation is but one articulation of the *Map of Fashion*. It is a specific piece that refers to the phenomenological connections between object and individual wearer. It is a map that is drawn through the cartographic movements of everyday life. These movements are realised through the pressings of the body against cloth, they are also the associations created through occasions of wear. In this way this map has connection to the tattooed maps on people's skin. These maps extend the garment from being an external object to being a covering and a narrative placed directly against the skin. For some, the tattoo garment is but a one off comment, a single statement or decoration. For others it is a lifelong narrative, an evolving pictorial landscape of aesthetics, associations and artistry. This is a connected yet different to the proposition of *Site for Transformation*. The tattooed garment is a permanent map that, without much pain and effort, cannot be changed. The map of *Site for Transformation* accepts the transience of connection between the body and object. The fragility of cloth reminds us that the object may hold the memory, a memory that has personal meaning, but it does not hold all memories of all occasions. There are multiple garments, multiple maps, wardrobes of experiences and meanings.

The collective *Map of Fashion* is the map of associations that each garment of everyday life makes within the street and in our wardrobes. This is not the hyper-garment of *Site for Transformation,* with its particular meanings, and yet it is. For it is the narrative of each of the garments with their embodied marks placed together in relation to each other. In this way the hyper-garment, leads us to contemplate the meaning of one and all of the items of identity. What they mean alone, for us and in relation to others.

25.4 Conclusions

It is proposed that through this discussion of one particular object, we are able to engage in a discourse about the collective. If every garment contains meaning (whether it be less literal, or as abstract as this particular piece) then how rich are the streets of the city, the stadium or the mall? The inhabited landscape, typically is a clothed landscape made of cloth or skin, marked with image, line, text and texture. This symphony of communication methods results in a dynamic and evolving topography that has multiple viewing points. Through this discussion of the object of fashion as a map unto itself (an intimate map created through its relationship to the body, moulded through use and context), new conversations and possibilities about the relationship between the self, place, narrative, cartography and the body emerge. *Site for Transformation* leads us to see the *Map of Fashion,* a map that we engage with consciously or not.

References

Krygier J (2006) Jake Barton's Performance Maps: An Essay, cartographic perspectives, Number 53, Winter, pp. 41–50

Malpas JE (1999) Place and Experience: A Philosophical Topography, Oxford University Press, London

Perec G (1997) Species of Spaces and Other Pieces, Penguin Books, London

Turchi P (2004) Maps of the Imagination: The Writer as Cartographer, Trinity University Press, Texas

26 Maps of What Might Be: A Dozen Works on Ideas and Possibilities

Peter Downton

School of Architecture and Design, RMIT University, Melbourne, Australia

Abstract

The aim here is to investigate some of the relations between ways of ordering and representing knowledge in two fields – architecture and cartography. This is undertaken by examining some of the ideas generated in the making of 12 small model pavilions. I call them 'epistemological pavilions' as they are an effort to map knowledge of space and place in a physical form.

As with cartographic maps these models embody knowledge. In this case their production is not often concerned with embodying knowledge of extant places, although sometimes they endeavor to illuminate relationships and conditions of built objects. The pavilion models embody the knowledge of their designer-maker, but do so in a projective manner, as designing concerns what might be, in the context of what is. These models thus occupy a territory that has overlaps with paper maps; they can also be read as speculative maps of what could be. Selected issues dealt with in the models that resonate with issues in mapping are explored in an effort to relate the two knowledge domains.

26.1 Introduction

An explorer bends over a table to consult a map. A viewer peers at a small model in a museum display case. Both explorer and viewer physically look down; hopefully neither is figuratively looking down on 'just a map' or 'just a model', for each of these items is a complex work of epistemology and practice. The explorer looks down upon an area of paper that both relies on, and embodies, rich constructed knowledges of measurement, discernments of differences, and partitionings of the world, for a map is a way of ordering a complex of scientific, technical, social and political endeavors in a systematised representation. The model embodies similar realms of knowledge and is evidence of its maker's designerly knowledge (Cross 1982, 2000). Both mappings and models are mechanisms for ordering and explanation, but they can also function as aids to exploration, projection and speculation for both their makers and for others.

In 1996 I began building a series of small architectural models termed 'epistemological pavilions'. Currently there are twelve of them and each would fit on an A4 size sheet. I fill two roles in this undertaking: I am the designer-maker and also a researcher-reporter who analyses and reflects on this enterprise and reports the outcomes of different understandings of my own investigations through making. The project is ongoing and has enabled the study of researching through designing. The models have drawn on my background in architecture including past research into, and study of, placemaking and cognitive mapping. Each pavilion is seen as a work in its own right, but also as a vehicle for conducting a larger study of design epistemology with a focus on the roles and nature of models in inquiry. This has naturally centred on architectural inquiry – on the ways physical (and computer-based) models of intended physical outcomes are used in architecture. This research program has recently become concerned with the relations between precedents such as extant pieces of architecture or everyday objects in the world and the constructed forms in the models. How are they connected? Are elements in the models simply distorted representations of the originals? Is the process of getting from an original to a model one of translation that entails a particular mapping? Or does the translation involve many micro-mappings that are each shaped by judgment? Grappling with such questions brings the modelling research to the domain of mapping.

The concept of 'mapping' employed here fundamentally entails selecting a set of elements and the relations between them and then mirroring or representing them by a second set of elements and relations. A mapping is required to establish this second set; this set is a map.

With this in mind, there are at least two ways in which to understand the title 'Maps of What Might Be'. It could refer to maps of things that *could* be built, where models are understood as mappings giving orderings of knowledge that, in this case, would enable construction. Alternatively, the title may allude to maps of what might, may or could be the case – that is speculations, explorations of some ideas and the raising of questions or possibilities. I intend both readings, but will emphasize the second in this, the latest in a number of examinations of the models. Elsewhere I have considered them for audiences mostly in architecture and other design disciplines and focused predominantly on ideas concerned with design research (Downton 2004, 2007a, 2007b).

The aim here is to investigate some of the relations between ways of ordering and representing knowledge in two fields – architecture and cartography. In the sections that follow I want to select a number of the models and examine some ideas that appear in their making. My principle focus is on mapping in a generic sense; the hope is that by considering issues of interest that arise in scrutinizing my modelling research, I can indicate some parallels and differences between the two fields and provoke some new understandings.

26.2 Models and Knowledges

My group of models can be read as embodying the various knowledges of their maker and thus they potentially encompass knowledge ranging from anything accumulated over my lifetime, through historical, theoretical and practice knowledge of architecture and design, to knowledge about how to use and fashion various materials at a small scale. On interrogation the models reveal this knowledge; they can, and do, intentionally endeavour (with varying success) to represent items and aspects of knowledge explicitly drawn from architecture and other areas. There is evidence of mappings between diverse knowledge domains and the knowledge space of these models.

Any more-or-less completed design for a building is likely to be represented as a set of drawings, by computer models, and through other standardized representational systems. Its designers may well have used rough sketches and quick physical models as part of their designing process. When most decisions have been taken there is often a public revelation using a detailed presentation model. This is an accurate representation of what is intended by the designers. Effort will then be expended to make the final outcome very like this model. When the building is finished, the model will thus be a map of the building: aspects of it will represent the building, materials are likely to differ, but proportion will be maintained between the reduced scale of the model and the full-scale building. We can look back and say that the model was a map of what would be.

The models I have made operate differently, as the pavilions are not intended to be constructed as full-scale buildings. In many instances they could, with appropriate adjustments and additions, be expanded to become habitable structures and in this sense my works are not markedly different from presentation models. Their fundamental difference lies in the fact that there is little designing undertaken in my models away from the evolving models themselves – they do not represent what has been already designed; the designing and the modelling are one. Inevitably, I think about them elsewhere (predominantly on public transport) and sometimes record these thoughts through drawing, but mostly designing and making are melded into one process. As with presentation models, mine are, when finished, repositories of the decisions made in their designing; they record the processes of their making. Specifically in terms of the means of making the design decisions, they are like any sketches or sketch models made in usual architectural design processes. However, instead of being rapidly constructed from card or balsa wood, the pavilion models are produced slowly from timbers, copper, aluminium, plastics and a range of possibly surprising reclaimed objects and materials. Each took months not minutes for their construction. This slowness leads to constant enrichment and to evolution into realms not anticipated at the outset. Knowledge production is inevitably performative; the models exemplify this. Each work takes its part in a number of spatial narratives – the stories of their making, the tales of their involvement in spaces of ideas and evocations and, finally, an account with greater scope which concerns where the project fits in research into design knowledge.

26.3 Projective Characteristics

Designing is always projective in a temporal sense. One cannot design backwards – the best that can be hoped for when looking back is a convincing post-rationalization of what was designed; a design action now does not affect the past – except possibly by leading to its understanding or to a new social construction of aspects of it. Maps report the past. They must concern what was the case when the data mapped was obtained. Strictly, this applies even to maps of very current data – perhaps even that emerging from an ongoing experiment employing equipment delivering a data stream that is being mapped in real time.

Models, like other modes of designing, are projective of a future state if they model something to be built. We might also claim they project into the future if they can be read and understood in the future. I am wary of this claim, as the same observation can be made of even an everyday newspaper article concerned with description of past events. However, any mapping is concerned with the production of a knowledge space and ordered knowledge is usually projected into the future for use by others.

At any point in their production, my design models are projective in that they speak of what might unfold in the process of their making; they speak of what might be in the realm of ideas and built objects.

26.4 The Representation of Ideas, Especially 'Place'

If I am to try and produce a paper map of an island, I expect to replicate the shape of the island. The more sophisticated my understanding of this idea, the more I could concern myself with translations of a thing on a curved surface being projected onto a flat plane, with issues of measurement, or with issues of assumption – such as which tide level the mapping assumes. In the interior of the island I might want to include the ruins of an ancient building believed to house the treasure and would expect to describe its place and shape. To do this I have to establish or adopt rules for representational systems that can be understood by potential users of the map in a near or distant future.

Consider the issues in architecture. Once an architectural work is designed, there are standard systems of codification and representation to describe to others what is to be built. However, no such systems exist to delimit what is to be designed, or to guide the production of form or space in designing. For example, there are no rule-governed systems for giving form to ideas such as those I have been dealing with in the models. Larger cultural idea systems can be plundered to link idea to form, but they mostly offer too rich a range of possibilities from which to draw; they leave the designer to make complex choices. What shape or form might speak of love, comfort or hope? The designer has to concoct a tale which does some translation work, tell a story about love or comfort and find an expressive form that partakes of this story. The choices available are manifold. Suppose that my story involves a simple place to sit – comfortable or otherwise. How do I give a form for sitting in the circumstances of the design? Should it be a rectangular block, a free-form dish or something other? Its connection to other elements in the design may partially determine my decisions about form, materials, and finishes, as it must take its place within the 'game' I am playing in a particular model. Judgements about the overall game precede decisions about details such as seats. Each set of decisions requires a number of fine judgments. It is possible that consideration of the seat may change my intentions respecting the larger story; maybe a subtle adjustment will be called for, potentially, however, the whole may be rethought in response to a seat-related insight.

This is a kind of mapping with endless degrees of freedom. Typically mapping exercises are more concerned with achieving a mapping of some portion of reality onto some item that can carry the representational knowledge system over time. Through such a process we claim to be able to tame and transmit items of knowledge. The relation of the map to the 'reality' mapped is inevitably contestable; it cannot be fully determined, explicated and enumerated. To map the island, many judgments must be made and these can be analyzed in a number of ways. A range of social, moral, political and other constructions can be formed to explain or critique the decisions taken to produce a mapping. The knowledge and techniques employed may be extended to map an adjacent island just beyond our sight and understanding.

As with the map of the island, mapping knowledge spaces onto a model (itself another knowledge space) involves choices about conventions of representation, but beyond this there are possible provocations to read the outcome of the mapping in inventive ways. To explore this in a model, consider the first one I made: 'Prospect Raft (Refuge Raft)'. Its roof was originally derived from the graphed form of the cusp in Rene Thom's catastrophe theory (Woodcock and Davis 1978, pp. 56–63 or http://www.newscientist.com/data/images/archive/1957/19574101.jpg). As well as the sheltering offered by this, there were also means of offering protection to an occupant – a defensible or guarded path to the central raft space, a fence and a screen. In other words a set of the usual mechanisms historically employed by humans to position themselves and see who or what is coming and allow time to react. People sitting in a coffee shop against the wall with a view of the door display the residues of this behavior. This model was envisioned as a habitable place, a raft afloat on a sea of ideas. Here lay a plausible difficulty and, to avoid exposure to too many ideas, some could be framed, selected for scrutiny, while others could be held at bay – all through the agency of sliding a screen or a window frame. *(Figure 26.1)*

There are two issues for examination here. The first is the mapping of ideas from divergent fields of knowledge that are based in words (although using mathematics and observational procedures), onto a system of representation not employing words. In the model, I used an assortment of physical forms, but similar issues arise if employing two-dimensional symbols from a vocabulary in art or from a standardized graphical system. The second issue is enmeshed with this first one: the idea of selection in a mapping or in the production of any model. What is to be mapped? Not all mappings entail simplification, but they must entail selection. A mapper must choose what is to be mapped and how. Thus exclusions are as significant as inclusions; by their very nature they are often omitted from discussion and examination. It is rewarding to consider what was suppressed and why, what was never given any thought, or what was avoided as too difficult.

Changes originate from many sources in designing: when the plastic roof of this first model melted under photographic lights, it was rebuilt drawing on the formal

Fig. 26.1. Prospect Raft (Refuge Raft)"

vocabulary I had established in building subsequent models. Thus, a new form that I deemed to serve the interests being explored in the model replaced the original shape. It is a representation of very similar ideas utilizing forms that differ from those originally devised to be appropriate; it highlights the lack of a precise representational system connecting ideas and forms. Any such representation draws its strength from its place in an overall system of ideas and their representation.

26.5 Time

Model number two 'The Palimpsest Tower' concerns the passage of time, and does so through exploring the duration of habitation of a particular site. *(Figure 26.2.)* Hence place and time were considered. I wrote that:

> *"This has been an occupied site for a lengthy period. Endurance of physical occupation parallels endurance of occupation of an intellectual site or space, a node in an otherwise less-differentiated realm of concepts and concerns. In the past (perhaps more than the present) physical occupation embodied knowing (myths, religions and temples all attest to this).*
>
> *There are ancient steps down to buried knowledge kept contained by a ventilated trapdoor. An aged and wind-swept tree demarks the entry; perhaps it is a, or even the, tree of knowledge. Old paving stones are set into long-firmed ground. The pavilion sits above this asymmetrically; it has used the grounding offered, but has been positioned with little regard for what went before. The interior of the pavilion is not as old as the basic ideas upon which it stands, perhaps a century*

Fig. 26.2. The Palimpsest Tower

or so has passed since it was in its brick and timber prime; it is a bit battered and dirty, but the front has had a quick coat of green paint. It has been surrounded, colonized, re-wrapped and enfolded by a bright and shiny copper construction, supported or shrouded by a skeletal steel frame that affords, from its balcony, a fresh view over the world of ideas ... Perhaps there is machinery within, as large pipes connect to the ground from the pavilion. ... Is it no more than technological image forging? There are signs of additions that are more recent, colonizations, softenings and domestication. A strong and simple table below an awning suggests that food and wine might be brought from within, and it is even possible to imagine some distant Tuscan hillsides, a haze of golden light in a late afternoon. The cues perhaps contradict, or at least require an effort of accommodation. What do we assume and what do we know?" (Downton 2004, p. 43)

Traditional maps imply roughly their own present. As a user of a map representing a locale, I expect to find in the world what the map indicates. I know that an old map is likely to present outdated, maybe misleading information, or perhaps provide knowledge of what is now lost or hidden. This results from the passage of time; it is not an effort to incorporate time into the mapping. For that, more specialist mappings are required, ones employing some means to represent change in something that varies and is mapped over a selected and represented period of time. The intent in the mappings differs. The use of the map is variable. Time is dealt with in a more allusive manner in the model: there is evidence of the passing of time that is imprecise – some parts can be presumed to have preceded others, or at least that is how they are intended to be construed by a thoughtful viewer. There is also, according to the accompanying narrative, a sense of experiential time unfurling slowly – more slowly than clock time – on a benign afternoon. These matters can be evoked through cues in the model. In a map, overlayings and redrawings may similarly function in the manner of a palimpsest. In neither case is time genuinely portrayed, rather it is what we might term 'time's residue' – the things we see as evidence of the passage of time, of occupation of place and duration of physical elements. In models and maps it is not time in itself, but the imaging of time's residue that is made. In this particular model there is also an imagining of time.

26.6 Adjustment and Accommodation

The issues of adjustment and accommodation came to pervade my view of the sixth model, 'Parasite; Life; Adjustment', *(Figure 26.3).* These were not the issues that informed its initial making: I began by considering the representation of life through mechanical means and played with ideas such as the Fibonacci series as an indicator of growth in a (formerly) living thing, or cloisters as a physical and cultural expression of spiritual life. The way this model was made over an interrupted period of two years, with a self-imposed avoidance of any kind of drawing, meant that intentions became lost and furry; some already-built parts were consequentially abandoned. I also utilized ready-made elements – in this case plastic model kit motorbike frames. They required adjustment, as did various parts that needed to mesh with them. Although it had been a tendency in prior models, I intentionally forced my modelling to accommodate anything already built and this has continued to apply. It is a mode of operation that parallels designing for urban sites in which architects must accommodate to the existing fabric around any given site and adjust their designs to fit the strictures of the locale. As mentioned, the only permitted deviation from an acceptance of the already-built was abandonment; some things made previously were rejected to allow me to embrace an idea I deemed to be an improvement.

As a strategy in knowledge production, accommodation to the resistances provided by the existing is identified by Pickering (1995) as a key element in the behavior of scientists. They are forced to accept that either their working methods (equipment, experimental design) do not deliver what they anticipated, or they come to understand that the material world does not work as they expected. Revision of aspects of their work becomes necessary to accommodate these resistances. Mappings and orderings of knowledge do not behave differently to this. The intended schema does not absorb the existing items of knowledge perhaps, or does not adequately portray the relations deemed to exist. The mapping system may suggest items or relations that are not known and may thus suggest the direction of knowledge searches.

Maps of the known often suggest possibilities for filling in the unknown: the half-known continent or the as yet empty boxes in a classificatory mapping such as the Periodic Table of elements in chemistry. (The latter is an instance where the

Fig. 26.3. Parasite; Life; Adjustment

system helped direct the search for more elements to fill the places set up by the ordering system.) Of course in old maps, where they ventured beyond the known, speculation and fantasy reigned. I am happy to venture into this realm in these models too.

26.7 Contradictions, Allusions and Even Ambiguity

The potential of multiple readings was explored in the fifth pavilion 'Fragments of a Venice'. The title alludes to the possibility of a number of Venices of perception, construction, invention or experience *(Figure 26.4)*. This was a Venice of my devising, not a representation of 'reality'. I made an attempt to ensure that more-or-less simultaneous alternative readings of parts of the model were possible. The starting point was a hull-like shape held aloft on props. For many people it reads as a boat. The props supporting only one side allude to those in a dry dock. They raise the issue of a likely succumbing to gravity and sit on leveling chocks that could function – if only there were props on the other side of the timber hull. In this case they reinforce the absurdity. The hull shape has windows mimicking the average proportion of windows in Venetian palazzi on one side and odd chimneys on the top. The first architect to see the model declared it was a palazzo; most people have seen a boat, although some read it as both at once. The base could be read as the ground since it holds up both the props below the boat and a campanile. I also saw it as water in which there are the well-known poles for tethering boats and gondolas. No one has acknowledged to me that they have noted this joint condition.

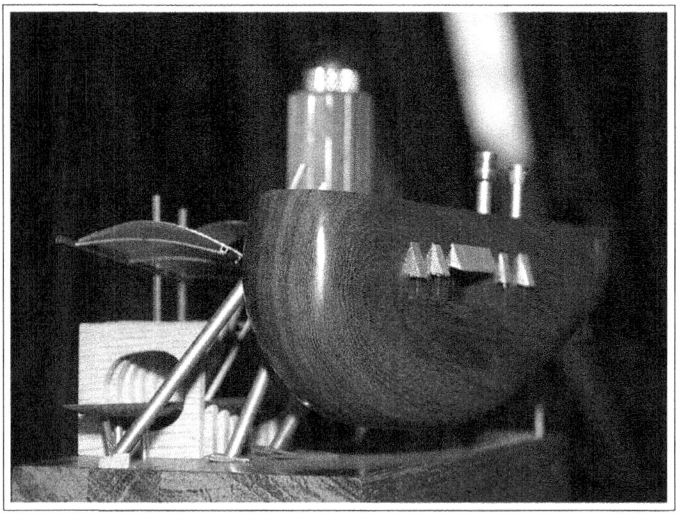

Fig. 26.4. Fragments of a Venice

The issues of multiple codings, contradictions and allusions are potential problems to be avoided in most mapping endeavours, but they permeate a number of the models. In pavilion 8, 'Music Jetty' such ideas are strongly in evidence, along with a range of allusions *(Figure 26.5)*. The upper layer of the pier is concerned with modern music with no desire for the tones and reverberation times suitable to Nineteenth century music provided by the lower level. A number of black rods project from the upper area. In my thinking they are enigmatically evolved from cross breeding microphone booms with fishing rods. Photos of this model were shown at a conference in the early days of the invasion of Iraq; to some of the audience they were guns. While I am happy with differing readings of my models and aware that it cannot be otherwise, attempting to extend a military reading to other elements of the model defeats me – they are an inexplicable oddity when incorporated with what I imagine to be possible alternative readings of the whole.

This raises the issues of judgment and evaluation by a designer-maker. On what grounds are decisions made when there are no 'right' answers? I cannot judge correctness as I might if I were making a model constrained solely by dictates of representation. I may need to judge if some element is enough like a 'real' thing to be read as such by an audience – will it be read as a staircase, seat or window? There is a degree of possible ambiguity here and sometimes I push this as far as possible: the literal-minded may not accept that I have modeled a window although others are happy to read the element as having ambiguity and possibly, thereby, humor. To the extent that mappings are involved, these are rather loose and fluid mappings of rather limited accuracy, but possibly desirable precision. In a way the idea of mapping is inverted; the map precedes the possible, ideas and relationships are explored and the explorations are projected into the realm of ideas and possible future constructions.

Fig. 26.5. Music Jetty

Not only do all mappings involve making decisions about what to both include and exclude, decisions about completeness and completion are also entailed. There are no reliable grounds for me to decide when a model is finished. I am not entirely certain that any of them are 'complete'. Several have been revisited. I declare them finished – or at least stopped for a long period – assign them to the cupboard and commence another one. I certainly cannot claim that the ideas I dealt with in any model have been fully explored. Perhaps they are sufficiently explored for the moment, but any and all of them provide enough material to continue the investigation in another model. Some themes, such as narrative, unavoidably reappear in a number of models. For many mapping exercises, I imagine that the process is stopped when the outcome is sufficient for the purpose at hand. This is probably true for any model if I knew what its purpose was.

26.8 Pattern

We seek and construct pattern. We expect to perceive patterns of events in the world around us and believe the events are thus more than coincidental occurrences. We impute relations between aspects of our world. We frequently employ and imply pattern as a mechanism in design and invent patterns to structure our stories and our buildings. Pattern goes hand-in-hand with meaning. Apparent lack of pattern is likewise imbued with 'meaning' for us, for that which we think is patternless is often dismissed as meaningless. 'Song Station', the seventh pavilion, explored aspects of this *(Figure 26.6).* It offers an acoustically augmented place from which to sing.

Fig. 26.6. Song Station

"Song and architecture have the concept of repetition and thus pattern as ordering devices in common. In this pavilion there is an A-B-A-B-A pattern to both the screens and the rear speaker elements behind the performance spaces. One of the 'B' spaces between the screens is centred on an existing, old and worn chimney structure; an enigmatic box punctuates the other space. The rear units, although conforming to the same A-B-A-B-A pattern, reverse the lengths of each element compared to the spacing of the screens, otherwise they share the same centres. Each 'A' element is also subdivided into an A-B-A-B-A pattern – in this case of equal widths. The same spacing pattern occurs in the arrangement of the red platforms if the vertical rods are taken as the 'B' elements. In contrast, there is an absence of ordering devices backstage. Pragmatic necessity rules." (Downton 2004, p. 77)

The simplest reading of a map is to assume that all markings on it have the same pattern of relations between them as those considered to exist between the elements of the world that are represented by the map. (Or at least that this is so within the constraints of legible map production.) This assertion of a similarity of relations in both the world and the map is an assertion that they share a pattern that connects them (Bateson 1979, pp. 8–11). The myriad assumptions employed to construct the relations in the world are not often questioned, for they are formed and filtered by the assembled history of Western technoscience that underlies our taken-for-granted sense of the way the world is – an isness that is too obvious to be questioned by most (Turnbull 2000, pp. 1–52). We name elements of the world and map them and their relations; naming and mapping are mutually dependent (Bateson 1979, p. 30). We devise a means of representing what we perceive as existing – we see what we believe and reify this into a map. In these models there is an effort to foreground some of these issues by making them into something that requires questioning to be at all intelligible. This is unlike those maps where the issues are submerged below an intelligibility that is formed from the instrumental value of the map.

26.9 Translation

The theme dominating my twelfth model 'The Pilgrim Temple of Canonic Desires' and frequently at play in earlier works, was the relation between an original and an element in a model. *(Figure 26.7)* The process of getting from a known piece of architecture to a piece of my model, I understand as a process of translation – held by dictionaries to be a language-based activity whereby something is turned into a new version of the original (Oxford English Dictionary 2007). This, by extension, can apply to representational systems. In the models a change of scale is involved, other characteristics such as proportion or form may be maintained, materials often change, but some aspects such as color or texture are arbitrarily varied.

Only one of many translations is abstracted for examination here. The model uses the roof plan shape of La Chapelle-Notre-Dame-du-Haut at Ronchamp by Le Corbusier (1950–1955) twice – once in timber as a floor plan where shape but not curvature is maintained and once in aluminium as a suspended awning where it is reversed and punctured by holes in the form and pattern of some of the windows in the original building. Architects recognize the games being played. Others may, depending on their knowledge and the degree to which they are helped by descriptions. For this to be a valid translation, many acts of judgment must be made. As the designer-maker I had to make selections, decide on amounts of mutation and abstraction to satisfy my own ends and also to enable the models to speak to an audience. The translated work has to satisfy in its own right; there is no measure of correctness (no matter how debatable) to which I might aspire as there might be in the translation of a novel for example. However, in those places where I wanted some elements of the model to be recognisable translations, similar criteria applied to those that affect translators between written languages – there was an effort to perform mappings that preserved essences or culture-dependent meaning through appropriate choices. Thus the stability of an idea is sought even at the expense of superficial difference. To the extent that significant aspects of my chosen originals were maintained in the pavilion models as recognizable borrowings or clear representations of these originals, then this might be seen as an act of transformation of the original.

This suggests both transformation and translation are involved, but in addition there are entirely new elements to be found. The three modes are mixed, stirred and shaken into a new cocktail. To do this any designer-maker must continuously select and choose, evaluate and decide.

Fig. 26.7. The Pilgrim Temple of Canonic Desires

26.10 Conclusions

To establish a model of something it is necessary to perform a mapping from the original to the model. At its simplest, this only requires one to say that this thing is 'like' that thing. The assertion of likeness may draw on many subtle judgments or be concerned with very few. With more care, instead of a mapping of characteristics between physical things, a mapping of non-physical ideas onto a physical entity may be performed. A paper map appears to be a specific instance of this that combines objects, relations and ideas. If we follow Bateson's epistemology (1979), then what is recorded in the map are the differences perceived and deemed to make a difference – in the case of the island map, above, in characteristics such as altitude, position, size, and length. They are characteristics for which we have metrics and modes of indicating them. We map what is not yet extant when we produce a model of a future building and ask the physical model, or the computer model, to reverse the process above and project an array of ideas of what the built outcome will be like. If these ideas become reality, there is a mapping from the original model to the constructed architecture. If, however, there is no intention to build a larger version of a model in the world, as is the case with my models, then there is no mapping forward in time and the model is a model of itself, a map of no other place.

In demonstrating some of those issues that pertain in the models and are also present in mapping, I have been performing a mapping of a set of ideas in one domain onto another domain. To this extent parts of either the domain of architecture or cartography could serve as a model of at least part of the other domain. By examining one, we learn something about the other – one of the normal uses of any model. The two knowledge domains are brought close enough to suggest a working relationship. This bringing together could lead to mutually constructing a third knowledge space – an idea that is central to Turnbull's (2000, pp. 224–228) definition of the 'transmodern'.

References

Bateson G (1979) Mind and Nature: a necessary unity. New York: EP Dutton

Downton P (2004) Studies in Design Research: Ten Epistemological Pavilions. Melbourne: RMIT University Press

Downton P (2007a) Making Dust, Acquiring Knowledge, in Downton P, Ostwald M, Mina A and Fairley A (eds) Homo Faber: Modelling Architecture, Archadia Press, Sydney, 2007

Downton P (2007b) From ideas to Models: an introspective report in Fairley A, Mina A and Downton P (eds) Homo Faber: Modelling Ideas, Melbourne: CD catalogue for an exhibition at the Melbourne Museum, pp.17–23

Cross N (1982) Designerly ways of knowing, Design Studies 3 (4), 221–227

Cross N (2000) Designerly ways of knowing: Design discipline versus design science, in Pizzocaro S, Arruda A, de Moraes D (eds), Design plus Research: Proceedings of the Politecnico di Milano Conference May 18–20, 2000, Milan: PhD programme in Industrial Design, Politecnico di Milano, 43–48. This paper is also at http://pcsiwa12.rett.polimi.it/~phddi/uk/01/dpr00/thoward.htm (Accessed June 11, 2008)

Oxford English Dictionary Online. http://www.oed.com. (Accessed June 11, 2008)

Pickering A (1995) The Mangle of Practice: Time, Agency and Science, Chicago: Chicago University Press

Turnbull D (2000) Masons, Tricksters and Cartographers, London: Routledge

Woodcock A, Davis M (1978) Catastrophe Theory, Harmondsworth: Penguin

www.newscientist.com/data/images/archive/1957/19574101.jpg (Accessed June 11, 2008)

27 On Sustenance: Operating Landscapes, Land Spaces and Land's Paces

Titusz Tarnai

Academy of Fine Arts Vienna, Austria

Abstract

This chapter portraits three projects of the Academy of Fine Arts Vienna: *Landscapes, Land's Paces* and *Land Spaces.*

27.1 Landscapes: Analogue and Digital

In fall 2007, the Platform for Analogue and Digital Production was installed at the Academy of Fine Arts Vienna with the purpose to analyze and search for possible redefinitions in working- and production methods in the realms of Architecture, Design, Engineering and Sculpting, and deriving pedagogical models. A key role is located in the potential of enabling an interaction of digital and analogue practices, especially the interaction of physical and virtual models. In the system of the curriculum, the adjustment of the architectural practice to an explosion of tools on the intersection of the analogue and the digital is being exercised. Expanding the field of analogue and digital transcriptions, subject-object relationships are generated and reconstructed. The tools, gadgets are apparati of observation and description with revolutionary potential that need to be examined to their capacity of informing the architectural practice. In the practice of the platform, application perspectives of design tools to which one has become comfortable with are destabilized, perverted and reinvented. The most (trans-)active place is the osmotic membrane delimiting analogue and digital. It is a border of intense traffic and volatile transactions: The place to be. As the less permeable walls of our cities, this delimitation needs to be examined for its validity. One of the goals of the ADP is to unfold the dichotomy wired into its own label: Being a platform of analogue and digital production, the transaction space between these two realms is charted, inscribing routes of (spatial) production. Current research is focusing on the exploration of experimental measuring and mapping techniques on a given territory. The so produced territorial models are deployed and become operative in the articulation of (urban) building strategies.

27.2 Land's Paces: Working the Dynamics of a Site

Nature and the city have traditionally been the binary states in which architecture would be sited. These two archetypical categories of environment form conditions to which in an architectural project relations are clarified and made explicit. In the course of the architectural intervention, nature is sustained or reinstalled as a heroic act of reverence and paralleled by or confused with declared insertions into the contested epistemological field of the city. Construction is at once a cultural, ecological, technological inscription into a given context. The notion of the site emerges as the instant of extracted and prioritized conditions that are defined as relevant to the act of building. Insisting on the principle of sustenance and expanding (folding) it from the occupant outward onto the site, buildings as interventions must seek, as a strategy of self-preservation to serve its environs. Understanding the responsibility of the building lying as much in the sheltering of the inhabitants as the sheltering of the site, the challenge of research in the studio is set in identifying, modeling and simulating ecological processes on the site and to adaptively insert an intervention that operates upon an inside-outside condition in architecture on the condition that an insertion into a site has to be a prosthetic device to a found local ecology: Assuming that the architectural program has to be amended with symbiotic operations to the host. This will determine what architecture is or can be, once sited. Buildings are life-support systems because buildings contain life-support systems.

Investigation in the studio departed from inquiring into the subtle differences between site, territory and zone, activating generative concepts triggered by the tension between the cultural and mechanic qualities of these spatial categories. Urban space is fragmented by administrative, legislative and other cultural and functional boundaries, most of which have a persuasive character in designating boundaries to sites of an intervention. As opposed to these constitutional boundaries, the city is carrying a multiplicity and simultaneity of operational (ecological, systematic, infrastructural) zones such as watersheds, which remain unreflected in the formation of intervention zones, thus neglecting to react to the organic needs of the city. In the method of the studio the field of action, the site, was designated by one point coordinate on the map of the city that by definition formed part of the field. Researching and revealing possible boundaries of the site went simultaneously with identifying the active (performing) characteristics of the site and envisioning its potential. In the specification of the research program, the classical demand of incorporating a given program on the site has been dropped, presenting a territory with initially no need for intervention *(Handlungsbedarf)*. Programmatic demand and need for intervention had to arise from identifying constraints, potentials and capacities diagnosed in situ. Architecture as an extension or prosthesis of a constructed local ecology, had to insert itself into the existing continuum of the zone.

Scheduling sustenance to the agenda of building, the first question to respond to, is what to sustain. Is sustenance a form of preservation: a conservative stance towards the condition of the landscape? Questioning preconceived ideas of nature, landscape, purpose, ecological integrity and not least conservation, a productive definition of a local sustainability has to follow. Departing from a pragmatic point of view, where architecture has always been and should continue to be subservient to the survival of mankind, makes evident that one deals with an extended ecosystem, defined by processes and transactions taking place on its domain. Thus where preservation is aiming to sustain a preferred image, sustenance of [desired] processes in turn urges the preservation of an ecosystem as such. Preference and desire then need to be detached from any moral or aesthetic pre-figuration:

Imagine a site without monuments. Imagine a project without preconceptions. Imagine a zone free of ideology. Imagine a city of free exchange. Imagine a landscape in transience. Imagine a place that you don't know. Of course first you have to free yourself, allow your perception to advance. Allow your body to be part of the site. The studio constructed interfaces, machines and devices allowing for engagement with the site beyond an urban pragmatism dominated by consolidated political forces, foregrounding geological conditions, geographical, climatic, dynamic, elastic, transactional conditions. Long enough cities have been devised to fulfill first the needs and then the desires of its citizens. The error committed by modern city planning was to aspire to rationalize and quantify these desires. Since this illusion is to fail, a renewed paradigm in territorial programming has to emerge.

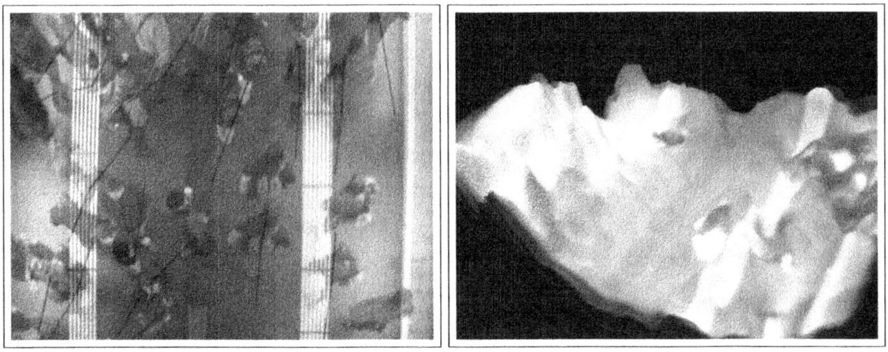

Fig. 27.1. Synaptic site model – Christian Tonko, ADP design studio, Academy of Fine Arts Vienna

Lately the focus of attention shifted on the transactional property of capital, of goods, bodies and ideas. This view on the mechanics of cities still remains disjoint from the also emerging view on the cycles and transformations in nature. Nature as an ecosystem, an evolutionary process, stands in conflict with civilization as an accelerated apparatus based on mutation and change. Can we bridge the gap? Architecture is pollution *(Umweltbelastung)*. Nature re-coined territory is next to its geometric and tectonic existence, a dynamic, rhythmic process. In looking for a language *(Sprachlichkeit)*, a means of concern and description for this emerging aspect of the territory, it becomes certain that the model of a territory has to be a model in motion. Next to the coordinates of a place, a valid model of a territory needs to account for the evolving and dynamic aspects of a site. It describes and reacts simultaneously. One could also speak of performative models, of models which are able to contribute a certain productive and diagnostic work *(Eigenleistung)* in respect to a given load *(Anforderung)*. As the capacity of models is not exhausted in representing but now comes equipped with a sensory and motored agency, the role of the model in turn needs to be investigated.

27.3 Land Spaces: Sustenance and Intervention – On the Search for Models of Territorial Reconstruction

Searching for an adequate mode of operation, an immanent system of growth and directive authority acting on a territorial basis has to be unraveled: Landscape is a territory on which man maps his own actions and extracts his intricate relation to it. All our actions are grounded, in some cases rooted – there is a datum on which we all confer. Thus landscape is a territorial medium carrying and recording the discourse on the state of collective society. The accessible subset of this datum is designated as a territory. The territory is the field of actions yielding relevance in their capacity to result in transaction and distribution. Actions unfold on the ground, concurring descriptive and inscribing operations. Later as a consequence of the division of planning and action, the datum of the ground is duplicated in maps carrying a catalog of prescriptive operations. Ideologies are constructed claiming authority to prognosis as a base of prescriptive deeds. Monopolies on the theorization of time, on the construction of history and projections historically used the establishment of valid (acceptable or enforced) models to designate their license of authority. The role of precedent is a persuasive one, as our cognition works through metaphoric transference. All concepts point to an original model that works as a precedent. Models are distilled from an observed entity, are made portable through systematic generalization and abstraction. Transplanting a model is to transfer a logical apparatus onto a distinct domain, similar to a metaphoric transference of meaning. Yet using models as an interlocutor, operational assets are fore-grounded and meaning is emerging through the process of transplantation[1]. Comparative investigations, the method of observation and abstraction of models lead to a model based thinking, which came to dominate mode of conduct in science. Scientific models are precise scripted abstractions – the scripts themselves follow the logic canonical to science itself – constituting and designating the phenomenon in question. (The object of science is the phenomenon, its objective the production of models that amount to the legitimization of a phenomenon.) A phenomenon in general is an isolated subset of the entirety of universal becoming. Once distilled, phenomena are exercised in an abstract temporality, partially reacting onto, distorting the concept of time. In this sense, the authority embedded in the model as a mechanism able to translate phenomenon to a system of causalities does generate a responsibility for the architect intervener to understand his action not so much in the installment of an alteration, to say a jump from one solid state to an other solid state, on the site, but in the successful implementation of a territorial logical engine that is able to produce a set of desired vibrations and/or able to handle motion and transaction on its domain.

The role of the architect thus is to understand the intricacies in the relationship between model space, territory and the public realm and to insert himself as a binding force. If cartography is a set of scripted operations with the aim to perform a systematic description of a specific aspect of a territory – condensing in the recording and encoding of qualitative differences in documents, every project that seeks an accountable insertion into the site catalyzes the production of charts and maps.[2] On the field of architectonic intervention two biographies collide: designating the datum of the site. Biography of nature, embedded in geological time, and the biography of the architect embedded in historical time. Due to the constructive responsibility towards the site, the architect can not withdraw and sublimate his subject in a scientific fashion. He needs to transform, not only to describe, and thus finds himself in a field of tension between empirical subjectivity and deterministic objectified reason. Only through the synthesis of new charts can qualities be incorporated into the envisioning and ratification of future architectures that aspire to overcome predominant conventional programmatic planning ideologies. Reformation of a site restrained by political and cultural anachronisms requires a shift of perspective leading to explorative documents and site descriptions. In this case the authority of the architect, supported by scientific systems of recording (observation) procures for the validity and reality of the emerging local topography of the site.

Facing a system of territorial potentials and kinetics, in which drifting conditions oppose needs of change and needs of stabilization, in turn constituted by states of

[1] Thus the transference of models always triggers the production of theory: the manuals of transference, conceptualizing and legitimizing at once.

[2] On the other hand, charts and maps contain and convey an intentionality: becoming purposeful justifications of action. The best example is the mapping of oilfields prior to extraction procedures.

arrestments as solid states interwoven with production and transaction cycles, the struggle lies in the exposure and overcoming of the resistance of a status quo. In an effort to unblock anachronistic contingencies in situ, one is confronted with the question of how an urban epistemology asserts itself. Historically any site bears a condition of the existing, its reality. Reality comes to being through all the extrapolations that have to be discarded as such, proving unviable to transgress. Thus reality at first exhibits a set of constraining markers, which solidify the territory and immobilize its substance. Immobilizations leading to accumulation build up an arrested inert mass which comes to being as a potential. Search for and recognition of historically arrested potentials emerges as an act of revealing and constructing. Landscape viewed as a meta-organism, foregrounds categories of becoming as opposed to being. In searching to expose the intricacies in the relationship between practice (as action, production), society (as an emergent system) and nature in the framework of unfolding time one finds

> *the fact that the forms of social production, like those of desiring-production, involve an unengendered nonproductive attitude, an element of antiproduciton coupled with the process, a full body that functions as a socius. This socius may be the body of the earth, that of the tyrant, or capital. (...) In a word, the socius as a full body forms a surface where all production is recorded, whereupon the entire surface seems to emanate from this recording surface. Society constructs its own delirium by recording the process of production; but it is not a conscious delirium, or rather is a true consciousness of a false movement, a true perception of an apparent objective movement, a true perception of the movement that is produced on the recording surface.*[3]

Allowing for the process of recording and observation to enter the practice of the architect as a tool, which is then harnessed engaged consciously, and nevertheless allowing for the observation process to remain a destabilizing element in the act of planning, the architectural model emerges as an apparatus that extends in both directions to on the one hand become the monopolized tool in describing the reality of a locality as territory, able to function as a gauge, measuring its current condition up to the minute, and on the other hand to transact the such received information, and feed it back onto the site through, and in the form of the intervention as device and procurement at once. Setting loose from epistemologies, the use of models becomes crucial in the provision of intervention proposals and strategies. In order to act within the unstable field of the landscape-ecosystem the architect as a spatially organizing authority has to develop models with seismic, calculating and prognostic abilities. Binding the model space with the territory and transferring precedents through sampling, the architect-planner forms part of a sensorial assemblage through

---

[3] Deleuze, Guattari – The Desiring Machines, Anti Oedipus, University of Minnesota Press, Minneapolis, 1993

his modeled observations. His practice oscillates between embodied sensuality and scientific rigor. Utopias are not invented, but deducted and constructed up to the minute. They can only work within confines and grow out of the present. Landscape is fixed as the totality of all that is in flux of which architecture as site programming is inserted as a permanent regulating element. ...[4]

Fig. 27.2. Johannes Flatz: Performative model – the responsive system within an avalanche as a measuring and scanning tool

[4] ...emitters emit, consumers consume, transformers transform, instances commission, events trigger, faces look, cars park, attractors attract, controllers dispense, planners devise, ears hear, curves bend, inflation and entropy increase, schedules delimit and activate, conductors conduct, coils induce, funnels funnel, wheels turn, water flows, objects deteriorate, tools bland, ideas transmute, records cross-fade, power forces and persuades, history records, science explains, simulations predict, man intervenes...

Fig. 27.3. Oliver Maerki: Urban intervention model – proposing a site massing in 4D, diffusing programs in unstable and amorphous building conditions

Fig. 27.4. Peter Jellitsch: Tectonic model – Based on a dynamic model of an earthquake, main point in the project is an engine that simulates dematerialization on given sight conditions by the application of tectonic drift sequences

28 Social Tapestries

Alice Angus, Giles Lane, Orlagh Woods and Karen Martin

Proboscis, London, United Kingdom

Abstract

Social Tapestries illustrates the work included in the Art and Cartography exhibition that traces the territories, research and outcomes from Proboscis *Social Tapestries* research programme.

28.1 Social Tapestries

Over the past five years many of Proboscis' artworks and projects have explored and developed the concept of 'public authoring', the everyday mapping and sharing of knowledge and experience by people about the places they live, work and play in. Central to this is *Social Tapestries* (2004–2008), a research programme of challenging and playful artworks, projects and experiments exploring how public authoring can pervade everyday life in different situations and contexts. It built upon Proboscis' earlier project *Urban Tapestries* (2002–2004) – which investigated the potential for grassroots knowledge mapping and sharing (public authoring) using a fusion of mobile and wireless technologies with geographic information systems.

For *Urban Tapestries* Proboscis created a software platform enabling people to annotate and share words, sounds and pictures about places – weaving their personal threads of knowledge to co-create the rich tapestry of the city. Two public trials were run to test the platform and begin to understand the social and cultural implications of public authoring. However the trials attracted participants who were, in the main, already familiar with the concepts and aims of spatial annotation and mobile technologies. *Social Tapestries* was designed to enable Proboscis to engage with other communities – not convened groups, but actual communities in real world situations. This project sought to build relationships in a number of contexts that would allow exploration of the concepts behind public authoring. *Social Tapestries* included discrete works and public collaborations with specific communities addressing education, social housing, community arts and local government. These included Havelock Housing Estate, St Marks Housing Co-operative, Jenny Hammond Primary School, users of London Fields and the Institute for International Visual Arts (inIVA) in London. With play and scavenging as two essential ingredients in the work, the work manouvered from inventing technology for software applications (in collaboration with Birkbeck College, London) to using paper based technologies and found materials. This in turn opened up novel arenas for working with other sectors such as community development and education, employing the tools, techniques and formats such as Diffusion eBooks and StoryCubes; as well as undertaking experiments combining adaptations of carnival costumes, toy robots and cheap home electronics with GPS positioning, environmental sensors, wireless data and online mapping technologies.

28.2 Conclusions

The Proboscis exhibit for Art and Cartography traced the territories and unfolding outcomes presenting processes, research and outcomes from *Social Tapestries*. The maps from *An Atlas of Enquiry* detail our collaborations, methods, tools, techniques and aims; the Storycubes playfully illustrate activities, artworks, interfaces, communities, partners and concepts of public authoring and the film *Play to Invent* is a playful exploration of some of the *Social Tapestries* projects and techniques. The display aimed to evoke the collaborative nature of our processes by inviting visitors to construct their own Storycube landscapes and share their thoughts.

URL

proboscis.org.uk
socialtapestries.net

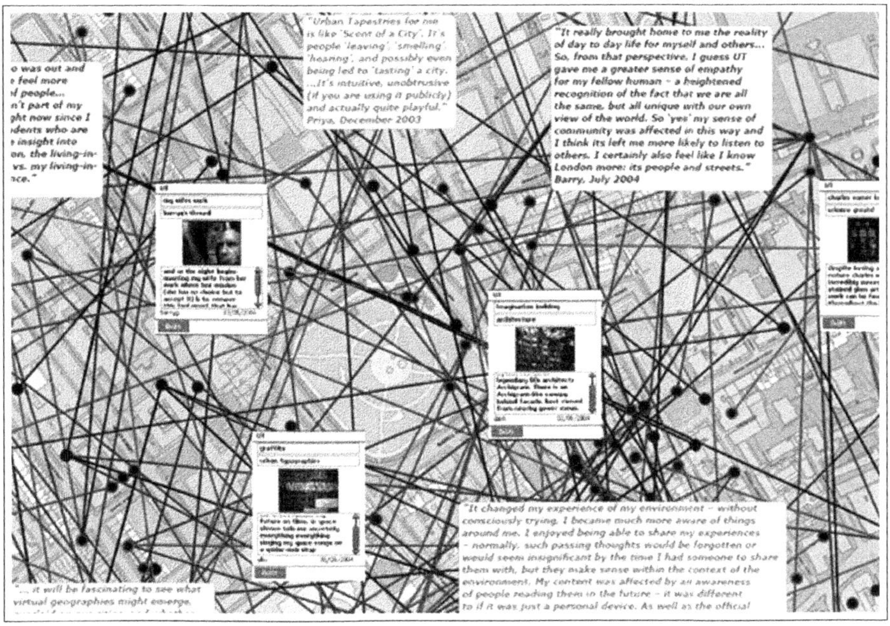

Fig. 28.1. Detail from *An Atlas of Enquiry* by Proboscis. Image Proboscis 2006

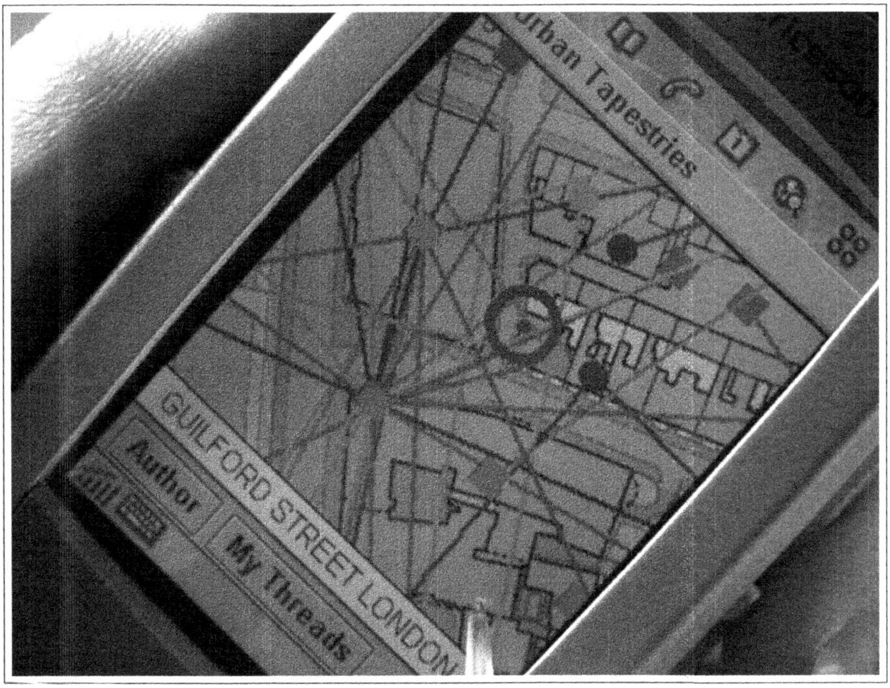

Fig. 28.2. *Urban Tapestries* mobile phone client. Image John Paul Bichard 2004

Fig. 28.3. *Snout*, detail of the Plague Doctor costume with environmental sensors, London. Photograph Thierry Bal, 2006

Fig. 28.4. Urban Tapestries bodystorming. Photograph Proboscis 2004

Fig. 28.5. *Social Tapestries Storybox*, set of 27 StoryCubes. Photograph Proboscis 2007

29 Community Karlsplatz, Vienna Video Screen Projection

Theresa Anna Häfele

Vienna, Austria

> *"The key thing may be to create vacuoles of noncommunication, circuit breakers, so we can elude control."*
>
> Deleuze 1995

Abstract

Karlsplatz is a square in downtown Vienna without clearly defined borders. This chapter deals with the diverse aspects and areas above ground and underground and portraits *Community Karlsplatz,* a project which aims to develop a cultural network for this urban area.

29.1 Introduction

In January 1997, on the initiative of the Künstlerhaus, the joint project *"Kunstplatz Karlsplatz"* was launched, whose specific objective was the coordination of and cooperation among the entire spectrum of cultural institutions situated around the Karlsplatz.

Fig. 29.1. Left: *Third Man,* Carol Reed (1949). UK, Arthaus. Right: *Scorpio,* Michael Winner (1973). USA, MGM Entertainment

A unifying framework within the given spatial parameters is not a solution – as has been proven by the rejection of several such project designs. The Karlsplatz has never been a "square" in the classic sense (from the perception of a river bank landscape up to the subterranean channeling of the Vienna River). It has no clearly defined borders, the surface area is transected by streets and extends diffusely far out past the Ressel Park. Trying to define the "Karlsplatz" as a specific locality is so difficult that it is almost impossible to identify it as a meeting place. As is frequently the case, decades of failed municipal planning has – due to the complexity of the matter – created its own interesting internal logic. A logic which involves the separation of what is above from what is below, a dissociation of surface and underground worlds.

29.2 Subterritory

Thousands of people pass through the underground pedestrian passageway every day. Tunnels and conduits extend almost 40 meters down underground. The four-level subterranean Karlsplatz parking garage lies buried beneath the area in front of the Karlskirche. The emergency exits are nestled almost imperceptibly up next to the oval water basin, in a kind of Baroque symmetry. An Italian café in the passageway also serves as an entrance to a nightclub. There are DJ's putting on their music in the Künstlerhauspassage. Not far away from all this the Musikverein has expanded its premises underground. The Opernpassage houses the "Operntoilette", complete with waltz music in the background.

29.3 Kunstplatz Karlsplatz Versus Museumsquartier (MQ)

The *MuseumsQuartier,* an international showcase of various museum facilities, defies comparison with the complexities of the Karlsplatz. The MuseumsQuartier has been called a "claustrophobic concept", primarily because of the clearly assigned interior spaces defined by the Baroque premises of the former Court

Stables. The external walls furnish the location with its own identity. Everything inside is *MuseumsQuartier.* The administration headquarters are located at the *Museumsquartier Errichtungs- und Betriebs-GmbH* which, as a kind of all-encompassing umbrella organisation, dominates and presides over a group of renowned cultural institutions. Less well-established younger artists are given additional support by the organisation *quartier 21.*

The Karlsplatz, with all its complexities, does not lend itself to a one-size-fits-all planning scheme in the traditional sense. The *Wien Museum,* which functions as an exhibition venue for art, architecture and design within the context of Viennese history, is ideally located on the Karlsplatz – at the pulsating centre of modern Viennese history. Despite its relocation to the *MuseumsQuartier,* the *Kunsthalle* has, with its *ProjectSpace,* maintained a vestigial presence at the Karlsplatz. In addition to the *Technical University,* the *Musikverein, Secession* and the *Künstlerhaus* have become renowned institutions in the course of time, although in essence they were originally associations founded by groups of autonomous artists in the spirit of the beginning 19th century. Their interests are, or were, of an ideological nature – in the broadest sense a "community".

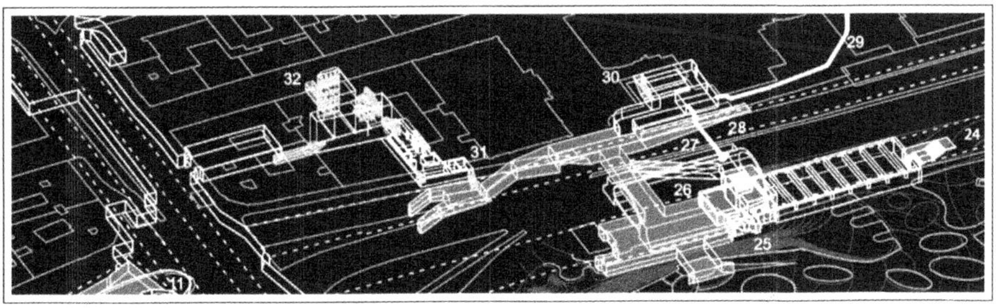

29.4 Network/Community

"Community" has acquired a new meaning due to its use on the internet: There we find forums, chatrooms, newsboards, social sharing networks, matchmaking sites, etc. These are all concerned with the exchange of information within specific topic areas. A topic area reflects the main interests of the Community, its name as well as its identity. This configuration is based on the infrastructure of the internet itself. The totality of these interconnections – the internet – provides a structure which enables users to communicate via all nodes and all channels. The internet nodes are the interfaces between computer networks. The computers are networked and each of them is connected to a smaller network of computers, and in turn these are connected to even smaller regional networks, etc.

In the *Community Karlsplatz* such types of networks are to be implemented in a spatial sense. The network nodes consist of eleven platforms which provide the

Communities with a sense of identity. Any person can become part of the Community. Anyone can determine the subject matter, everyone has the opportunity to participate creatively and to communicate. To draw an analogy to the virtual community platform MySpace: The user himself becomes the active "player". He maps out his area and invites others to pay him a visit there. These guests introduce their friends, to whom he can also pay a visit. The artist is the user and his "friends" become part of the creative network. There is no single creative "pool", but rather several of them. These are networked among one another. They are also "friends" and are responsible for organising collaborative projects and activities. At each platform there are Community members who also contribute their own networks. Every user can also become an active "player" and thus an integral part of the *Community Karlsplatz.*

The innovative concept of a *Community Karlsplatz* attempts to bypass the brand-name management and marketing of art as it is practiced by certain well-known institutions. There is a number of established art institutions in Vienna which, through the use of their names, claim to "guarantee" the absolute exclusivity of certain exhibition objects or artists. These institutions are in effect lobbies, and their selection criteria amount more or less to a form of censorship.

The *Community Karlsplatz* has neither a curator nor a fixed lobbying group. Its spatial concept relies on the given area already in and around the subterranean passage, although there is also a broad variety of resources (sewage system, Vienna River, empty spaces above the underground train shafts, the Karlsplatz parking garage, private cellars, etc.) which are normally not perceived as part of the urban/park setting. These are to be slightly adapted using connective elements. The resulting structure will resemble the spatial configuration of a network. The entrances are temporary access points, which thus enables a flexible and continuous re-organisation of the network.

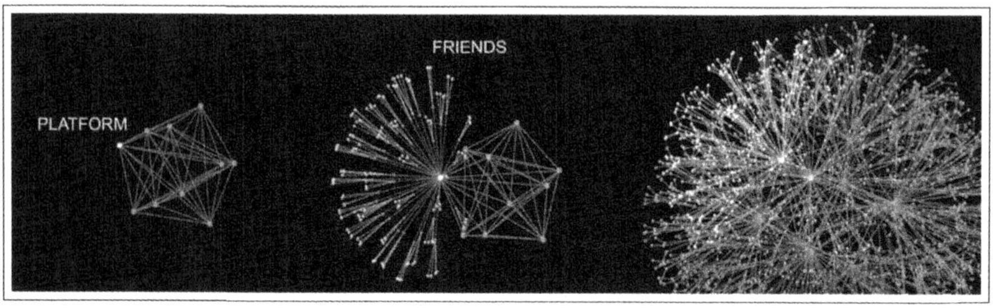

References

Deleuze G (1995) Negotiations: 1972–1990. New York: Columbia
Scorpio (1973) Michael Winner. USA, MGM Entertainment
Third Man (1949) Carol Reed. UK, Arthaus

30 4816 – Urban Expeditions in Vienna Along the GPS Matrix

Evamaria Trischak

Vienna, Austria

Abstract

Vienna lies 48° north of the equator and 16° east of Greenwich, which is located on the prime meridian. For this reason all GPS coordinates in Vienna begin with N 48° and E 16° and the project's name 4816 is a reference to this fact.

The focus of interest are the precise points where minutes intersect within Vienna's city limits. The point N 48° 11.000' E 16° 19.000' is just one example of these 185 intercept points, which are approximately 1.9 kilometers apart on the north-south axis and about 1.25 on the east-west axis. The intention of 4816 is to locate these intersections with the aid of a GPS tool and document them in photographs. Four photos, one for each point of the compass, are required for 4816.

Project Description

Traversing one of these intercept points will take you on an expedition through Vienna. Robert Eder compared 4816 to the psychogeographic method of "dérive" of the Situationists associated with Guy Debord (Paris 1950s). The term psychogeography denotes exploration of the immediate effect of geographic environment on individual emotional behavior. One of the methods employed was dérive, or drifting, in cities. In this context the term dérive refers to an experimental procedure in the city, specifically moving through various urban environments (Eder 2004).

The object of 4816 is exploration of the city independently of the street systems that have developed over time. The process by which participants find the intercept points takes priority over the results, field research is the focus. The Situationists go so far as to say, "Everything that alters our perception of the street is more important than what alters our perception of painting" (Ohrt 1995, p. 43).

On the other hand, Erik Meinharter called an article about 4816 in dérive, the magazine of urban research, "Stichprobe Stadt" (The City as Sample). In it he compares 4816's method of exploring the city and experiments in the natural sciences, for which samples of vegetation are collected (Meinharter 2007). Vienna becomes the focus of an expedition, an area being surveyed, a playground and an abstract matrix. The intercept points are connected by straight lines that have nothing to do with conventional pathways, streets and roads. A systematic and conceptual study of the city of Vienna is inspired by the matter-of-fact strategy of registration and realisation.

"Looking through the project's database will provide a sense of the fascinating nature of such a 'mechanical' approach to good old Vienna" (Möchel 2007).

The exploratory approach shows both a macro and a micro perspective of the city. The project provides both an overview of the city and samples with photographs taken at the points where the minutes intersect. Details are made visible. Each documented point is a zoom toward the reality of the abstract grid. The individual observations are juxtaposed, functioning as building blocks, which can be used to obtain a new general view of the terrain. This represents an attempt to organize the record of a geographic area which is subjected to constant change.

The samples show images that go beyond the narrow canon of dominant gazes and the stereotypical view of Vienna. In addition to offensively communicated symbolic images, a different reality is shown, one of intermediate spaces. After examining 4816's photographs Neil Freeman, who performed a similar project in Chicago, stated, "I didn't recognize any of Vienna" (Freeman 2007).

By now 90 % of the intercept points were found. The points still needed include one in the gardens of a Castle and on the grounds of the power station in Simmering.

Acknowledgement

The author thanks Steve Wilder for the translation of the chapter.

URL

http://4816.nsew.at

References

Eder R (2004) 4816 dem Bedürfnis auf der Spur. http://4816.nsew.at/4816/public/home/theorie_reder, 08/2008

Freeman N (2008) Chicago Mile by Mile. http://fakeisthenewreal.org/milexmile/, 08/2008

Ohrt R (ed) (1995) Der Beginn einer Epoche. Texte der Situationisten. Hamburg: Nautilus

Meinharter E (2007) Stichprobe Stadt. Das Projekt >> 4816, in dérive. Zeitschrift für Stadtforschung, 26. http://www.nextroom.at/article.php?article_id=26075, 08/2008

Möchel E (2007) Das GPS-Weichbild von Wien. In: futureZone. http://futurezone.orf.at/it/stories/214514/, 08/2008

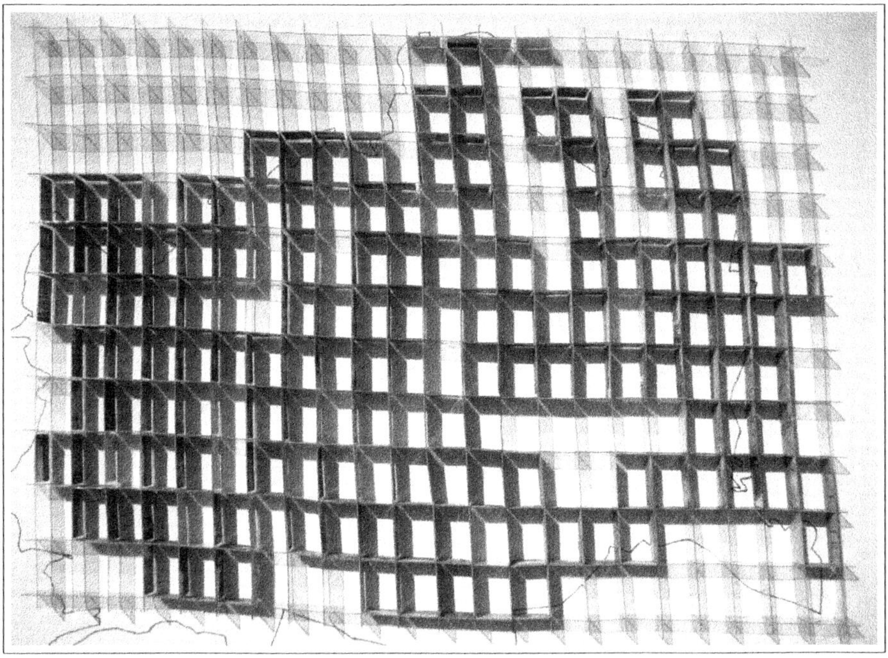

Fig. 30.1. The 4816-model of Vienna (1:50.000, plexiglas). Evamaria Trischak (© 2008)

Fig. 30.2. The 4816-model shows the gps-minute-grid and the pictures already taken. Evamaria Trischak (© 2008)

Fig. 30.3. Some of the 4816-points are still waiting for their discoverers. Evamaria Trischak (© 2008)

Fig. 30.4. The result of the follow-up project '5107' in Dortmund/Phoenixhalle. http://www.nsew.at/5107/ Thomas Wucherpfennig (© 2008)

Section VII: Installation Art

31 Dream but Conflict

Wolfgang Fiel

Vienna, Austria

Abstract

A shift from the restricted administration of a controlled and passive environment towards multiple bottom-sideway processes through the dynamic empowerment of a wider public, aims at active use of electronic networks; a micro polity as affirmative topography of desire (Hardt and Negri 2000).

With this chapter I attempt to summarise the basic conceptual elements for what I call "World Maps of Dissipative Resources". Much has been said and written about maps in general, its manifold political, socio-economic, ethical or moral implications. Maps have been drawn as tools of domination, deception or means for the interpretation of history. But yet even the most obvious objective doesn't seem to prevent the habitual assertion for scientific objectivity. The enduring quest for unity and normative regulations falls short of addressing the complexity of processes that are assumed to be highly dynamic. Hidden momentum, invisible desire or secrecy, are sources for sudden eruption, prone to change a seemingly stable arrangement within an instant. Hence, rendering these 'ghost-like' and subjective entities visible is the quest I attempt to set forth.

31.1 Introduction

In his book entitled "What is life?" Schrödinger (1944) has speculated about the basic principle at work that keeps the entropy of a living organism relatively constant and would comply with the second law of thermodynamics. (The total entropy of any isolated thermodynamic system tends to increase over time, approaching a maximum value.) He coined the term "negative entropy" that each living organism would have to detract from the environment continually. Subsequently for him the key to life was: "Order by Order".

Complementally to Schrödingers concept von Foerster (1960) proposed what he called "second key to life: Order by disturbance".

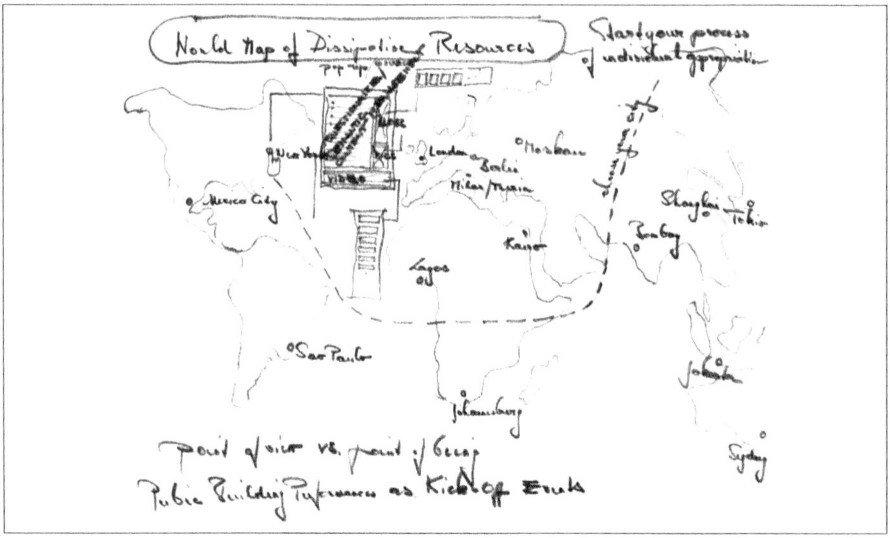

Fig. 31.1. World Map of Dissipative Resources

31.2 Form and Structure: The Operational Paradigm

If the concept of 'random' disturbance as genetic imperative is appropriate, the foundation for a better understanding of the evolutionary dynamics of cognitive processing might be provided. In how far this concept bears artistic relevance is inextricably linked with a proper understanding of the dialog between an organism and its environment and the implicit necessity for a strict definition of the term 'form' qua 'structure' with respect to the following assumptions.

If – using an analogy to the etymological roots of the term structure and contrary to a common understanding – form isn't understood as the outer appearance of a physical entity (for which the term 'Gestalt' seems more appropriate), the main focus is shifted towards the regularities that determine the outer appearance. Subsequently we gain a comprehensive insight into the act of formalising a problem. Once a concrete problem is put on a level with one of the above mentioned random disturbances, creative (intuitive) problem analysis effectively is the modification and re-organisation of already existing 'internal models' in order to gain a sufficient (operable) but transitory arrangement. For a useful description of the processes at work the tools offered by Analytical Linguistics fall short, since the conjunction of signifier and signified offers no operational aspect for a process that is assumed to be on the whole dynamic (organism – environment interaction). Having specified the term form as the 'tentative' process of analysing the structure of a physical entity presented to the perceiving organism one will often experience, that different objects can share the same aspects or a single object may be seen from different angles (association). It is exactly this process within the framework of the subject – object interaction that is of specific interest. A decisive quality of internal models is the nature of their heterarchical organisation (Wiener 1996). Accordingly the capacity of the entire system cannot be reduced to the mere sum of the information contained in each of the models. The elements are connected among one another by a large number of functional links and therefore are relational by nature. Relations between objects and/or events suddenly become manifest because of the unexpected emergence of structural congruence and bring about the important moment of surprise. The specific challenge is how to organise the discrete elements at hand in such a way, that within the manifold interplay of given context and actor (observer) a reference can be drawn to the already known, leaving room though for 'interstitial spaces', that necessitate individual structuring intervention. Through these interventions individual meaning is constituted. Already indicated by the term actor, her/his part is rendered active and is of integral bearing for the anticipated chain of events. Subsequently the process of affirmative appropriation affects the present condition of the 'field of action'.

Fig. 31.2. Structural framework of objects, its ontologies, and relations

Fig. 31.3. Morphological features of the database

31.3 Database

The characteristics of this dynamic heterarchy of individually constituted internal models are transposable into the operational properties of web-communities, which embody a form of connectedness beyond the sheer multiplicity of their users.

1. The 'collective objects' of the Database provide the framework for a web-community to engage in a process that provides non-linear access to the individual items contained in a collaborative exchange, and forms of retrieval.
2. The collective (shared) set of objects, presented within a Semantic Content Management System, is undergoing a process of individual attribution, and valuation. These annotations can be based on formal ontologies of concepts and relationships that provide a formal – and hopefully widely accepted – vocabulary for the particular domain.
3. In accordance with the concept of 'random' disturbance as genetic imperative the structure of this database has to be receptive for sudden change (events) that occur outside the system (T_{n+1} (n = 0, 1, 2, 3...)).
4. Transformation of language based knowledge into a formal system that can be interpreted and reasoned about by applications.
5. Applications based on ISO-Code (operable machine code), effectively can control a rapid prototyping process tuned along a set of parameters.

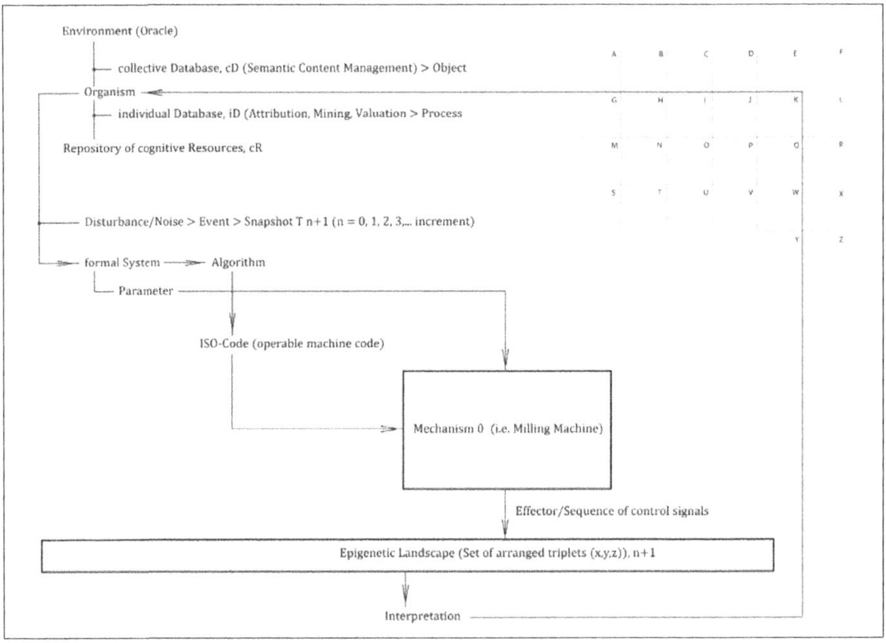

Fig. 31.4. Model for Dissipative Mapping

6. This Mechanism, i.e. a milling machine produces a sequence of control signals that eventually run the tools that produce a physical 'snapshot' (set of arranged triplets (x, y, z) of the system state at a given time (T_n).
7. The resulting model (Epigenetic Landscape) is subject to interpretation, information that is fed back into the system on the level of individual attribution.

Looking at information systems, one recognizes the additional need for explicit operators that construct and update the model itself, as the subject evolves or more information is discovered about it.

31.4 Model, Epigenetic Landscape

Virtual forms are real (not symbolic, nor ideal) 'folds' in a real and n-dimensional space and they can trigger indefinite morphogenetic occurrences in n+1 space (the space of the next higher dimension). In this n+1 space, where forms become dynamic or unfold, unfettered freedom and indefiniteness reigns since numbers, quality, and combinations of real energy contained within the folds become recognisable unexpectedly – almost as a coincidence.

One of the most important geometric terms coined to describe the relation between phenomenal forms (phenotypes) and morphogenetic fields in which they emerge, is Conrad Waddington's term "Epigenetic Landscape". An Epigenetic Landscape is an undulating topological surface in a phase space (i.e., a descriptive model without explanation). Its many valleys correspond with possible trajectories (forms) of a body evolving (emerging) on it.

The epigenetic landscape is merely a template or a virtual form that was constructed in a different dimension as multiplicity produced by an extremely complex energy field. The forms that

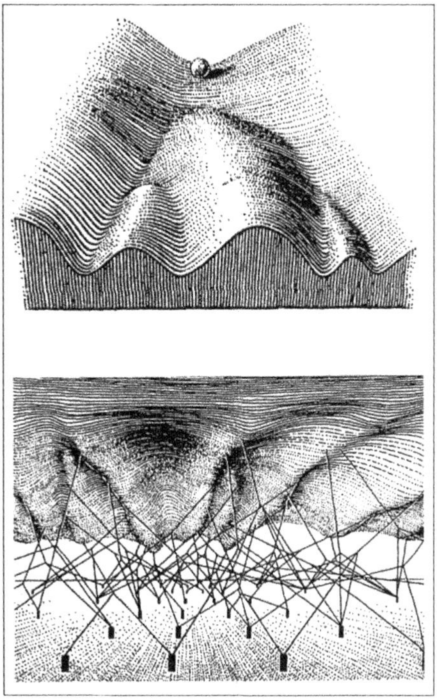

Fig. 31.5. The complex system of interactions underlying the epigenetic landscape. The pegs in the ground represent genes; the strings leading from them the chemical tendencies which the genes produce. The modelling of the epigenetic landscape, which slopes down from above one's head towards the distance, is controlled by the pull of these numerous guy-ropes which are ultimately anchored to the genes.

were developed in the course of this exercise are to be seen as evolving forms within a dissipative physical system. This means that the curves represented on 'flat planes' by functional graphs or diagrams, (e.g., statistical information of urban development) serve to generate formations that are subject to numerous modifications. Layering, local density (as basis for spatiality) of a coherent probability (specific use) for these modifications is implicit. By the term intelligence, applied to one of these temporary conditions, is meant the potential for their virtual folds, which are modulated by singularity and serve to trigger morphogenetic occurrences. Such a dissipative system acts dynamically on effective energies (regulative energies) and then progresses to the next state of evolutive succession.

31.5 Phantom Public?

"This is what we wish to attempt: Where matters-of-fact have failed, let's try what I have called matters-of-concern. What we are trying to register is a huge sea change in our conceptions of science, our grasps of facts, our understanding of objectivity. For too long, objects have been wrongly portrayed as matters-of-fact. This is unfair to them, unfair to science, unfair to objectivity, unfair to experience. They are much more interesting, variegated, uncertain, complicated, far reaching, heterogeneous, risky, historical, local, material and networked than the pathetic version offered for

Fig. 31.6. Public Building Performance, "Triumphal Arch"

too long by philosophers. Rocks are not simply there to be kicked at, desks to be thumped at. 'Facts are facts are facts'? Yes, but they are also a lot of other things in addition." (Latour 2005)

As indicated with the above quote by Bruno Latour, besides the technological implications to the concept there is yet another dimension to the mapping of hidden agencies in the process of urban and collective spatial appropriation. For too long the emphasis on participative processes has been obstructed by a sturdy definition of the social as determined by structural necessity, notions of assumed collective identities, and the prevailing attitude of constructing cohesive meaning for the social body as a whole, a "Phantom Public" (Lippmann 1927). The particularities have been neglected for the common good or the pursuit of an all-encompassing consensus in the process of dialectic reasoning. However, not only the contrast between the local and the global but also among the various constituencies on all levels of investigation sheds new light on the question that is at the heart of political determination: the very definition of what composes the social. It is crucial to emphasise the necessity of looking into the uncertainties of groups, actions, objects and matters of fact in order to shape the process of localising the global, and distributing the local.

References

Althusser L (1982) Sul pensiero marxista. In Sul materialismo aleatorio. Milan, Edizioni Unicopli

Foerster H (1960) Über selbst-organisierende Systeme und ihre Umwelten. In: Sicht und Einsicht, Heidelberg, Carl-Auer-Systeme Verlag, pp.115–130

Hardt M, Negri A (2000) Empire. London, Harvard University Press

Schrödinger E (1944) Ordnung, Unordnung und Entropie. In Was ist Leben?, München, Piper, pp.120–132

Latour B (2005) From Realpolitik to Dingpolitik. In Latour B, Weibel P (eds) Making Things Public. Cambridge, MA, MIT Press, pp. 14–41

Lippmann W (1927) The Phantom Public, New Brunswick and London, Transaction Publishers (Reprint 2002)

Waddington CH (1956) Principles of Embryology, New York, Macmillan

Wiener O (1996) Form und Inhalt. In Herken R (ed) Schriften zur Erkenntnistheorie. Vienna, SpringerWienNewYork, pp.113–144

32 Pressure Today

Peter Dykhuis

Halifax, NS, Canada

Abstract

Pressure Today collates the 'pressure-full' mail of my every-day life with daily, continental barometric pressure maps traced onto business envelopes that previously contained bills, invoices and professional announcements addressed to me. The result is an on-going record of meteorological phenomenon layered over an idiosyncratic map of my personal, professional, economic and social networks.

2004 – Present

Pressure Today began on 14 November 2004 and collates the 'pressure-full' mail of my every-day life with daily, continental barometric pressure maps traced onto business envelopes that previously contained bills, invoices and professional announcements addressed to me. The result is an on-going record of meteorological phenomenon layered over an idiosyncratic map of my personal, professional, economic and social networks. As a project, *Pressure Today* expands the concept of mapping beyond the construction of geophysical territory and develops new metaphors where transitory high-and-low pressure systems are superimposed on waves, cycles and flows of personal and public mail.

Pressure Today is also a self-portrait of my place on the earth since my postal addresses at home and work are specific destinations for a variety of personal and professional materials. The project sometimes reveals too much information about me: who I bank with, charities that I contribute to, political parties that I endorse, friends and family who send personal notes or colleagues that forward professional material to me. Throughout this, an outwardly spiraling web is constructed from mail sent to me, one where I am, at least by name and address, the centre of a world of my own making and imagination.

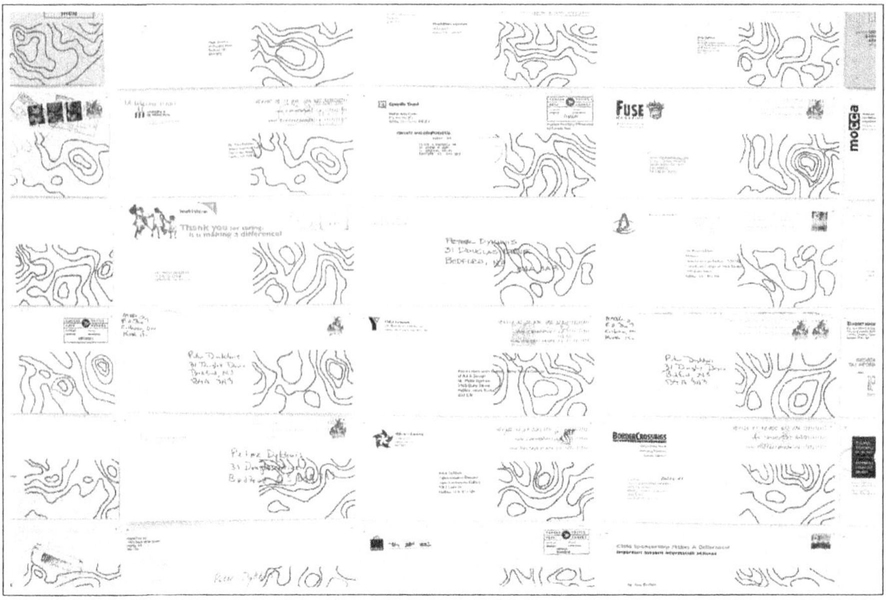

Fig. 32.1. Sharpie marker on used business envelopes. Average envelope 10.5 x 24.0 cm

33 Early-Modern Portuguese Maps: Metaphors of Space and Spaces of Metaphor

Manuela Mourão

Department of English, Old Dominion University, Norfolk, VA, USA

Abstract

This project uses early-modern Portuguese maps as backgrounds of paintings that explore metaphors of expansion, colonialism and race.

Work Description

That maps 'construct' space as surely as they represent it is now a commonplace in the discipline of Cartography. It is not surprising, therefore, that the metaphoric possibilities of maps inspire writers and visual artists, even as the scientific impulse behind map-making taunts any one interested in rigor.

Early maps, often so exquisitely illustrated, were (and continue to be) used as decorative pieces. They figure so prominently in the paintings of Dutch interiors by Jean Vermeer precisely because they were, from early on, often displayed in homes as decorative art. But more than a touch of realism, the maps in Vermeer's paintings provide ample illustration of their potential to be used non-referentially.

As a scholar and a visual artist, I am deeply compelled by the dual possibility of treating maps not only as the metaphors of space they typically stand for, but also as spaces of metaphor. Indeed, it is the map as a space of metaphor that interests me primarily, so this project explores the connections between art and cartography by making art with maps. The paintings I am showing are part of a wider project (entitled *Whitewash)* which analyses the impact of the sixteenth-century Portuguese maritime expansion on the Portuguese concept of race, as well as the contemporary ramifications of this historic legacy.

Starting in the late 1400s with the arrival of Vasco da Gama's ships in India, and continuing well into the 16th century with the establishment of commercial outposts along a number of coastal areas in the Indian Ocean, the interaction between the

Portuguese and the non-Western world had a significant impact on the cultures of all nations involved. With this project I wish to reflect on the implications of present-day interpretations of the 15th and 16th century Portuguese maritime "discoveries." While they had a major impact in bringing the West and the East closer together, they also facilitated the slave trade and colonization. The recent celebration, in Europe, of the 500th anniversary of these "discoveries" led me to reflect on how this historical period is represented today and on how these representations may continue to shape present-day understanding of colonialism and race – particularly the European construction of whiteness that was generated by, and underpins, these discoveries.

Whitewash seeks to contribute to this reflection. These paintings do so in particular by bringing together three central (and inter-related) elements of the Portuguese 'discoveries' – the ocean (the way to expansion), the maps (the records of that expansion), and whiteness (the racial formation resulting from that expansion). The paintings are done on wood panels with copies of sixteenth-century maps selected from the work of early-modern Portuguese cartographers as the background. These maps, documenting the contemporary Portuguese presence in the oceans of the world, are painted over in different values of white in order to problematize the very concept of whiteness. Through specific technical and compositional choices dramatizing the process of erosion – e.g., the white surface is eroded and reveals the maps underneath – the paintings use the maps' inherent metaphorical possibilities and work, in their turn, as metaphors for the failure of the colonizing enterprise and of its glorification in historical representations.

In treating maps as spaces of metaphor that can illuminate the world in ways that far surpass their referentiality, this project takes up, albeit in a very particular way, Antje Lehn's challenge to look over the borders of Cartography as a discipline in order to "learn how to represent the world much better than a 'perfect' representation ... would do."

Acknowledgements

My thanks to Zoltan Biederman for his help with identifying the maps by Bartolomeu Lasso and João Teixeira.

Fig. 33.1. *Mare Arabicum a*, mixed media, 3'x2' (2006); background: Atlas Catalan by Abraão Cresques (1375)

Fig. 33.2. *Mare Arabicum b*, mixed media, 3'x2' (2006); background: Map of Portugal by Álvaro Seco (1561?) and Ethiopia, Arabia, Persia and India by Bartolomeu Lasso (1596)

Fig. 33.3. *Mare Atlanticum 1*, mixed media, 16" x 12" (2005); background: Atlas Catalan by Abraão Cresques (1375)

Fig. 33.4. *Mare Atlanticum 2*, mixed media, 24" x 18" (2007); background: Atlantic Chart by Jorge Reinel (c. 1534)

Fig. 33.5. *Mare Indicum 3*, mixed media, 24" x 18" (2007); background: Miller Atlas (folio 2) by Pedro Reinel, Jorge Reinel, and Lopo Homem (c. 1519)

Fig. 33.6. *Mare Leucorum*, mixed media, 6,5' x 4' (2007); background: World Atlas (folio 10) by Diogo Homem (c. 1564)

Fig. 33.7. *Mare Rubrum 1*, mixed media, 2'x2' (2007); background: World Atlas (chart 4) by João Teixeira (1643)

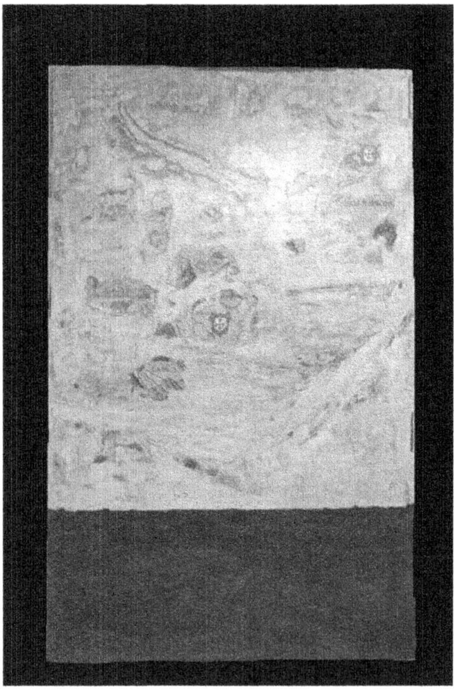

Fig. 33.8. *Mare Rubrum 2*, mixed media, 6,5x4' (2007); background: World Atlas (folio 8) by Diogo Homem (c. 1564)

34 This Is Not a Map

Sabine Müller-Funk and Maria Christine Holter

Vienna, Austria

Abstract

In the project THIS IS NOT A MAP the word MAP overwrites the map itself and the landscape described by this map. The word MAP lies as a superimposed pattern on both, the actual map and the the Rosa Mayreder Park, Vienna, thereby depicting itself and the filter which it constitutes.

34.1 Introduction

THIS IS NOT A MAP is the title of Sabine Müller-Funk's contribution to the symposium *Art and Cartography – Cartography and Art.* The German born Müller-Funk, now living and working in Vienna and in a small community in Lower Austria, can be briefly described as a multimedia artist who creates objects, installations, land art and performances out of her deep interest in text, script and the process of her own handwriting. In realizing her work she always uses site-specific materials as the background or as a part of her – sometimes very large-scale – inscriptions. The provocative title, alluding to Magritte's *This is not a pipe* written onto a picture of a pipe, is a good point of departure for a conversation about the artist's strategies in THIS IS NOT A MAP and her art in general.

34.2 Interview

Holter: *What did you have in mind for the installation THIS IS NOT A MAP (Figure 34.1) at the Rosa Mayreder Park, which is part of the large area generally referred to as Karlsplatz in Vienna?*

Müller-Funk: In this project I see maps – and more specifically city maps – as a field characterized by different semiotic systems: the icon as a sign for houses,

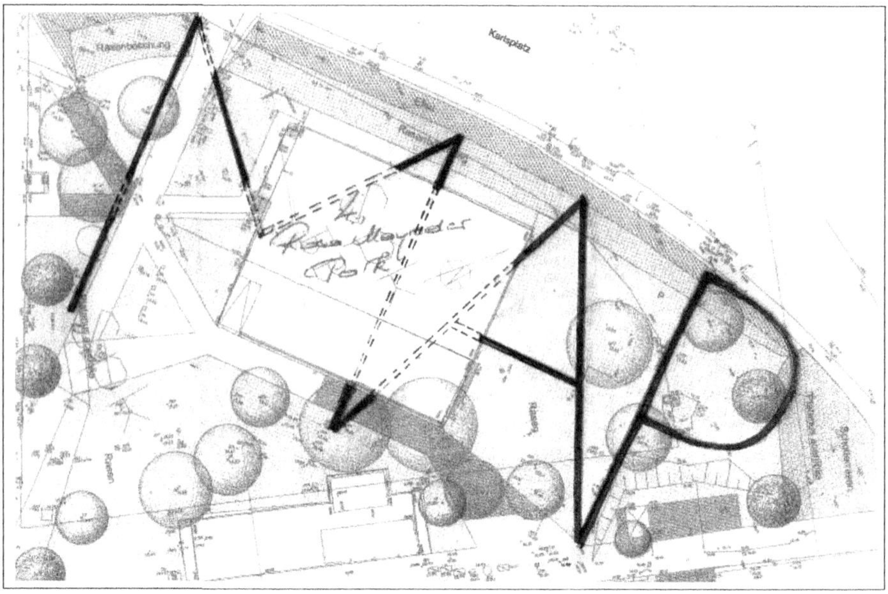

Fig. 34.1. Conception of THIS IS NOT A MAP, 2007, Vienna. Map of the Rosa Mayreder Park marked with the planned intervention: the word MAP written by means of red and white construction site ribbons

churches, courses of rivers, ...; the symbol, whose letters form names and thereby marks villages or urban areas; and last but not least the image, in which the landscape is presented as a painting due to the bird's eye perspective of the map. These different systems always overlap and mix in a map. Maps constitute a survey or measuring of the landscape (or the city), which lets the absent landscape arise before the mind's eye and thereby reaches out to pull the absent landscape into presence. Obviously a map never depicts a landscape 1:1, and it always incorporates a filter – that of the cartographer, and also that of the viewer, the user of the map. In my project THIS IS NOT A MAP the word MAP overwrites the map itself and the landscape described by this map. The word MAP lies as a superimposed pattern on both, the actual map and the landscape (the Rosa Mayreder Park), thereby depicting itself and the filter which it constitutes.

Holter: *Unfortunately the project could not be executed as planned. What were the obstacles to its 1:1 realization?*

Müller-Funk: Since THIS IS NOT A MAP would have been an intervention in pubic space, I had to submit the project to a municipal department of the City of Vienna. It was turned down, because they argued that the installation would disappear under the January snow. They did not want to put up snow poles to fix the plastic ribbons onto them. Ironically there is no snow in Vienna at the

Fig. 34.2. ASPHALT WRITING, 2004, Drosendorf (Austria). Artist working with asphalt sealing gun

moment that would have covered my installation, even if I had fixed the red-and-white-striped construction site ribbons to the ground.

Holter: *Let's go back to a few projects that also revolved around your main theme of combining landscape with text – as such not only describing the landscape, but actually writing on it. For the symposium ON THE ROAD. STILL in 2004, your piece ASPHALT WRITING (Figure 34.2) was executed directly on a country road leading to and from a small village. You had to use quite an unusual writing tool. Can you tell us a bit more about that performance with an asphalt sealing gun?*

Müller-Funk: The symposium dealt with the topics of roads and writing, which are both engraved traces of attempts to communicate. My contribution in this symposium was to connect thoughts about communication and roads with the material which physically forms the road, namely asphalt. I wrote my text with liquid asphalt on a 4.5-meter-wide and 450-meter-long stretch of asphalt road using a machine that is normally employed for street repair. So first I had to learn how to write all over again, because it is not easy to handle such a big and clumsy writing tool. The pump itself was connected to a kind of tractor, which pulled the entire equipment including myself at the end. But after some practice, writing with the material asphalt on asphalt resulted in a very rewarding experience of being literally *within* the text. The asphalt sealing gun moved along a trace of reflection of the now.

Fig. 34.3. IN/DESCRIPTION OF A RIVERBED, 2004, Italy, latex paint on stones, 20 x 300 cm

Holter: *Whereas handling the asphalt pump as a giant fountain pen was difficult in terms of the process of writing, the surface you wrote on in the project IN/ DESCRIPTION OF A RIVERBED (Figure 34.3) – stones in a riverbed – must have surely been a challenge for you. Do you also expect a certain amount of effort from your audience?*

Müller-Funk: In this project I equalized the different latitudes of the stones by making the letters of the text thicker or thinner, which was not easy, as you can imagine. As to the challenge for the viewer, I think decoding an artist's work always requires a bit of an effort, since there are usually different layers of understanding and meaning.

Holter: *Scale, often very vast scale, seems to interest you. In Japan you cut the giant letters of the word HERE into a bamboo forest (Figure 34.4), and – as we will discuss a bit later – in a project executed in the countryside of Lower Austria you raked many cubic meters of hay to leave your verbal trace in the landscape. What is it that draws you to such large-scale performances?*

Müller-Funk: It is probably the urge to become familiar with a special area of land-scape. One could also say it is a way of conquering the space by doing something very physical with it. In Japan I was cutting the word HERE out of the bamboo and thereby, so to speak, *carving the forest*. Each letter was about 20 meters long and 5 meters wide, so the word HERE was written/inscribed very physically into

Fig. 34.4. HERE, 2006, Japan, bamboo forest. Each cutout letter ca. 20 m x 10 m, 400 m red plastic string leading through the land art piece

the forest. A 400 m-long red string guided the visitor through the forest along the cutout footpath forming the word HERE. Labels attached to the string indicated the visitor's momentary position in a particular moment of time. Thus walking and forming the HERE were the same: while moving along the red string the observer was always (T)HERE.

Holter: *For your performance In/DeSCRIPTION of a market (Figure 34.5) at Vienna's Naschmarkt, a big open food market in the vicinity of Karlsplatz, you used an approach similar to that of Asphalt writing. Again your writing tool was a spraying device – you wrote on long wooden boards with a paint gun. Nevertheless, this project seemed to focus on the participation of the viewer. The audience was invited to bring Naschmarkt-related text sequences, and in return, at the end of the performance, people were allowed to take fragments of the*

Fig. 34.5. NASCHMARKT/ASCHENMARKT: IN/DESCRIPTION OF A MARKET, 2006, Vienna. Writing performance with a latex paint spray gun on 36 wooden boards each 25 x 250 cm

installation home. How important is the audience to you, and how do you usually chose the texts for your work?

Müller-Funk: I like to work with opposites: in my smaller art objects I deal with the themes of handwriting and privacy, but on the other hand I like to open up this intimacy, to extend my horizon and work in these large scales. In the performance In/DeSCRIPTION of a market my private handwriting and the public texts were interwoven: food market themes, symbolic and real exchange, (hand) writing and sign. The text described the market, the performance and itself.

Holter: *We have just a little time left to look at UMBRUCH (Figure 34.6), the project I already mentioned above when talking about your work for Japan. In UMBRUCH you raked the letters of the word "Umbruch" into a rural landscape. It is important to understand the double meaning of Umbruch in German. The term is used in printing – in English it translates as "makeup" – and means the way the text is divided into pages in a book. Umbruch also means "upheaval", here used in terms of an ongoing social change visible also in the transformation of the rural landscape. Please tell us more about UMBRUCH and about life at the Austrian-Czech border.*

Müller-Funk: When I conceived the work, I was also thinking of a third meaning of the word: Umbruch meaning to break up the soil with a plough pulled by a

Fig. 34.6. UMBRUCH, 2006, Lower Austria. Letters of the word Umbruch raked into mown grass, land art piece ca. 100 m x 500 m.

tractor. So my initial idea was to plow the word UMBRUCH into the soil, but the owner of the meadow, a farmer, was so directly involved in the social upheaval – the problems caused by the EU market, his kids not wanting to take over the farm for understandable reasons – that he did not want to see his future scratched into the soil. He would not allow or could not bear the plowing, and agreed only to the performance of raking the word into the freshly mown grass.

This artist/critic talk took place at the *project space* of the *Vienna Kunsthalle* on February 1, 2008 within the framework of the art presentations for the exhibition *zoomandscale*.

GPSR Compliance

The European Union's (EU) General Product Safety Regulation (GPSR) is a set of rules that requires consumer products to be safe and our obligations to ensure this.

If you have any concerns about our products, you can contact us on ProductSafety@springernature.com

In case Publisher is established outside the EU, the EU authorized representative is:

Springer Nature Customer Service Center GmbH
Europaplatz 3
69115 Heidelberg, Germany

Batch number: 10755244

Printed by Printforce, the Netherlands